Sample Preparation Techniques in Analytical Chemistry

CHEMICAL ANALYSIS

A SERIES OF MONOGRAPHS ON ANALYTICAL CHEMISTRY AND ITS APPLICATIONS

Editor
J. D. WINEFORDNER

VOLUME 162

A complete list of the titles in this series appears at the end of this volume.

Sample Preparation Techniques in Analytical Chemistry

Edited by

SOMENATH MITRA

Department of Chemistry and Environmental Science
New Jersey Institute of Technology

A JOHN WILEY & SONS, INC., PUBLICATION

Published by John Wiley & Sons, Inc., Hoboken, New Jersey.
Published simultaneously in Canada.

For general information on our other products and services please contact our Customer Care Department within the U.S. at 877-762-2974, outside the U.S. at 317-572-3993 or fax 317-572-4002.

Wiley also publishes its books in a variety of electronic formats. Some content that appears in print, however, may not be available in electronic format.

Library of Congress Cataloging-in-Publication Data:

Sample preparation techniques in analytical chemistry / edited by Somenath Mitra.
     p.   cm. — (Chemical analysis; v. 162)
Includes index.
  ISBN 0-471-32845-6 (cloth : acid-free paper)
  1. Sampling. 2. Chemistry, Analytic—Methodology. I. Mitra, S.
(Somenath), 1959–   II. Series.
  QD75.4.S24S26 2003
  543—dc21

                                  2003001379

Printed in the United States of America

10 9 8 7 6 5 4 3 2 1

To the hands in the laboratory
and
the heads seeking information

CONTENTS

CHAPTER 3 **EXTRACTION OF SEMIVOLATILE ORGANIC COMPOUNDS FROM SOLID MATRICES** **139**
Dawen Kou and Somenath Mitra

**SECTION C   SAMPLE PREPARATION IN MICROSCOPY
               AND SPECTROSCOPY**

**CHAPTER 9   SAMPLE PREPARATION FOR
               MICROSCOPIC AND SPECTROSCOPIC
               CHARACTERIZATION OF SOLID
               SURFACES AND FILMS   377**
               Sharmila M. Mukhopadhyay

CONTRIBUTORS

Roman Brukh, Department of Chemistry and Environmental Science, New Jersey Institute of Technology, Newark, NJ 07102

Zafar Iqbal, Department of Chemistry and Environmental Science, New Jersey Institute of Technology, Newark, New Jersey 07102

Mahesh Karwa, Department of Chemistry and Environmental Science, New Jersey Institute of Technology, Newark, NJ 07102

Barbara B. Kebbekus, Department of Chemistry and Environmental Science, New Jersey Institute of Technology, Newark, NJ 07102

Dawen Kou, Department of Chemistry and Environmental Science, New Jersey Institute of Technology, Newark, NJ 07102

Somenath Mitra, Department of Chemistry and Environmental Science, New Jersey Institute of Technology, Newark, NJ 07102

Sharmila M. Mukhopadhyay, Department of Mechanical and Materials Engineering, Wright State University, Dayton, OH 45435

Bhama Parimoo, Department of Pharmaceutical Chemistry, Rutgers University College of Pharmacy, Piscataway, NJ 08854

Satish Parimoo, Aderans Research Institute, Inc., 3701 Market Street, Philadelphia, PA 19104

Gregory C. Slack, Department of Chemistry, Clarkson University, Potsdam, NY 13676

Nicholas H. Snow, Department of Chemistry and Biochemistry, Seton Hall University, South Orange, NJ 07079

Martha J. M. Wells, Center for the Management, Utilization and Protection of Water Resources and Department of Chemistry, Tennessee Technological University, Cookeville, TN 38505

PREFACE

There has been unprecedented growth in measurement techniques over the last few decades. Instrumentation, such as chromatography, spectroscopy and microscopy, as well as sensors and microdevices, have undergone phenomenal developments. Despite the sophisticated arsenal of analytical tools, complete noninvasive measurements are still not possible in most cases. More often than not, one or more pretreatment steps are necessary. These are referred to as *sample preparation*, whose goal is enrichment, cleanup, and signal enhancement. Sample preparation is often the bottleneck in a measurement process, as they tend to be slow and labor-intensive. Despite this reality, it did not receive much attention until quite recently. However, the last two decades have seen rapid evolution and an explosive growth of this industry. This was particularly driven by the needs of the environmental and the pharmaceutical industries, which analyze large number of samples requiring significant efforts in sample preparation.

Sample preparation is important in all aspects of chemical, biological, materials, and surface analysis. Notable among recent developments are faster, greener extraction methods and microextraction techniques. Specialized sample preparations, such as self-assembly of analytes on nanoparticles for surface enhancement, have also evolved. Developments in high-throughput workstations for faster preparation–analysis of a large number of samples are impressive. These use 96-well plates (moving toward 384 wells) and robotics to process hundreds of samples per day, and have revolutionized research in the pharmaceutical industry. Advanced microfabrication techniques have resulted in the development of miniaturized chemical analysis systems that include microscale sample preparation on a chip. Considering all these, sample preparation has evolved to be a separate discipline within the analytical/measurement sciences.

The objective of this book is to provide an overview of a variety of sample preparation techniques and to bring the diverse methods under a common banner. Knowing fully well that it is impossible to cover all aspects in a single text, this book attempts to cover some of the more important and widely used techniques. The first chapter outlines the fundamental issues relating to sample preparation and the associated quality control. The remainder of the book is divided into three sections. In the first we describe various extraction and enrichment approaches. Fundamentals of extraction, along with specific details on the preparation of organic and metal analytes, are presented. Classical methods such as Soxhlett and liquid–liquid extraction are described, along with recent developments in widely accepted methods such as SPE, SPME, stir-bar microextraction, microwave extraction, supercritical extraction, accelerated solvent extraction, purge and trap, headspace, and membrane extraction.

The second section is dedicated to the preparation for nucleic acid analysis. Specific examples of DNA and RNA analyses are presented, along with the description of techniques used in these procedures. Sections on high-throughput workstations and microfabricated devices are included. The third section deals with sample preparation techniques used in microscopy, spectroscopy, and surface-enhanced Raman.

The book is intended to be a reference book for scientists who use sample preparation in the chemical, biological, pharmaceutical, environmental, and material sciences. The other objective is to serve as a text for advanced undergraduate and graduate students.

I am grateful to the New Jersey Institute of Technology for granting me a sabbatical leave to compile this book. My sincere thanks to my graduate students Dawen Kou, Roman Brukh, and Mahesh Karwa, who got going when the going got tough; each contributed to one or more chapters.

New Jersey Institute of Technology SOMENATH MITRA
Newark, NJ

CHAPTER

1

SAMPLE PREPARATION: AN ANALYTICAL PERSPECTIVE

SOMENATH MITRA AND ROMAN BRUKH

*Department of Chemistry and Environmental Science,
New Jersey Institute of Technology, Newark, New Jersey*

1.1. THE MEASUREMENT PROCESS

The purpose of an analytical study is to obtain information about some object or substance. The substance could be a solid, a liquid, a gas, or a biological material. The information to be obtained can be varied. It could be the chemical or physical composition, structural or surface properties, or a sequence of proteins in genetic material. Despite the sophisticated arsenal of analytical techniques available, it is not possible to find every bit of information of even a very small number of samples. For the most part, the state of current instrumentation has not evolved to the point where we can take an instrument to an object and get all the necessary information. Although there is much interest in such noninvasive devices, most analysis is still done by taking a part (or portion) of the object under study (referred to as the *sample*) and analyzing it in the laboratory (or at the site). Some common steps involved in the process are shown in Figure 1.1.

The first step is *sampling*, where the sample is obtained from the object to be analyzed. This is collected such that it represents the original object. Sampling is done with variability within the object in mind. For example, while collecting samples for determination of Ca^{2+} in a lake, it should be kept in mind that its concentrations can vary depending on the location, the depth, and the time of year.

The next step is *sample preservation*. This is an important step, because there is usually a delay between sample collection and analysis. Sample preservation ensures that the sample retains its physical and chemical characteristics so that the analysis truly represents the object under study. Once

Sample Preparation Techniques in Analytical Chemistry, Edited by Somenath Mitra
ISBN 0-471-32845-6 Copyright © 2003 John Wiley & Sons, Inc.

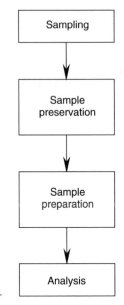

Figure 1.1. Steps in a measurement process.

the sample is ready for analysis, *sample preparation* is the next step. Most samples are not ready for direct introduction into instruments. For example, in the analysis of pesticides in fish liver, it is not possible to analyze the liver directly. The pesticides have to be extracted into a solution, which can be analyzed by an instrument. There might be several processes within sample preparation itself. Some steps commonly encountered are shown in Figure 1.2. However, they depend on the sample, the matrix, and the concentration level at which the analysis needs to be carried out. For instance, trace analysis requires more stringent sample preparation than major component analysis.

Once the sample preparation is complete, the analysis is carried out by an instrument of choice. A variety of instruments are used for different types of analysis, depending on the information to be acquired: for example, chromatography for organic analysis, atomic spectroscopy for metal analysis, capillary electrophoresis for DNA sequencing, and electron microscopy for small structures. Common analytical instrumentation and the sample preparation associated with them are listed in Table 1.1. The sample preparation depends on the analytical techniques to be employed and their capabilities. For instance, only a few microliters can be injected into a gas chromatograph. So in the example of the analysis of pesticides in fish liver, the ultimate product is a solution of a few microliters that can be injected into a gas chromatograph. Sampling, sample preservation, and sample preparation are

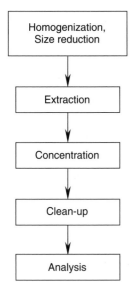

Figure 1.2. Possible steps within sample preparation.

all aimed at producing those few microliters that represent what is in the fish. It is obvious that an error in the first three steps cannot be rectified by even the most sophisticated analytical instrument. So the importance of the prior steps, in particular the sample preparation, cannot be understressed.

1.1.1. Qualitative and Quantitative Analysis

There is seldom a unique way to design a measurement process. Even an explicitly defined analysis can be approached in more than one ways. Different studies have different purposes, different financial constraints, and are carried out by staff with different expertise and personal preferences. The most important step in a study design is the determination of the purpose, and at least a notion of the final results. It should yield data that provide useful information to solve the problem at hand.

The objective of an analytical measurement can be qualitative or quantitative. For example, the presence of pesticide in fish is a topic of concern. The questions may be: Are there pesticides in fish? If so, which ones? An analysis designed to address these questions is a *qualitative analysis*, where the analyst screens for the presence of certain pesticides. The next obvious question is: How much pesticide is there? This type of analysis, *quantitative analysis*, not only addresses the presence of the pesticide, but also its concentration. The other important category is *semiqualitative analysis*. Here

Table 1.1. Common Instrumental Methods and the Necessary Sample Preparation Steps Prior to Analysis

Analytes	Sample Preparation	Instrument[a]
Organics	Extraction, concentration, cleanup, derivatization	GC, HPLC, GC/MS, LC/MS
Volatile organics	Transfer to vapor phase, concentration	GC, GC-MS
Metals	Extraction, concentration, speciation	AA, GFAA, ICP, ICP/MS
Metals	Extraction, derivatization, concentration, specia-tion	UV-VIS molecular absorp-tion spectrophotometry, ion chromatography
Ions	Extraction, concentration, derivatization	IC, UV-VIS
DNA/RNA	Cell lysis, extraction, PCR	Electrophoresis, UV-VIS, florescence
Amino acids, fats carbohydrates	Extraction, cleanup	GC, HPLC, electrophoresis
Microstructures	Etching, polishing, reac-tive ion techniques, ion bombardments, etc.	Microscopy, surface spectros-copy

[a]GC, gas chromatography; HPLC, high-performance liquid chromatography; MS, mass spectroscopy; AA, atomic absorption; GFAA, graphite furnace atomic absorption; ICP, inductively coupled plasma; UV-VIS, ultraviolet–visible molecular absorption spectroscopy; IC, ion chromatography.

the concern is not exactly how much is there but whether it is above or below a certain threshold level. The prostate specific antigen (PSA) test for the screening of prostate cancer is one such example. A PSA value of 4 ng/L (or higher) implies a higher risk of prostate cancer. The goal here is to determine if the PSA is higher or lower then 4 ng/L.

Once the goal of the analyses and target analytes have been identified, the methods available for doing the analysis have to be reviewed with an eye to accuracy, precision, cost, and other relevant constraints. The amount of labor, time required to perform the analysis, and degree of automation can also be important.

1.1.2. Methods of Quantitation

Almost all measurement processes, including sample preparation and analysis, require calibration against chemical standards. The relationship between a detector signal and the amount of analyte is obtained by recording the response from known quantities. Similarly, if an extraction step is involved, it is important to add a known amount of analyte to the matrix and measure its recovery. Such processes require standards, which may be prepared in the laboratory or obtained from a commercial source. An important consideration in the choice of standards is the matrix. For some analytical instruments, such as x-ray fluorescence, the matrix is very important, but it may not be as critical for others. Sample preparation is usually matrix dependent. It may be easy to extract a polycyclic aromatic hydrocarbon from sand by supercritical extraction but not so from an aged soil with a high organic content.

Calibration Curves

The most common calibration method is to prepare standards of known concentrations, covering the concentration range expected in the sample. The matrix of the standard should be as close to the samples as possible. For instance, if the sample is to be extracted into a certain organic solvent, the standards should be prepared in the same solvent. The calibration curve is a plot of detector response as a function of concentration. A typical calibration curve is shown in Figure 1.3. It is used to determine the amount of analyte in the unknown samples. The calibration can be done in two ways, best illustrated by an example. Let us say that the amount of lead in soil is being measured. The analytical method includes sample preparation by acid extraction followed by analysis using atomic absorption (AA). The stan-

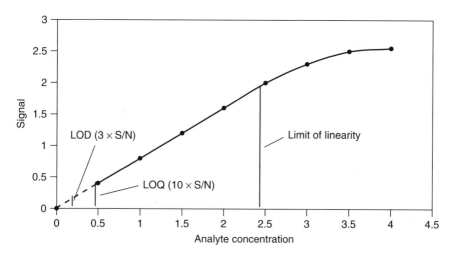

Figure 1.3. Typical calibration curve.

dards can be made by spiking clean soil with known quantities of lead. Then the standards are taken through the entire process of extraction and analysis. Finally, the instrument response is plotted as a function of concentration. The other option assumes quantitative extraction, and the standards are used to calibrate only the AA. The first approach is more accurate; the latter is simpler. A calibration method that takes the matrix effects into account is the method of standard addition, which is discussed briefly in Chapter 4.

1.2. ERRORS IN QUANTITATIVE ANALYSIS: ACCURACY AND PRECISION

All measurements are accompanied by a certain amount of error, and an estimate of its magnitude is necessary to validate results. The error cannot be eliminated completely, although its magnitude and nature can be characterized. It can also be reduced with improved techniques. In general, errors can be classified as random and systematic. If the same experiment is repeated several times, the individual measurements cluster around the mean value. The differences are due to unknown factors that are stochastic in nature and are termed *random errors*. They have a Gaussian distribution and equal probability of being above or below the mean. On the other hand, *systematic errors* tend to bias the measurements in one direction. Systematic error is measured as the deviation from the true value.

1.2.1. Accuracy

Accuracy, the deviation from the true value, is a measure of systematic error. It is often estimated as the deviation of the mean from the true value:

$$\text{accuracy} = \frac{\text{mean} - \text{true value}}{\text{true value}}$$

The true value may not be known. For the purpose of comparison, measurement by an established method or by an accredited institution is accepted as the true value.

1.2.2. Precision

Precision is a measure of reproducibility and is affected by random error. Since all measurements contain random error, the result from a single measurement cannot be accepted as the true value. An estimate of this error is necessary to predict within what range the true value may lie, and this is done by repeating a measurement several times [1]. Two important parameters, the *average value* and the *variability of the measurement*, are obtained from this process. The most widely used measure of average value is the arithmetic mean, $\bar{x}$:

$$\bar{x} = \frac{\sum x_i}{n}$$

where $\sum x_i$ is the sum of the replicate measurements and n is the total number of measurements. Since random errors are normally distributed, the common measure of variability (or precision) is the standard deviation, σ. This is calculated as

$$\sigma = \sqrt{\frac{\sum(x_i - \bar{x})^2}{n}} \qquad (1.1)$$

When the data set is limited, the mean is often approximated as the true value, and the standard deviation may be underestimated. In that case, the unbiased estimate of σ, which is designated s, is computed as follows:

$$s = \sqrt{\frac{\sum(x_i - \bar{x})^2}{n - 1}} \qquad (1.2)$$

As the number of data points becomes larger, the value of s approaches that of σ. When n becomes as large as 20, the equation for σ may be used. Another term commonly used to measure variability is the *coefficient of variation* (CV) or *relative standard deviation* (RSD), which may also be expressed as a percentage:

$$\text{RSD} = \frac{s}{\bar{x}} \quad \text{or} \quad \% \, \text{RSD} = \frac{s}{\bar{x}} \times 100 \qquad (1.3)$$

Relative standard deviation is the parameter of choice for expressing precision in analytical sciences.

Precision is particularly important when sample preparation is involved. The variability can also affect accuracy. It is well known that reproducibility of an analysis decreases disproportionately with decreasing concentration [2]. A typical relationship is shown in Figure 1.4, which shows that the uncertainty in trace analysis increases exponentially compared to the major and minor component analysis. Additional deviations to this curve are expected if sample preparation steps are added to the process. It may be prudent to assume that uncertainty from sample preparation would also increase with decrease in concentration. Generally speaking, analytical

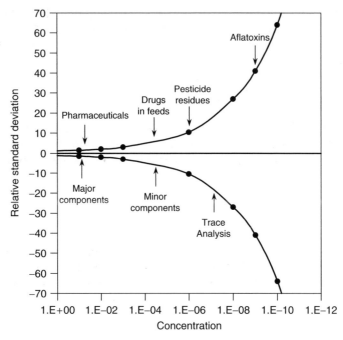

Figure 1.4. Reproducibility as a function of concentration during analytical measurements. (Reproduced from Ref. 3 with permission from LC-GC North America.)

instruments have become quite sophisticated and provide high levels of accuracy and precision. On the other hand, sample preparation often remains a rigorous process that accounts for the majority of the variability. Going back to the example of the measurement of pesticides in fish, the final analysis may be carried out in a modern computer-controlled gas chromatograph/mass spectrograph (GC-MS). At the same time, the sample preparation may involve homogenization of the liver in a grinder, followed by Soxhlett extraction, concentration, and cleanup. The sample preparation might take days, whereas the GC-MS analysis is complete in a matter of minutes. The sample preparation also involves several discrete steps that involve manual handling. Consequently, both random and systematic errors are higher during sample preparation than during analysis.

The relative contribution of sample preparation depends on the steps in the measurement process. For instance, typically two-thirds of the time in an analytical chromatographic procedure is spent on sample preparation. An example of the determination of olanzapine in serum by high-performance liquid chromatography/mass spectroscopy (HPLC-MS) illustrates this point [3]. Here, samples were mixed with an internal standard and cleaned up in a solid-phase extraction (SPE) cartridge. The quantitation was done by a cali-bration curve. The recovery was 87 ± 4% for three assays, whereas repeat-ability of 10 replicate measurements was only 1 to 2%. A detailed error analysis [3] showed that 75% of the uncertainty came from the SPE step and the rest came from the analytical procedure. Of the latter, 24% was attrib-uted to uncertainty in the calibration, and the remaining 1% came from the variation in serum volume. It is also worth noting that improvement in the calibration procedure can be brought about by measures that are signifi-cantly simpler than those required for improving the SPE. The variability in SPE can come from the cartridge itself, the washing, the extraction, the drying, or the redissolution steps. There are too many variables to control.

Some useful approaches to reducing uncertainty during sample prepara-tion are given below.

Minimize the Number of Steps

In the example above, the sample preparation contributed 75% of the error. When multiple steps such as those shown in Figure 1.2 are involved, the uncertainty is compounded. A simple dilution example presented in Figure 1.5 illustrates this point. A 1000-fold dilution can be performed in one step: 1 mL to 1000 mL. It can also be performed in three steps of 1 : 10 dilutions each. In the one-step dilution, the uncertainty is from the uncertainty in the volume of the pipette and the flask. In the three-step dilution, three pipettes and three flasks are involved, so the volumetric uncertainty is compounded that many times. A rigorous analysis showed [3] that the uncertainty in the one-step dilution was half of what was expected in the three-step process.

If and when possible, one or more sample preparation steps (Figure 1.2) should be eliminated. The greater the number of steps, the more errors there are. For example, if a cleanup step can be eliminated by choosing a selective extraction procedure, that should be adapted.

Use Appropriate Techniques

Some techniques are known to provide higher variability than others. The choice of an appropriate method at the outset can improve precision. For example, a volume of less than 20 mL can be measured more accurately and precisely with a syringe than with a pipette. Large volumes are amenable to precise handling but result in dilution that lowers sensitivity. The goal should be to choose a combination of sample preparation and analytical instrumentation that reduces both the number of sample preparative steps and the RSD. Automated techniques with less manual handling tend to have higher precision.

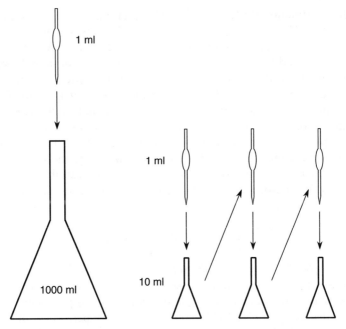

Figure 1.5. Examples of single and multiple dilution of a sample. (Reproduced from Ref. 3 with permission from LC-GC North America.)

1.2.3. Statistical Aspects of Sample Preparation

Uncertainty in a method can come from both the sample preparation and the analysis. The total variance is the sum of the two factors:

$$\sigma_T^2 = \sigma_s^2 + \sigma_a^2 \tag{1.4}$$

The subscript T stands for the total variance; the subscripts s and a stand for the sample preparation and the analysis, respectively. The variance of the analytical procedure can be subtracted from the total variance to estimate the variance from the sample preparation. This could have contribution from the steps shown in Figure 1.2:

$$\sigma_s^2 = \sigma_h^2 + \sigma_{ex}^2 + \sigma_c^2 + \sigma_{cl}^2 \tag{1.5}$$

where σ_h relates to homogenization, σ_{ex} to extraction, σ_c to concentration, and σ_{cl} to cleanup. Consequently, the overall precision is low even when a high-precision analytical instrument is used in conjunction with low-precision sample preparation methods. The total variance can be estimated by repeating the steps of sample preparation and analysis several times.

Usually, the goal is to minimize the number of samples, yet meet a specific level of statistical certainty. The total uncertainty, E, at a specific confidence level is selected. The value of E and the confidence limits are determined by the measurement quality required:

$$E = \frac{z\sigma}{\sqrt{n}} \tag{1.6}$$

where σ is the standard deviation of the measurement, z the percentile of standard normal distribution, depending on the level of confidence, and n the number of measurements. If the variance due to sample preparation, σ_s^2, is negligible and most of the uncertainty is attributed to the analysis, the minimum number of analysis per sample is given by

$$n_a = \left(\frac{z\sigma_a}{E_a}\right)^2 \tag{1.7}$$

The number of analyses can be reduced by choosing an alternative method with higher precision (i.e., a lower σ_a) or by using a lower value of z, which means accepting a higher level of error. If the analytical uncertainty is negligible ($\sigma_a \to 0$) and sample preparation is the major issue, the minimum number of samples, n_s, is given by

$$n_s = \left(\frac{z\sigma_s}{E_s}\right)^2 \tag{1.8}$$

Again, the number of samples can be reduced by accepting a higher uncertainty or by reducing σ_s. When σ_a and σ_s are both significant, the total error E_T is given by

$$E_T = z\left(\frac{\sigma_s^2}{n_s} + \frac{\sigma_a^2}{n_a}\right)^{1/2} \tag{1.9}$$

This equation does not have an unique solution. The same value of error, E_T, can be obtained by using different combinations of n_s and n_a. Combinations of n_s and n_a should be chosen based on scientific judgment and the cost involved in sample preparation and analysis.

A simple approach to estimating the number of samples is to repeat the sample preparation and analysis to calculate an overall standard deviation, s. Using Student's t distribution, the number of samples required to achieve a given confidence level is calculated as

$$n = \left(\frac{ts}{e}\right)^2 \tag{1.10}$$

where t is the t-statistic value selected for a given confidence level and e is the acceptable level of error. The degrees of freedom that determine t can first be chosen arbitrarily and then modified by successive iterations until the number chosen matches the number calculated.

Example

Relative standard deviation of repeat HPLC analysis of a drug metabolite standard was between 2 and 5%. Preliminary measurements of several serum samples via solid-phase extraction cleanup followed by HPLC analyses showed that the analyte concentration was between 5 and 15 mg/L and the standard deviation was 2.5 mg/L. The extraction step clearly increased the random error of the overall process. Calculate the number of samples required so that the sample mean would be within ± 1.2 mg/L of the population mean at the 95% confidence level.

Using equation (1.10), assuming 10 degrees of freedom, and referring to the t-distribution table from a statistics textbook, we have $t = 2.23$, $s = 2.5$, and $e = 1.2$ mg/L, so $n = (2.23 \times 2.5/1.2)^2 = 21.58$ or 22. Since 22 is significantly larger than 10, a correction must be made with the new value of t corresponding to 21 degrees of freedom ($t = 2.08$): $n = (2.08 \times 2.5/1.2)^2 = 18.78$ or 19. Since 19 and 22 are relatively close, approximately that many samples should be tested. A higher level of error, or a lower confidence level, may be accepted for the reduction in the number of samples.

1.3. METHOD PERFORMANCE AND METHOD VALIDATION

The criteria used for evaluating analytical methods are called *figures of merit*. Based on these characteristics, one can predict whether a method meets the needs of a certain application. The figures of merit are listed in Table 1.2. Accuracy and precision have already been discussed; other important characteristics are sensitivity, detection limits, and the range of quantitation.

Table 1.2. Figures of Merit for Instruments or Analytical Methods

No.	Parameter	Definition
1	Accuracy	Deviation from true value
2	Precision	Reproducubility of replicate measurements
3	Sensitivity	Ability to discriminate between small differences in concentration
4	Detection limit	Lowest measurable concentration
5	Linear dynamic range	Linear range of the calibration curve
6	Selectivity	Ability to distinguish the analyte from interferances
7	Speed of analysis	Time needed for sample preparation and analysis
8	Throughput	Number of samples that can be run in a given time period
9	Ease of automation	How well the system can be automated
10	Ruggedness	Durability of measurement, ability to handle adverse conditions
11	Portability	Ability to move instrument around
12	Greenness	Ecoefficiency in terms of waste generation and energy consumption
13	Cost	Equipment cost + cost of supplies + labor cost

1.3.1. Sensitivity

The *sensitivity* of a method (or an instrument) is a measure of its ability to distinguish between small differences in analyte concentrations at a desired confidence level. The simplest measure of sensitivity is the slope of the calibration curve in the concentration range of interest. This is referred to as the *calibration sensitivity*. Usually, calibration curves for instruments are linear and are given by an equation of the form

$$S = mc + s_{bl}$$
(1.11)

where S is the signal at concentration c and s_{bl} is the blank (i.e., signal in the absence of analyte). Then m is the slope of the calibration curve and hence the sensitivity. When sample preparation is involved, recovery of these steps has to be factored in. For example, during an extraction, only a fraction proportional to the extraction efficiency r is available for analysis. Then equation (1.11) reduces to

$$S = mrc + s_{tbl}$$
(1.12)

Now the sensitivity is mr rather than m. The higher the recovery, the higher the sensitivity. Near 100% recovery ensures maximum sensitivity. The blank is also modified by the sample preparation step; s_{tbl} refers to the blank that arises from total contribution from sample preparation and analysis.

Since the precision decreases at low concentrations, the ability to distinguish between small concentration differences also decreases. Therefore, sensitivity as a function of precision is measured by *analytical sensitivity*, which is expressed as [4]

$$a = \frac{mr}{s_s} \tag{1.13}$$

where s_s is the standard deviation based on sample preparation and analysis. Due to its dependence on s_s, analytical sensitivity varies with concentration.

1.3.2. Detection Limit

The *detection limit* is defined as the lowest concentration or weight of analyte that can be measured at a specific confidence level. So, near the detection limit, the signal generated approaches that from a blank. The detection limit is often defined as the concentration where the signal/noise ratio reaches an accepted value (typically, between 2 and 4). Therefore, the smallest distinguishable signal, S_m, is

$$S_m = X_{tbl} + k s_{tbl} \tag{1.14}$$

where, X_{tbl} and s_{tbl} are the average blank signal and its standard deviation. The constant k depends on the confidence level, and the accepted value is 3 at a confidence level of 89%. The detection limit can be determined experimentally by running several blank samples to establish the mean and standard deviation of the blank. Substitution of equation (1.12) into (1.14) and rearranging shows that

$$C_m = \frac{S_m - S_{tbl}}{m} \tag{1.15}$$

where C_m is the minimum detectable concentration and s_m is the signal obtained at that concentration. If the recovery in the sample preparation step is factored in, the detection limit is given as

$$C_m = \frac{S_m - S_{tbl}}{mr} \tag{1.16}$$

Once again, a low recovery increases the detection limit, and a sample preparation technique should aim at 100% recovery.

1.3.3. Range of Quantitation

The lowest concentration level at which a measurement is quantitatively meaningful is called the *limit of quantitation* (LOQ). The LOQ is most often defined as 10 times the signal/noise ratio. If the noise is approximated as the standard deviation of the blank, the LOQ is $(10 \times s_{tbl})$. Once again, when the recovery of the sample preparation step is factored in, the LOQ of the overall method increases by $1/r$.

For all practical purposes, the upper limit of quantitation is the point where the calibration curve becomes nonlinear. This point is called the *limit of linearity* (LOL). These can be seen from the calibration curve presented in Figure 1.3. Analytical methods are expected to have a *linear dynamic range* (LDR) of at least two orders of magnitude, although shorter ranges are also acceptable.

Considering all these, the recovery in sample preparation method is an important parameter that affects quantitative issues such as detection limit, sensitivity, LOQ, and even the LOL. Sample preparation techniques that enhance performance (see Chapters 6, 9, and 10) result in a recovery (r) larger that 1, thus increasing the sensitivity and lowering detection limits.

1.3.4. Other Important Parameters

There are several other factors that are important when it comes to the selection of equipment in a measurement process. These parameters are items 7 to 13 in Table 1.2. They may be more relevant in sample preparation than in analysis. As mentioned before, very often the bottleneck is the sample preparation rather than the analysis. The former tends to be slower; consequently, both measurement speed and sample throughput are determined by the discrete steps within the sample preparation. Modern analytical instruments tend to have a high degree of automation in terms of autoinjectors, autosamplers, and automated control/data acquisition. On the other hand, many sample preparation methods continue to be labor-intensive, requiring manual intervention. This prolongs analysis time and introduces random/systematic errors.

A variety of portable instruments have been developed in the last decade. Corresponding sample preparation, or online sample preparation methods, are being developed to make integrated total analytical systems. Many sample preparation methods, especially those requiring extraction, require solvents and other chemicals. Used reagents end up as toxic wastes, whose disposal is expensive. Greener sample preparation methods generate less spent reagent. Last but not the least, cost, including the cost of equipment, labor, and consumables and supplies, is an important factor.

1.3.5. Method Validation

Before a new analytical method or sample preparation technique is to be implemented, it must be validated. The various figures of merit need to be determined during the validation process. Random and systematic errors are measured in terms of precision and bias. The detection limit is established for each analyte. The accuracy and precision are determined at the concentration range where the method is to be used. The linear dynamic range is established and the calibration sensitivity is measured. In general, method validation provides a comprehensive picture of the merits of a new method and provides a basis for comparison with existing methods.

A typical validation process involves one or more of the following steps:

- *Determination of the single operator figures of merit.* Accuracy, precision, detection limits, linear dynamic range, and sensitivity are determined. Analysis is performed at different concentrations using standards.
- *Analysis of unknown samples.* This step involves the analysis of samples whose concentrations are unknown. Both qualitative and quantitative measurements should be performed. Reliable unknown samples are obtained from commercial sources or governmental agencies as certified reference materials. The accuracy and precision are determined.
- *Equivalency testing.* Once the method has been developed, it is compared to similar existing methods. Statistical tests are used to determine if the new and established methods give equivalent results. Typical tests include Student's *t*-test for a comparison of the means and the *F*-test for a comparison of variances.
- *Collaborative testing.* Once the method has been validated in one laboratory, it may be subjected to collaborative testing. Here, identical test samples and operating procedures are distributed to several laboratories. The results are analyzed statistically to determine bias and interlaboratory variability. This step determines the ruggedness of the method.

Method validation depends on the type and purpose of analysis. For example, the recommended validation procedure for PCR, followed by capillary gel electrophoresis of recombinant DNA, may consist of the following steps:

1. Compare precision by analyzing multiple (say, six) independent replicates of reference standards under identical conditions.
2. Data should be analyzed with a coefficient of variation less than a specified value (say, 10%).

3. Validation should be performed on three separate days to compare precision by analyzing three replicates of reference standards under identical conditions (once again the acceptance criteria should be a prespecified coefficient of variation).

4. To demonstrate that other analysts can perform the experiment with similar precision, two separate analysts should make three independent measurements (the acceptance criterion is once again a prespecified RSD).

5. The limit of detection, limit of quantitation, and linear dynamic range are to be determined by serial dilution of a sample. Three replicate measurements at each level are recommended, and the acceptance criterion for calibration linearity should be a prespecified correlation coefficient (say, an r^2 value of 0.995 or greater).

6. The molecular weight markers should fall within established migration time ranges for the analysis to be acceptable. If the markers are outside this range, the gel electrophoresis run must be repeated.

1.4. PRESERVATION OF SAMPLES

The sample must be representative of the object under investigation. Physical, chemical, and biological processes may be involved in changing the composition of a sample after it is collected. Physical processes that may degrade a sample are volatilization, diffusion, and adsorption on surfaces. Possible chemical changes include photochemical reactions, oxidation, and precipitation. Biological processes include biodegradation and enzymatic reactions. Once again, sample degradation becomes more of an issue at low analyte concentrations and in trace analysis.

The sample collected is exposed to conditions different from the original source. For example, analytes in a groundwater sample that have never been exposed to light can undergo significant photochemical reactions when exposed to sunlight. It is not possible to preserve the integrity of any sample indefinitely. Techniques should aim at preserving the sample at least until the analysis is completed. A practical approach is to run tests to see how long a sample can be held without degradation and then to complete the analysis within that time. Table 1.3 lists some typical preservation methods. These methods keep the sample stable and do not interfere in the analysis.

Common steps in sample preservation are the use of proper containers, temperature control, addition of preservatives, and the observance of recommended sample holding time. The holding time depends on the analyte of interest and the sample matrix. For example, most dissolved metals are

Table 1.3. Sample Preservation Techniques

Sample	Preservation Method	Container Type	Holding Time
pH	—	—	Immediately on site
Temperature	—	—	Immediately on site
Inorganic anions			
Bromide, chloride fluoride	None	Plastic or glass	28 days
Chlorine	None	Plastic or glass	Analyze immediately
Iodide	Cool to 4°C	Plastic or glass	24 hours
Nitrate, nitrite	Cool to 4°C	Plastic or glass	48 hours
Sulfide	Cool to 4°C, add zinc acetate and NaOH to pH 9	Plastic or glass	7 days
Metals			
Dissolved	Filter on site, acidify to pH 2 with HNO_2	Plastic	6 months
Total	Acidify to pH 2 with HNO_2	Plastic	6 month
Cr(VI)	Cool to 4°C	Plastic	24 hours
Hg	Acidify to pH 2 with HNO_2	Plastic	28 days
Organics			
Organic carbon	Cool to 4°C, add H_2SO_2 to pH 2	Plastic or brown glass	28 days
Purgeable hydrocarbons	Cool to 4°C, add 0.008% $Na_2S_2O_3$	Glass with Teflon septum cap	14 days
Purgeable aromatics	Cool to 4°C, add 0.008% $Na_2S_2O_3$ and HCl to pH 2	Glass with Teflon septum cap	14 days
PCBs	Cool to 4°C	Glass or Teflon	7 days to extraction, 40 days after
Organics in soil	Cool to 4°C	Glass or Teflon	As soon as possible
Fish tissues	Freeze	Aluminum foil	As soon as possible
Biochemical oxygen demand	Cool to 4°C	Plastic or glass	48 hours
Chemical oxygen demand	Cool to 4°C	Plastic or glass	28 days

(Continued)

Table 1.3. *(Continued)*

Sample	Preservation Method	Container Type	Holding Time
DNA	Store in TE (pH 8) under ethanol at $-20°C$; freeze at -20 or $-80°C$		Years
RNA	Deionized formamide at $-80°C$		Years
Solids unstable in air for surface and spectroscopic characterization	Store in argon-filled box; mix with hydrocarbon oil		

stable for months, whereas Cr(VI) is stable for only 24 hours. Holding time can be determined experimentally by making up a spiked sample (or storing an actual sample) and analyzing it at fixed intervals to determine when it begins to degrade.

1.4.1. Volatilization

Analytes with high vapor pressures, such as volatile organics and dissolved gases (e.g., HCN, SO_2) can easily be lost by evaporation. Filling sample containers to the brim so that they contain no empty space (headspace) is the most common method of minimizing volatilization. Solid samples can be topped with a liquid to eliminate headspace. The volatiles cannot equilibrate between the sample and the vapor phase (air) at the top of the container. The samples are often held at low temperature ($4°C$) to lower the vapor pressure. Agitation during sample handling should also be avoided. Freezing liquid samples causes phase separation and is not recommended.

1.4.2. Choice of Proper Containers

The surface of the sample container may interact with the analyte. The surfaces can provide catalysts (e.g., metals) for reactions or just sites for irreversible adsorption. For example, metals can adsorb irreversibly on glass surfaces, so plastic containers are chosen for holding water samples to be analyzed for their metal content. These samples are also acidified with HNO_3 to help keep the metal ions in solution. Organic molecules may also interact with polymeric container materials. Plasticizers such as phthalate esters can diffuse from the plastic into the sample, and the plastic can serve as a sorbent (or a membrane) for the organic molecules. Consequently, glass containers are suitable for organic analytes. Bottle caps should have Teflon liners to preclude contamination from the plastic caps.

Oily materials may adsorb strongly on plastic surfaces, and such samples are usually collected in glass bottles. Oil that remains on the bottle walls should be removed by rinsing with a solvent and be returned to the sample. A sonic probe can be used to emulsify oily samples to form a uniform suspension before removal for analysis.

1.4.3. Absorption of Gases from the Atmosphere

Gases from the atmosphere can be absorbed by the sample during handling, for example, when liquids are being poured into containers. Gases such as O_2, CO_2, and volatile organics may dissolve in the samples. Oxygen may oxidize species, such as sulfite or sulfide to sulfate. Absorption of CO_2 may change conductance or pH. This is why pH measurements are always made at the site. CO_2 can also bring about precipitation of some metals. Dissolution of organics may lead to false positives for compounds that were actually absent. Blanks are used to check for contamination during sampling, transport, and laboratory handling.

1.4.4. Chemical Changes

A wide range of chemical changes are possible. For inorganic samples, controlling the pH can be useful in preventing chemical reactions. For example, metal ions may oxidize to form insoluble oxides or hydroxides. The sample is often acidified with HNO_3 to a pH below 2, as most nitrates are soluble, and excess nitrate prevents precipitation. Other ions, such as sulfides and cyanides, are also preserved by pH control. Samples collected for NH_3 analysis are acidified with sulfuric acid to stabilize the NH_3 as NH_4SO_4.

Organic species can also undergo changes due to chemical reactions. Storing the sample in amber bottles prevents photooxidation of organics (e.g., polynuclear aromatic hydrocarbons). Organics can also react with dissolved gases; for example, organics can react with trace chlorine to form halogenated compounds in treated drinking water samples. In this case, the addition of sodium thiosulfate can remove the chlorine.

Samples may also contain microorganisms, which may degrade the sample biologically. Extreme pH (high or low) and low temperature can minimize microbial degradation. Adding biocides such as mercuric chloride or pentachlorophenol can also kill the microbes.

1.4.5. Preservation of Unstable Solids

Many samples are unstable in air. Examples of air-sensitive compounds are alkali metal intercalated C_{60}, carbon nanotubes, and graphite, which are usually prepared in vacuum-sealed tubes. After completion of the intercalation reaction in a furnace, the sealed tubes may be transferred directly to a Raman spectrometer for measurement. Since these compounds are photosensitive, spectra need to be measured using relatively low laser power densities. For x-ray diffraction, infrared, and x-ray photoelectron spectroscopy (XPS), the sealed tubes are transferred to an argon-filled dry box with less than 10 parts per million (ppm) of oxygen. The vacuum tubes are cut open in the dry box and transferred to x-ray sampling capillaries. The open ends of the capillaries are carefully sealed with soft wax to prevent air contamination after removal from the dry box. Samples for infrared spectroscopy are prepared by mixing the solid with hydrocarbon oil and sandwiching a small amount of this suspension between two KBr or NaCl plates. The edges of the plates are then sealed with soft wax. For the XPS measurements, the powder is spread on a tape attached to the sample holder and inserted into a transfer tube of the XPS spectrometer, which had previously been introduced into the dry box. Transfer of unstable compounds into the sampling chamber of transmission and scanning electron microscopes are difficult. The best approaches involve preparing the samples in situ for examination.

1.5. POSTEXTRACTION PROCEDURES

1.5.1. Concentration of Sample Extracts

The analytes are often diluted in the presence of a large volume of solvents used in the extraction. This is particularly true when the analysis is being done at the trace level. An additional concentration step is necessary to increase the concentration in the extract. If the amount of solvent to be removed is not very large and the analyte is nonvolatile, the solvent can be vaporized by a gentle stream of nitrogen gas flowing either across the surface or through the solution. This is shown in Figure 1.6. Care should be taken that the solvent is lost only by evaporation. If small solution droplets are lost as aerosol, there is the possibility of losing analytes along with it. If large volume reduction is needed, this method is not efficient, and a rotary vacuum evaporator is used instead. In this case, the sample is placed in a round-bottomed flask in a heated water bath. A water-cooled condenser is attached at the top, and the flask is rotated continually to expose maximum liquid surface to evaporation. Using a small pump or a water aspirator, the pressure inside the flask is reduced. The mild warming, along with the lowered pressure, removes the solvent efficiently, and the condensed solvent distills into a separate flask. Evaporation should stop before the sample reaches dryness.

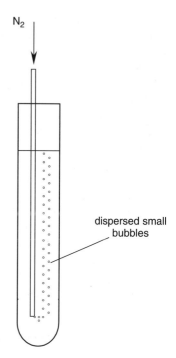

Figure 1.6. Evaporation of solvent by nitrogen.

For smaller volumes that must be reduced to less than 1 mL, a Kuderna–Danish concentrator (Figure 1.7) is used. The sample is gently heated in a water bath until the needed volume is reached. An air-cooled condenser provides reflux. The volume of the sample can readily be measured in the narrow tube at the bottom.

1.5.2. Sample Cleanup

Sample cleanup is particularly important for analytical separations such as GC, HPLC, and electrophoresis. Many solid matrices, such as soil, can contain hundreds of compounds. These produce complex chromatograms, where the identification of analytes of interest becomes difficult. This is especially true if the analyte is present at a much lower concentration than the interfering species. So a cleanup step is necessary prior to the analytical measurements. Another important issue is the removal of high-boiling materials that can cause a variety of problems. These include analyte adsorption in the injection port or in front of a GC-HPLC column, false positives from interferences that fall within the retention window of the analyte, and false negatives because of a shift in the retention time window.

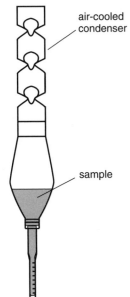

Figure 1.7. Kuderna–Danish sample concentrator.

In extreme cases, instrument shut down may be necessary due to the accumulation of interfacing species.

Complex matrices such as, soil, biological materials, and natural products often require some degree of cleanup. Highly contaminated extracts (e.g., soil containing oil residuals) may require multiple cleanup steps. On the other hand, drinking water samples are relatively cleaner (as many large molecules either precipitate out or do not dissolve in it) and may not require cleanup [5].

The following techniques are used for cleanup and purification of extracts.

Gel-Permeation Chromatography

Gel-permeation chromatography (GPC) is a size-exclusion method that uses organic solvents (or buffers) and porous gels for the separation of macromolecules. The packing gel is characterized by pore size and exclusion range, which must be larger than the analytes of interest. GPC is recommended for the elimination of lipids, proteins, polymers, copolymers, natural resins, cellular components, viruses, steroids, and dispersed high-molecular-weight compounds from the sample. This method is appropriate for both polar and nonpolar analytes. Therefore, it is used for extracts containing a broad range of analytes. Usually, GPC is most efficient for removing high-boiling materials that condense in the injection port of a GC or the front of the GC column [6]. The use of GPC in nucleic acid isolation is discussed in Chapter 8.

Acid–Base Partition Cleanup

Acid–base partition cleanup is a liquid–liquid extraction procedure for the separation of acid analytes, such as organic acids and phenols from base/ neutral analytes (amines, aromatic hydrocarbons, halogenated organic compounds) using pH adjustment. This method is used for the cleanup of petroleum waste prior to analysis or further cleanup. The extract from the prior solvent extraction is shaken with water that is strongly basic. The basic and neutral components stay in the organic solvent, whereas the acid analytes partition into the aqueous phase. The organic phase is concentrated and is ready for further cleanup or analysis. The aqueous phase is acidified and extracted with an organic solvent, which is then concentrated (if needed) and is ready for analysis of the acid analytes (Figure 1.8).

Solid-Phase Extraction and Column Chromatography

The solvent extracts can be cleaned up by traditional column chromatography or by solid-phase extraction cartridges. This is a common cleanup method that is widely used in biological, clinical, and environmental sample preparation. More details are presented in Chapter 2. Some examples include the cleanup of pesticide residues and chlorinated hydrocarbons, the separation of nitrogen compounds from hydrocarbons, the separation of aromatic compounds from an aliphatic–aromatic mixture, and similar applications for use with fats, oils, and waxes. This approach provides efficient cleanup of steroids, esters, ketones, glycerides, alkaloids, and carbohydrates as well. Cations, anions, metals, and inorganic compounds are also candidates for this method [7].

The column is packed with the required amount of a sorbent and loaded with the sample extract. Elution of the analytes is effected with a suitable solvent, leaving the interfering compounds on the column. The packing material may be an inorganic substance such as Florisil (basic magnesium silicate) or one of many commercially available SPE stationary phases. The eluate may be further concentrated if necessary. A Florisil column is shown in Figure 1.9. Anhydrous sodium sulfate is used to dry the sample [8].

These cleanup and concentration techniques may be used individually, or in various combinations, depending on the nature of the extract and the analytical method used.

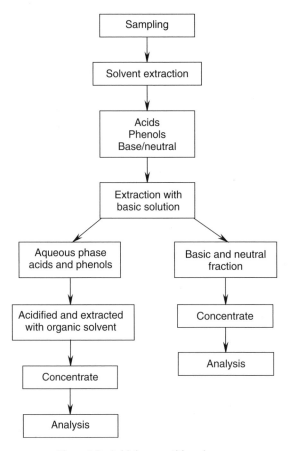

Figure 1.8. Acid–base partition cleanup.

1.6. QUALITY ASSURANCE AND QUALITY CONTROL DURING SAMPLE PREPARATION

As mentioned earlier, the complete analytical process involves sampling, sample preservation, sample preparation, and finally, analysis. The purpose of quality assurance (QA) and quality control (QC) is to monitor, measure, and keep the systematic and random errors under control. QA/QC measures are necessary during sampling, sample preparation, and analysis. It has been stated that sample preparation is usually the major source of variability in a measurement process. Consequently, the QA/QC during this step is of utmost importance. The discussion here centers on QC during sample preparation.

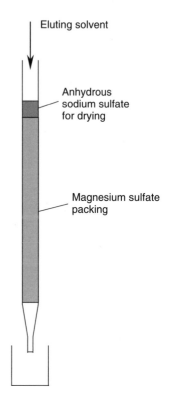

Figure 1.9. Column chromatography for sample cleanup.

Quality assurance refers to activities that demonstrate that a certain quality standard is being met. This includes the management process that implements and documents effective QC. *Quality control* refers to procedures that lead to statistical control of the different steps in the measurement process. So QC includes specific activities such as analyzing replicates, ensuring adequate extraction efficiency, and contamination control.

Some basic components of a QC system are shown in Figure 1.10. Competent personnel and adequate facilities are the most basic QC requirements. Many modern analytical/sample preparation techniques use sophisticated instruments that require specialized training. *Good laboratory practice* (GLP) refers to the practices and procedures involved in running a laboratory. Efficient sample handling and management, record keeping, and equipment maintenance fall under this category. *Good measurement practices* (GMPs) refer to the specific techniques in sample preparation and analysis. On the other hand, GLPs are independent of the specific techniques and refer to general practices in the laboratory. An important QC step is to have formally documented GLPs and GMPs that are followed carefully.

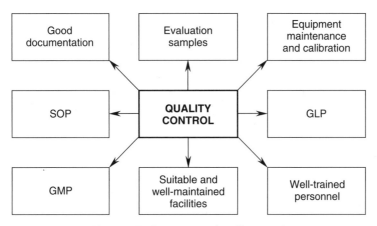

Figure 1.10. Components of quality control.

Standard operating procedures (SOPs) are written descriptions of procedures of methods being followed. The importance of SOPs cannot be understated when it comes to methods being transferred to other operators or laboratories. Strict adherence to the SOPs reduces bias and improves precision. This is particularly true in sample preparation, which tends to consist of repetitive processes that can be carried out by more than one procedure. For example, extraction efficiency depends on solvent composition, extraction time, temperature, and even the rate of agitation. All these parameters need to be controlled to reduce variability in measurement. Changing the extraction time will change the extraction efficiency, which will increase the relative standard deviation (lower precision). The SOP specifies these parameters. They can come in the form of published standard methods obtained from the literature, or they may be developed in-house. Major sources of SOPs are protocols obtained from organizations, such as the American Society for Testing and Materials and the U.S. Environmental Protection Agency (EPA).

Finally, there is the need for proper documentation, which can be in written or electronic forms. These should cover every step of the measurement process. The sample information (source, batch number, date), sample preparation/analytical methodology (measurements at every step of the process, volumes involved, readings of temperature, etc.), calibration curves, instrument outputs, and data analysis (quantitative calculations, statistical analysis) should all be recorded. Additional QC procedures, such as blanks, matrix recovery, and control charts, also need to be a part of the record keeping. Good documentation is vital to prove the validity of data. Analyt-

Table 1.4. Procedures in Quality Control

QC Parameters	Procedure
Accuracy	Analysis of reference materials or known standards
Precision	Analysis of replicate samples
Extraction efficiency	Analysis of matrix spikes
Contamination	Analysis of blanks

ical data that need to be submitted to regulatory agencies also require detailed documentation of the various QC steps.

The major quality parameters to be addressed during sample preparation are listed in Table 1.4. These are accuracy, precision, extraction efficiency (or recovery), and contamination control. These quality issues also need to be addressed during the analysis that follows sample preparation. Accuracy is determined by the analysis of evaluation samples. Samples of known concentrations are analyzed to demonstrate that quantitative results are close to the true value. The precision is measured by running replicates. When many samples are to be analyzed, the precision needs to be checked periodically to ensure the stability of the process. Contamination is a serious issue, especially in trace measurements such as environmental analysis. The running of various blanks ensures that contamination has not occurred at any step, or that if it has, where it occurred. As mentioned before, the detection limits, sensitivity, and other important parameters depend on the recovery. The efficiency of sample preparation steps such as extraction and cleanup must be checked to ensure that the analytes are being recovered from the sample.

1.6.1. Determination of Accuracy and Precision

The levels of accuracy and precision determine the quality of a measurement. The data are as good as random numbers if these parameters are not specified. Accuracy is determined by analyzing samples of known concentration (evaluation samples) and comparing the measured values to the known. Standard reference materials are available from regulatory agencies and commercial vendors. A standard of known concentration may also be made up in the laboratory to serve as an evaluation sample.

Effective use of evaluation samples depends on matching the standards with the real-world samples, especially in terms of their matrix. Take the example of extraction of pesticides from fish liver. In a real sample, the pesticide is embedded in the liver cells (intracellular matter). If the calibration standards are made by spiking livers, it is possible that the pesticides will be absorbed on the outside of the cells (extracellular). The extraction of extracellular pesticides is easier than real-world intracellular extractions. Consequently, the extraction efficiency of the spiked sample may be significantly higher. Using this as the calibration standard may result in a negative bias. So matrix effects and matrix matching are important for obtaining high accuracy. Extraction procedures that are powerful enough not to have any matrix dependency are desirable.

Precision is measured by making replicate measurements. As mentioned before, it is known to be a function of concentration and should be determined at the concentration level of interest. The *intrasample variance* can be determined by splitting a sample into several subsamples and carrying out the sample preparation/analysis under identical conditions to obtain a measure of RSD. For example, several aliquots of homogenized fish liver can be processed through the same extraction and analytical procedure, and the RSD computed. The *intersample variance* can be measured by analyzing several samples from the same source. For example, different fish from the same pond can be analyzed to estimate the intersample RSD.

The precision of the overall process is often determined by the extraction step rather than the analytical step. It is easier to get high-precision analytical results; it is much more difficult to get reproducible extractions. For example, it is possible to run replicate chromatographic runs (GC or HPLC) with an RSD between 1 and 3%. However, several EPA-approved methods accept extraction efficiencies anywhere between 70 and 120%. This range alone represents variability as high as 75%. Consequently, in complex analytical methods that involve several preparative steps, the major contributor to variability is the sample preparation.

1.6.2. Statistical Control

Statistical evidence that the precision of the measurement process is within a certain specified limit is referred to as *statistical control*. Statistical control does not take the accuracy into account. However, the precision of the measurement should be established and statistical control achieved before accuracy can be estimated.

Control Charts

Control charts are used for monitoring the variability and to provide a graphical display of statistical control. A *standard*, a reference material of known concentration, is analyzed at specified intervals (e.g., every 50 samples). The result should fall within a specified limit, as these are replicates. The only variation should be from random error. These results are plotted on a control chart to ensure that the random error is not increasing or that a

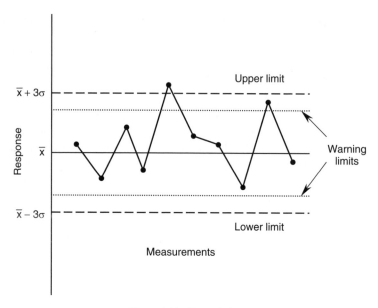

Figure 1.11. Control chart.

systematic bias is not taking place. In the control chart shown in Figure 1.11, replicate measurements are plotted as a function of time. The centerline is the average, or expected value. The upper (UCL) and lower (LCL) control limits are the values within which the measurements must fall. Normally, the control limits are $\pm 3\sigma$, within which 99.7% of the data should lie. For example, in a laboratory carrying out microwave extraction on a daily basis, a standard reference material is extracted after a fixed number of samples. The measured value is plotted on the control chart. If it falls outside the control limit, readjustments are necessary to ensure that the process stays under control.

Control charts are used in many different applications besides analytical measurements. For example, in a manufacturing process, the control limits are often based on product quality. In analytical measurements, the control limits can be established based on the analyst's judgment and the experimental results. A common approach is to use the mean of select measurements as the centerline, and then a multiple of the standard deviation is used to set the control limits. Control charts often plot regularly scheduled analysis of a standard reference material or an audit sample. These are then tracked to see if there is a trend or a systematic deviation from the centerline.

Control Samples

Different types of control samples are necessary to determine whether a measurement process is under statistical control. Some of the commonly used control standards are listed here.

1. *Laboratory control standards* (LCSs) are certified standards obtained from an outside agency or commercial source to check whether the data being generated are comparable to those obtained elsewhere. The LCSs provide a measure of the accuracy and can be used as audits. A source of LCSs is *standard reference materials* (SRMs), which are certified standards available from the National Institute of Standards and Testing (NIST) in the United States. NIST provides a variety of solid, liquid, and gaseous SRMs which have been prepared to be stable and homogeneous. They are analyzed by more than one independent methods, and their concentrations are certified. Certified standards are also available from the European Union Community Bureau of Reference (BCR), government agencies such as the EPA, and from various companies that sell standards. These can be quite expensive. Often, samples are prepared in the laboratory, compared to the certified standards, and then used as secondary reference materials for daily use.

2. *Calibration control standards* (CCSs) are used to check calibration. The CCS is the first sample analyzed after calibration. Its concentration may or may not be known, but it is used for successive comparisons. A CCS may be analyzed periodically or after a specified number of samples (say, 20). The CCS value can be plotted on a control chart to monitor statistical control.

1.6.3. Matrix Control

Matrix Spike

Matrix effects play an important role in the accuracy and precision of a measurement. Sample preparation steps are often sensitive to the matrix. Matrix spikes are used to determine their effect on sample preparation and analysis. Matrix spiking is done by adding a known quantity of a component that is similar to the analyte but not present in the sample originally. The sample is then analyzed for the presence of the spiked material to evaluate the matrix effects. It is important to be certain that the extraction recovers most of the analytes, and spike recovery is usually required to be at least 70%. The matrix spike can be used to accept or reject a method.

For example, in the analysis of chlorophenol in soil by accelerated solvent extraction followed by GC-MS, deuterated benzene may be used as the matrix spike. The deuterated compound will not be present in the original sample and can easily be identified by GC-MS. At the same time, it has chemical and physical properties that closely match those of the analyte of interest.

Often, the matrix spike cannot be carried out at the same time as the analysis. The spiking is carried out separately on either the same matrix or on one that resembles the samples. In the example above, clean soil can be spiked with regular chlorophenol and then the recovery is measured. However, one should be careful in choosing the matrix to be spiked. For instance, it is easy to extract different analytes from sand, but not so if the analytes have been sitting in clay soil for many years. The organics in the soil may provide additional binding for the analytes. Consequently, a matrix spike may be extracted more easily than the analytes in real-world samples. The extraction spike may produce quantitative recovery, whereas the extraction efficiency for real samples may be significantly lower. This is especially true for matrix-sensitive techniques, such as supercritical extraction.

Surrogate Spike

Surrogate spikes are used in organic analysis to determine if an analysis has gone wrong. They are compounds that are similar in chemical composition and have similar behavior during sample preparation and analysis. For example, a deuterated analog of the analyte is an ideal surrogate during GC-MS analysis. It behaves like the analyte and will not be present in the sample originally. The surrogate spike is added to the samples, the standards, the blanks, and the matrix spike. The surrogate recovery is computed for each run. Unusually high or low recovery indicates a problem, such as contamination or instrument malfunction. For example, consider a set of samples to be analyzed for gasoline contamination by purge and trap. Deuterated toluene is added as a surrogate to all the samples, standards, and blanks. The recovery of the deuterated toluene in each is checked. If the recovery in a certain situation is unusually high or low, that particular analysis is rejected.

1.6.4. Contamination Control

Many measurement processes are prone to contamination, which can occur at any point in the sampling, sample preparation, or analysis. It can occur in the field during sample collection, during transportation, during storage, in the sample workup prior to measurement, or in the instrument itself. Some

Table 1.5. Sources of Sample Contamination

Measurement Step	Sources of Contamination
Sample collection	Equipment Sample handling and preservation Sample containers
Sample transport and storage	Sample containers Cross-contamination from other samples
Sample preparation	Sample handling, carryover in instruments Dilutions, homogenization, size reduction Glassware and instrument Ambient contamination
Sample analysis	Carryover in instrument Instrument memory effects Reagents Syringes

common sources of contamination are listed in Table 1.5. Contamination becomes a major issue in trace analysis. The lower the concentration, the more pronounced is the effect of contamination.

Sampling devices themselves can be a source of contamination. Contamination may come from the material of construction or from improper cleaning. For example, polymer additives can leach out of plastic sample bottles, and organic solvents can dissolve materials from surfaces, such as cap liners of sample vials. Carryover from previous samples is also possible. Say that a sampling device was used where the analyte concentration was at the 1 ppm level. A 0.1% carryover represents a 100% error if the concentration of the next sample is at 1 part per billion (ppb).

Contamination can occur in the laboratory at any stage of sample preparation and analysis. It can come from containers and reagents or from the ambient environment itself. In general, contamination can be reduced by avoiding manual sample handling and by reducing the number of discrete processing steps. Sample preparations that involve many unautomated manual steps are prone to contamination. Contaminating sources can also be present in the instrument. For instance, the leftover compounds from a previous analysis can contaminate incoming samples.

Blanks

Blanks are used to assess the degree of contamination in any step of the measurement process. They may also be used to correct relatively constant,

Table 1.6. Types of Blanks

Blank Type	Purpose	Process
System or instrument blank	Establishes the baseline of an instrument in the absence of sample	Determine the background signal with no sample present
Solvent or calibration blank	To measure the amount of the analytical signal that arises from the solvents and reagents; the zero solution in the calibration series	Analytical instrument is run with solvents/reagents only
Method blank	To detect contamination from reagents, sample handling, and the entire measurement process	A blank is taken through the entire measurement procedure
Matched-matrix blank	To detect contamination from field handling, transportation, or storage	A synthetic sample that matches the basic matrix of the sample is carried to the field and is treated in the same fashion as the sample
Sampling media	To detect contamination in the sampling media such as filters and sample adsorbent traps	Analyze samples of unused filters or traps to detect contaminated batches
Equipment blank	To determine contamination of equipment and assess the efficiency or equipment cleanup procedures	Samples of final equipment cleaning rinses are analyzed for contaminants

unavoidable contamination. Blanks are samples that do not contain any (or a negligible amount of) analyte. They are made to simulate the sample matrix as closely as possible. Different types of blanks are used, depending on the procedure and the measurement objectives. Some common blanks are listed in Table 1.6. Blank samples from the laboratory and the field are required to cover all the possible sources of contamination. We focus here on those blanks that are important from a sample preparation perspective.

System or Instrument Blank. It is a measure of system contamination and is the instrumental response in the absence of any sample. When the background signal is constant and measurable, the usual practice is to consider that level to be the zero setting. It is generally used for analytical instruments but is also applicable for instruments for sample preparation.

The instrument blank also identifies memory effects or carryover from previous samples. It may become significant when a low-concentration sample is analyzed immediately after a high-concentration sample. This is especially true where preconcentration and cryogenic steps are involved. For example, during the purge and trap analysis of volatile organics, some components may be left behind in the sorbent trap or at a cold spot in the instrument. So it is a common practice to run a deionized water blank between samples. These blanks are critical in any instrument, where sample components may be left behind only to emerge during the next analysis.

Solvent/Reagent Blank. A solvent blank checks solvents and reagents that are used during sample preparation and analysis. Sometimes, a blank correction or zero setting is done based on the reagent measurement. For example, in atomic or molecular spectroscopy, the solvents and reagents used in sample preparation are used to provide the zero setting.

Method Blank. A method blank is carried through all the steps of sample preparation and analysis as if it were an actual sample. This is most important from the sample preparation prospective. The same solvents/reagents that are used with the actual samples are used here. For example, in the analysis of metals in soil, a clean soil sample may serve as a method blank. It is put through the extraction, concentration, and analysis steps encountered by the real samples. The method blank accounts for contamination that may occur during sample preparation and analysis. These could arise from the reagents, the glassware, or the laboratory environment.

Other types of blanks may be employed as the situation demands. It should be noted that blanks are effective only in identifying contamination. They do not account for various errors that might exist. Blanks are seldom used to correct for contamination. More often, a blank above a predetermined value is used to reject analytical data, making reanalysis and even resampling necessary. The laboratory SOPs should identify the blanks necessary for contamination control.

REFERENCES

1. D. Scoog, D. West, and J. Holler, *Fundamentals of Analytical Chemistry*, Saunders College Publishing, Philadelphia, 1992.
2. W. Horwitz, L. Kamps, and K. Boyer, *J. Assoc. Off. Anal. Chem.*, **63**, 1344–1354 (1980).
3. V. Meyer, *LC-GC North Am.*, **20**, 106–112, 2 (2002).

4. B. Kebbekus and S. Mitra, *Environmental Chemical Analysis*, Chapman & Hall, New York, 1998.

5. *Test Methods: Methods for Organic Chemical Analysis of Municipal and Industrial Wastewater*, U.S. EPA-600/4-82-057.

6. U.S. EPA method 3640A, *Gel-Permeation Cleanup*, 1994, pp. 1–15.

7. V. Lopez-Avila, J. Milanes, N. Dodhiwala, and W. Beckert, *J. Chromatogr. Sci.*, **27**, 109–215 (1989).

8. P. Mills, *J. Assoc. Off. Anal. Chem.*, **51**, 29 (1968).

CHAPTER

2

PRINCIPLES OF EXTRACTION AND THE EXTRACTION OF SEMIVOLATILE ORGANICS FROM LIQUIDS

MARTHA J. M. WELLS

Center for the Management, Utilization and Protection of Water Resources and Department of Chemistry, Tennessee Technological University, Cookeville, Tennessee

2.1. PRINCIPLES OF EXTRACTION

This chapter focuses on three widely used techniques for extraction of semi-volatile organics from liquids: liquid–liquid extraction (LLE), solid-phase extraction (SPE), and solid-phase microextraction (SPME). Other techniques may be useful in selected circumstances, but these three techniques have become the extraction methods of choice for research and commercial analytical laboratories. A fourth, recently introduced technique, stir bar sorptive extraction (SBSE), is also discussed.

To understand any extraction technique it is first necessary to discuss some underlying principles that govern all extraction procedures. The chemical properties of the analyte are important to an extraction, as are the properties of the liquid medium in which it is dissolved and the gaseous, liquid, supercritical fluid, or solid extractant used to effect a separation. Of all the relevant solute properties, five chemical properties are fundamental to understanding extraction theory: *vapor pressure, solubility, molecular weight, hydrophobicity*, and *acid dissociation*. These essential properties determine the transport of chemicals in the human body, the transport of chemicals in the air–water–soil environmental compartments, and the transport between immiscible phases during analytical extraction.

Extraction or separation of dissolved chemical component X from liquid phase A is accomplished by bringing the liquid solution of X into contact with a second phase, B, given that phases A and B are immiscible. Phase B may be a solid, liquid, gas, or supercritical fluid. A distribution of the com-

Sample Preparation Techniques in Analytical Chemistry, Edited by Somenath Mitra
ISBN 0-471-32845-6   Copyright © 2003 John Wiley & Sons, Inc.

ponent between the immiscible phases occurs. After the analyte is distributed between the two phases, the extracted analyte is released and/or recovered from phase B for subsequent extraction procedures or for instrumental analysis.

The theory of chemical equilibrium leads us to describe the reversible distribution reaction as

$$X_A \rightleftharpoons X_B \qquad (2.1)$$

and the equilibrium constant expression, referred to as the *Nernst distribution law* [1], is

$$K_D = \frac{[X]_B}{[X]_A} \qquad (2.2)$$

where the brackets denote the concentration of X in each phase at constant temperature (or the activity of X for nonideal solutions). By convention, the concentration extracted into phase B appears in the numerator of equation (2.2). The equilibrium constant is independent of the rate at which it is achieved.

The analyst's function is to optimize extracting conditions so that the distribution of solute between phases lies far to the right in equation (2.1) and the resulting value of K_D is large, indicating a high degree of extraction from phase A into phase B. Conversely, if K_D is small, less chemical X is transferred from phase A into phase B. If K_D is equal to 1, equivalent concentrations exist in each phase.

2.1.1. Volatilization

Volatilization of a chemical from the surface of a liquid is a partitioning process by which the chemical distributes itself between the liquid phase and the gas above it. Organic chemicals said to be *volatile* exhibit the greatest tendency to cross the liquid–gas interface. When compounds volatilize, the concentration of the organic analyte in the solution is reduced. *Semivolatile* and *nonvolatile* (or *involatile*) describe chemicals having, respectively, less of a tendency to escape the liquid they are dissolved in and pass into the atmosphere above the liquid.

As discussed in this book, certain sample preparation techniques are clearly more appropriate for volatile compounds than for semivolatile and nonvolatile compounds. In this chapter we concentrate on extraction methods for semivolatile organics from liquids. Techniques for extraction of volatile organics from solids and liquids are discussed in Chapter 4.

Henry's Law Constant

If the particular extracting technique applied to a solution depends on the volatility of the solute between air and water, a parameter to predict this behavior is needed to avoid trial and error in the laboratory. The volatilization or escaping tendency (fugacity) of solute chemical X can be estimated by determining the gaseous, G, to liquid, L, distribution ratio, K_D, also called the *nondimensional,* or *dimensionless, Henry's law constant, H'*.

$$H' = K_D = \frac{[X]_G}{[X]_L} \qquad (2.3)$$

The larger the magnitude of the Henry's law constant, the greater the tendency for volatilization from the liquid solvent into the gaseous phase [2–4].

According to equation (2.3), the Henry's law constant can be estimated by measuring the concentration of X in the gaseous phase and in the liquid phase at equilibrium. In practice, however, the concentration is more often measured in one phase while concentration in the second phase is determined by mass balance. For dilute neutral compounds, the Henry's law constant can be estimated from the ratio of vapor pressure, P_{vp}, and solubility, S, taking the molecular weight into consideration by expressing the molar concentration:

$$H = \frac{P_{vp}}{S} \qquad (2.4)$$

where P_{vp} is in atm and S is in mol/m^3, so H is in atm·m^3/mol.

Vapor Pressure

The vapor pressure, P_{vp}, of a liquid or solid is the pressure of the compound's vapor (gas) in equilibrium with the pure, condensed liquid or solid phase of the compound at a given temperature [5–9]. Vapor pressure, which is temperature dependent, increases with temperature. The vapor pressure of chemicals varies widely according to the degree of intermolecular attractions between like molecules: The stronger the intermolecular attraction, the lower the magnitude of the vapor pressure. Vapor pressure and the Henry's law constant should not be confused. *Vapor pressure* refers to the volatility from the pure substance into the atmosphere; the *Henry's law constant* refers to the volatility of the compound from liquid solution into the air. Vapor pressure is used to estimate the Henry's law constant [equation (2.4)].

Solubility

Solubility is also used to estimate the Henry's law constant [equation (2.4)]. *Solubility* is the maximum amount of a chemical that can be dissolved into another at a given temperature. Solubility can be determined experimentally or estimated from molecular structure [6,10–12].

The Henry's law constant, H, calculated from the ratio of vapor pressure and solubility [equation (2.4)] can be converted to the dimensionless Henry's law constant, H', [equation (2.3)] by the expression

$$H' = \frac{P_{vp}(MW)}{0.062ST} \tag{2.5}$$

where P_{vp} is the vapor pressure in mmHg, MW the molecular weight, S the water solubility in mg/L, T the temperature in Kelvin, and 0.062 is the appropriate universal gas constant [9].

For the analyst's purposes, it is usually sufficient to categorize the escaping tendency of the organic compound from a liquid to a gas as high, medium, or low. According to Henry's law expressed as equation (2.4), estimating the volatilization tendency requires consideration of both the vapor pressure and the solubility of the organic solute. Ney [13] ranks vapor pressures as

- *Low:* 1×10^{-6} mmHg
- *Medium:* between 1×10^{-6} and 1×10^{-2} mmHg
- *High:* greater than 1×10^{-2} mmHg

while ranking water solubilities as

- *Low:* less than 10 ppm
- *Medium:* between 10 and 1000 ppm
- *High:* greater than 1000 ppm

However, note that in Ney's approach, concentration expressed in parts per million (ppm) does not incorporate molecular weight. Therefore, it does not consider the identity or molecular character of the chemical.

Rearranging equation (2.4) produces

$$P_{vp} = HS \tag{2.6}$$

In this linear form, a plot (Figure 2.1) of vapor pressure (y-axis) versus solubility (x-axis) yields a slope representing the Henry's law constant at values

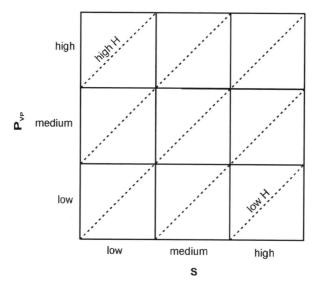

Figure 2.1. Henry's law constant at values of constant H conceptually represented by diagonal (dotted) lines on a plot of vapor pressure (P_{vp}) versus solubility, S.

of constant H. From this figure it can be deduced that low volatility from liquid solution is observed for organic chemicals with low vapor pressure and high solubility, whereas high volatility from liquid solution is exhibited by compounds with high vapor pressure and low solubility. Intermediate levels of volatility result from all other vapor pressure and solubility combinations. H is a ratio, so it is possible for compounds with low vapor pressure and low solubility, medium vapor pressure and medium solubility, or high vapor pressure and high solubility to exhibit nearly equivalent volatility from liquid solution.

The Henry's law constant can be used to determine which extraction techniques are appropriate according to solute volatility from solution. If the Henry's law constant of the analyte (solute) is less than the Henry's law constant of the solvent, the solute is nonvolatile in the solvent and the solute concentration will increase as the solvent evaporates. If the Henry's law constant of the analyte (solute) is greater than the Henry's law constant of the solvent, the solute is semivolatile to volatile in the solvent. In a solution open to the atmosphere, the solute concentration will decrease because the solute will evaporate more rapidly than the solvent.

Mackay and Yuen [2] and Thomas [4] provide these guidelines for organic solutes in water (Figure 2.2):

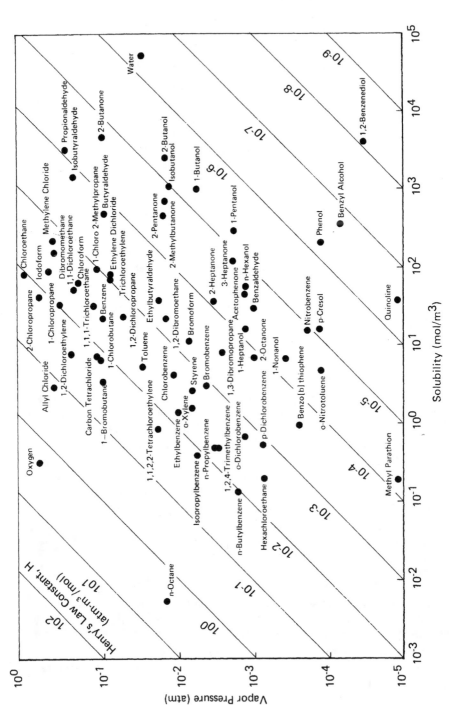

Figure 2.2. Solubility, vapor pressure, and Henry's law constant for selected chemicals [2,4]. (Reprinted with permission from Ref. 2. Copyright © 1980 Elsevier Science.)

42

- *Nonvolatile:* volatilization is unimportant for $H < 3 \times 10^{-7}$ atm·m^3/mol (i.e., H for water itself at 15°C)
- *Semivolatile:* volatilizes slowly for $3 \times 10^{-7} < H < 10^{-5}$ atm·m^3/mol
- *Volatile:* volatilization is significant in the range $10^{-5} < H < 10^{-3}$ atm·m^3/mol
- *Highly volatile:* volatilization is rapid if $H > 10^{-3}$ atm·m^3/mol

Schwarzenbach et al. [8] illustrate the Henry's law constant (Figure 2.3c) for selected families of hydrocarbons in relation to vapor pressure (Figure 2.3a) and solubility (Figure 2.3b). Vapor pressure (Figure 2.3a) and solubility (Figure 2.3b) tend to decrease with increasing molecular size. In Figure 2.3c, the Henry's law constant is expressed in units of atm·L/mol, whereas the Henry's law constant in Figure 2.2 is expressed in units of atm·m^3/mol. Applying Mackay and Yuen's, and Thomas's volatility guidelines to the units in Figure 2.3c, the Henry's law constant for semivolatile compounds in water lies between $3 \times 10^{-4} < H < 10^{-2}$ atm·L/mol (since 1 L = 0.001 m^3). Highly volatile compounds lie above a Henry's law constant of 1 atm·L/mol. For example, Figure 2.3c illustrates that a high-molecular-weight polycyclic aromatic hydrocarbon (PAH) such as benzo[a]pyrene ($C_{20}H_{12}$) is semivolatile in its tendency to escape from water according to the Henry's law constant, whereas a low-molecular-weight PAH, naphthalene ($C_{10}H_8$), is volatile.

The most common gas–liquid pair encountered in analytical extractions is the air–water interface. The extraction methods discussed in this chapter are most applicable to organic solutes that are considered nonvolatile and semivolatile. However, it is possible to extend these techniques to more volatile chemicals as long as careful consideration of the tendency of the solute to volatilize is made throughout the extraction process.

2.1.2. Hydrophobicity

Studies about the nature of the hydrophobic effect have appeared in the literature since the early work of Traube in 1891 [14]. According to Tanford, a hydrophobic effect arises when any solute is dissolved in water [15]. (*Hydrophobic effects, hydrophobic bonds,* and *hydrophobic interactions* are used synonymously in the literature.) *A hydrophobic bond* has been defined as one "which forms when non-polar groups in an aqueous solvent associate, thereby decreasing the extent of interaction with surrounding water molecules, and liberating water originally bound by the solutes" [16]. In the past, the hydrophobic effect was believed to arise from the attraction of nonpolar groups for each other [17]. Although a "like-attracts-like" interaction certainly plays a role in this phenomenon, current opinion views the strong

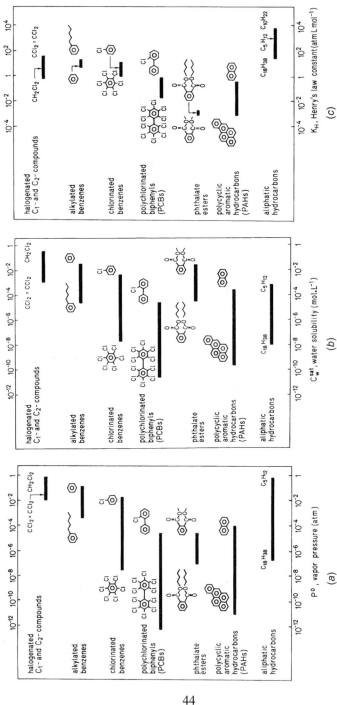

Figure 2.3. Ranges of (*a*) saturation vapor pressure (*P*°) values at 25°C, (*b*) water solubilities (C_w^{sat}), and (*c*) Henry's law constants (K_H) for some important classes of organic compounds. (Reprinted with permission from Ref. 8. Copyright © 1993 John Wiley & Sons, Inc.)

44

forces between water molecules as the primary cause of the hydrophobic effect. The detailed molecular structure of liquid water is complex and not well understood [18]. Many of the unusual properties of water are the result of the three-dimensional network of hydrogen bonds linking individual molecules together [19].

The attractive forces between water molecules are strong, and foreign molecules disrupt the isotropic arrangement of the molecules of water. When a nonpolar solute is dissolved in water, it is incapable of forming hydrogen bonds with the water, so some hydrogen bonds will have to be broken to accommodate the intruder. The breaking of hydrogen bonds requires energy. Frank and Evans [20] suggested that the water molecules surrounding a nonpolar solute must rearrange themselves to regenerate the broken bonds. Thermodynamic calculations indicate that when this rearrangement occurs, a higher degree of local order exists than in pure liquid water. Tanford [15] concludes that the water molecules surrounding a nonpolar solute do not assume one unique spatial arrangement, but are capable of assuming various arrangements, subject to changes in temperature and hydrocarbon chain length. The first layer of water molecules surrounding the solute cavity and subsequent layers are often termed *flickering clusters* [20–22].

An intruding hydrocarbon must compete with the tendency of water to re-form the original structure and is "squeezed" out of solution [23]. This hydrophobic effect is attributed to the high cohesive energy density of water because the interactions of water with a nonpolar solute are weaker than the interactions of water with itself [24]. Leo [22] notes that "part of the energy 'cost' of creating the cavity in each solvent is 'paid back' when the solvent interacts favorably with parts of the solute surface."

Recognizing that the *hydrophobic effect* (or more generally, a *solvophobic effect*) exists when solutes are dissolved in water leads to considering the influence of this property on the distribution of a solute between immiscible phases during extraction. A parameter that measures hydrophobicity is needed. This parameter is considered important to describe transport between water and hydrophobic biological phases (such as lipids or membranes), between water and hydrophobic environmental phases (such as organic humic substances), and between water and hydrophobic extractants (such as methylene chloride or reversed-phase solid sorbents). Although the earliest attempts to quantitate hydrophobicity used olive oil as the immiscible reference phase [25,26], since the 1950s, n-octanol has gained widespread favor as the reference solvent [27].

The general equilibrium constant expression in equation (2.2) can be rewritten to express the distribution of solute chemical X between water (W) and n-octanol (O) as

$$K_{OW} = K_D = \frac{[X]_O}{[X]_W} \qquad (2.7)$$

The n-octanol/water partition coefficient, K_{OW} (also referred to as P_{OW}, P, or P_{oct}), is a dimensionless, "operational" [21] or "phenomenological" [24] definition of hydrophobicity based on the n-octanol reference system [28]. The amount of transfer of a solute from water into a particular immiscible solvent or bulk organic matter will not be identical to the mass transfer observed in the n-octanol/water system, but K_{OW} is often directly proportional to the partitioning of a solute between water and various other hydrophobic phases [8]. The larger the value of K_{OW}, the greater is the tendency of the solute to escape from water and transfer to a bulk hydrophobic phase. When comparing the K_{OW} values of two solutes, the compound with the higher number is said to be the more hydrophobic of the two.

The n-octanol/water reference system covers a wide scale of distribution coefficients, with K_{OW} values varying with organic molecular structure (Figure 2.4). The magnitude of the n-octanol/water partition coefficient generally increases with molecular weight. The differences in K_{OW} cover several orders of magnitude, such that hydrophobicity values are often reported on a logarithmic scale (i.e., log K_{OW} or log P), in the range -4.0 to $+6.0$ [21].

The distribution coefficient refers to the hydrophobicity of the entire molecule. Within a family of organic compounds it is sometimes useful to deal with hydrophobic substituent constants that relate the hydrophobicity of a derivative, log P_X, to that of the parent molecule, log P_H. Therefore, a substituent parameter, π, has been defined [21] as

$$\pi_X = \log P_X - \log P_H \qquad (2.8)$$

where a positive value means the substituent is more hydrophobic (i.e., prefers n-octanol to water) relative to hydrogen, and a negative value indicates that the substituent prefers the water phase and is more hydrophilic than hydrogen (Table 2.1). The hydrophobic contribution of a substituent such as CH_3, Cl, OH, or NO_2 varies according to the molecular subenvironment of the substituent [21,30].

In order to use the value of the distribution coefficient between n-octanol and water as a guide for methodology to use when extracting organic compounds from water, the effect of variation in the degree of hydrophobicity must be considered. If a solute has low hydrophobicity, according to equation (2.7), it will prefer to remain in the aqueous phase relative to n-octanol. If a solute has very high hydrophobicity, it will prefer to be in the n-octanol phase. Intuitively, highly hydrophobic organic chemicals are easier to extract from water by a second immiscible, hydrophobic phase, but analyti-

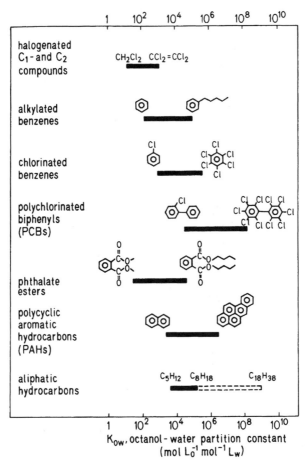

Figure 2.4. Ranges in octanol–water partition constants (K_{OW}) for some important classes of organic compounds. (Reprinted with permission from Ref. 8. Copyright © 1993 John Wiley & Sons, Inc.)

cally they can subsequently be difficult to remove from the immiscible phase. Ney [13] defines low K_{OW} as values less than 500 (log K_{OW} = 2.7), midrange values as $500 \leq K_{OW} \leq 1000$ ($2.7 \leq$ log $K_{OW} \leq 3.0$), and high K_{OW} values as greater than 1000 (log K_{OW} > 3.0). Others [31,32] found it useful to consider compounds with a log K_{OW} less than 1 as highly hydrophilic, and compounds with a log K_{OW} above 3 to 4 (depending on the nature of the immiscible phase) as highly hydrophobic.

The relationship between water solubility and the n-octanol/water partition coefficient must be addressed. Why are both parameters included in

Table 2.1. Substituent Constants Derived from Partition Coefficients

Functional Group	Aromatic Para-Substituted Systems (π)									
	Monobenzenes	Phenoxyacetic Acid	Phenylacetic Acid	Benzoic Acid	Benzyl Alcohol	Nitrobenzenes	Benzamides	Phenols	Anilines	Acetanilides
OCH$_3$	-0.02	-0.04	0.01	0.08	0	0.18	0.21	-0.12		-0.02
CH$_3$	0.56	0.52	0.45	0.42	0.48	0.52	0.53	0.48	0.49	0.24
NO$_2$	-0.28	0.24	-0.04	0.02	0.16	-0.39	0.17	0.50	0.49	0.50
Cl	0.71	0.70	0.70	0.87	0.86	0.54	0.88	0.93		0.71

Source: Data from Refs. 21, 29, and 30.

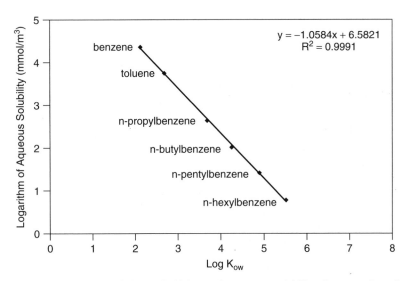

Figure 2.5. Comparison of hydrophobicity and aqueous solubility for a series of *n*-alkylbenzenes. (Data from Ref. 33.)

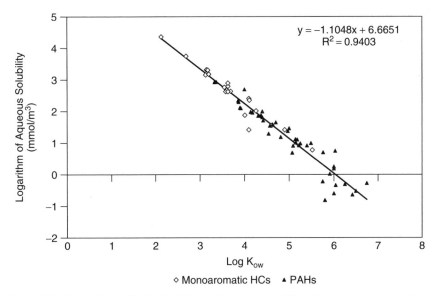

Figure 2.6. Comparison of hydrophobicity and aqueous solubility for monoaromatic hydrocarbons (HCs) and polycyclic aromatic hydrocarbons (PAHs). (Data from Ref. 6 and 33.)

the list of key chemical properties? In general, there is a trend toward an inverse relationship between these parameters such that high water solubility is generally accompanied by low hydrophobicity, and vice versa. Many authors use this relationship to estimate one of these parameters from the other. However, it is this author's opinion that the n-octanol/water partition coefficient and water solubility are not interchangeable (via inverse relationships) because they measure different phenomena. Water solubility is a property measured at maximum capacity or saturation. The n-octanol/water partition coefficient measures distribution across an interface. While the relationship between water solubility and the n-octanol/water partition coefficient may be highly correlated for closely related families of congeners (Figure 2.5), as the diversity of the compounds compared increases, the correlation between these two parameters decreases (Figure 2.6). However, solubility should remain on the list of essential chemical properties because if the value of the octanol–water partition coefficient is unavailable, water solubility can be used as a surrogate. Also, solubility is used to estimate the Henry's law constant.

2.1.3. Acid–Base Equilibria

The acid–base character of a chemical and the pH of the aqueous phase determine the distribution of ionized–nonionized species in solution. Starting from the equilibrium dissociation of a weak acid, HA,

$$HA \rightleftharpoons H^+ + A^- \tag{2.9}$$

the equilibrium constant for dissociation of a weak acid can be written as

$$K_a = \frac{[H^+][A^-]}{[HA]} \tag{2.10}$$

Analogously, the dissociation of the conjugate acid, BH^+, of a base, B, is described as

$$BH^+ \rightleftharpoons H^+ + B \tag{2.11}$$

and the related constant is

$$K_a = \frac{[H^+][B]}{[BH^+]} \tag{2.12}$$

Ionizable compounds' K_a values (Figure 2.7) have an orders-of-magnitude

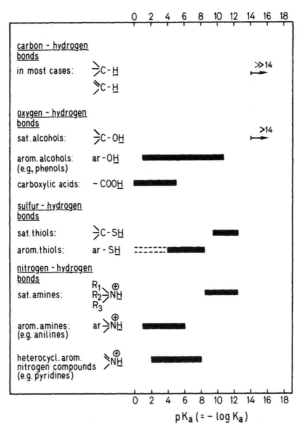

Figure 2.7. Ranges of acid dissociation constants (pK_a) for some important classes of organic compounds. (Reprinted with permission from Ref. 8. Copyright © 1993 John Wiley & Sons, Inc.)

range. This makes it useful to describe K_a values in terms of logarithms; that is, $pK_a = -\log K_a$.

Two graphical methods described here, a master variable (pC–pH) diagram and a distribution ratio diagram, are extremely useful aids for visualizing and solving acid–base problems. They help to determine the pH at which an extraction should be performed. Both involve the choice of a *master variable*, a variable important to the solution of the problem at hand. The obvious choice for a master variable in acid–base problems is $[H^+]$ [equations (2.9)–(2.12)], or pH when expressed as the negative logarithm of $[H^+]$.

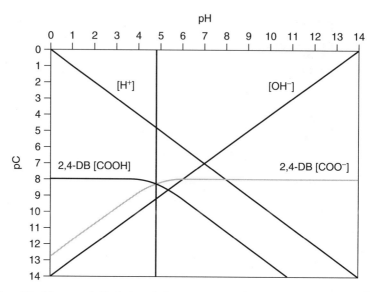

Figure 2.8. Master variable (pC–pH) diagram for 2,4-DB; pK_a = 4.8, C_T = 1 × 10^{-8} M.

To prepare a *pC–pH diagram*, the master variable, pH, is plotted on the
x-axis. On the y-axis, the concentration of chemical species is plotted as a
function of pH. The concentration, C, of each chemical species is expressed
as a logarithm (log C), or more often as the negative logarithm of its concentration, that is pC (analogous to pH). The pC–pH diagram (Figure 2.8)
for a representative acidic solute, 4-(2,4-dichlorophenoxy)butanoic acid or
2,4-DB, is prepared by first determining that the pK_a for this compound is
4.8. A reasonable concentration to assume for trace levels of this compound
in water is 2.5 ppm or 1 × 10^{-8} M, since the molecular weight of 2,4-DB is
249.1. Based on the molar concentration of 1 × 10^{-8}, pC has a value of 8.
By mass balance, the total concentration at any given pH value, C_T, is the
sum of all species. That is,

$$C_T = [\text{HA}] + [\text{A}^-] \qquad (2.13)$$

for a monoprotic acid, as in the example in Figure 2.8. The diagonal line
connecting pH, pC values $(0, 0)$ with $(14, 14)$ represents the hydrogen
ion concentration, and the diagonal line connecting pH, pC values $(0, 14)$
with $(14, 0)$ represents the hydroxide ion concentration, according to the
expression

$$[\text{H}^+][\text{OH}^-] = K_W = 10^{-14} \qquad (2.14)$$

where K_W is the ion product of water. The vertical line in Figure 2.8 indicates data at which the pH $=$ pK_a.

To graph the curves representing [HA] and [A$^-$], a mathematical expression of each as a function of [H$^+$] (a function of the master variable) is needed. The appropriate equation for [HA] is derived by combining the equilibrium constant for dissociation of a weak acid [equation (2.10)] with the mass balance equation [equation (2.13)] to yield

$$[HA] = \frac{[H^+]C_T}{[H^+] + K_a} \qquad (2.15)$$

Analogously, solving for [A$^-$] yields

$$[A^-] = \frac{K_a C_T}{[H^+] + K_a} \qquad (2.16)$$

Point-by-point plotting of equations (2.15) and (2.16) produces the curves for the nonionized, 2,4-DB[COOH], and ionized, 2,4-DB[COO$^-$], species in Figure 2.8. This approach can be expanded to generate master variable diagrams of more complex polyprotic systems (Figure 2.9) such as phosphoric

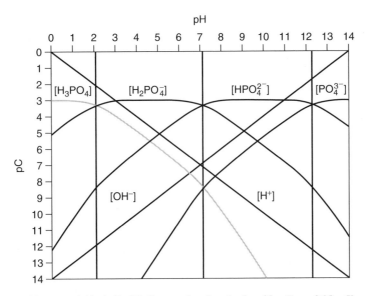

Figure 2.9. Master variable (pC–pH) diagram for phosphoric acid: p$K_{a1} = 2.15$, p$K_{a2} = 7.20$, and p$K_{a3} = 12.35$, $C_T = 1 \times 10^{-3}$ M.

acid. Figure 2.9 was generated by using the acid dissociation constants of phosphoric acid, $pK_{a1} = 2.15$, $pK_{a2} = 7.20$, and $pK_{a3} = 12.35$. Additionally, a total phosphate concentration of 0.001 M was assumed. In this case, $C_T = [H_3PO_4] + [H_2PO_4^-] + [HPO_4^{2-}] + [PO_4^{3-}]$. Figures 2.8 and 2.9 were produced using a free software package, EnviroLand version 2.50, available for downloading from the Internet [34]. Alternatively, equations (2.15) and (2.16) can be input to spreadsheet software to produce pC–pH diagrams.

A second graphical approach to understanding acid–base equilibria is preparation of a *distribution ratio diagram*. The fraction, α, of the total amount of a particular species is plotted on the y-axis versus the master variable, pH, on the x-axis, where

$$\alpha_{HA} = \frac{[HA]}{[A^-] + [HA]} \tag{2.17}$$

and

$$\alpha_{A^-} = \frac{[A^-]}{[A^-] + [HA]} \tag{2.18}$$

By combining equations (2.15), (2.16), and (2.18), a distribution diagram (Figure 2.10) for acetic acid can be prepared given that the acid dissociation constant is 1.8×10^{-5} with an assumed concentration of 0.01 M. The vertical line in Figure 2.10, positioned at $x = 4.74$, is a reminder that when the pH of the solution is equal to the pK_a of the analyte, the α value is 0.5, which signifies that the concentration of HA is equal to the concentration of A^-. The distribution diagram can be used to determine the fraction of ionized or nonionized acetic acid at any selected pH.

Another way of understanding the distribution of species as a function of pH is to apply the *Henderson–Hasselbach equation*:

$$pH = pK_a + \log \frac{[A^-]}{[HA]} \tag{2.19}$$

which is derived by taking the negative logarithm of both sides of equation (2.10). The Henderson–Hasselbach equation provides a useful relationship between system pH and acid–base character taking the ratio of ionized to nonionized species into consideration.

To calculate the relative amount of A^- present in a solution in which the pH is 1 unit above the pK_a (i.e., $pH = pK_a + 1$), apply the Henderson–

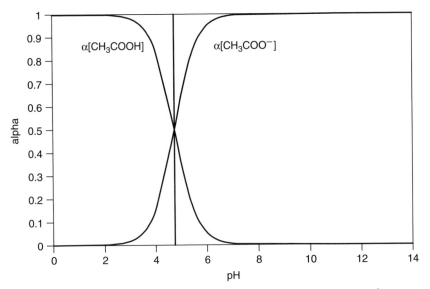

Figure 2.10. Distribution diagram for acetic acid; $pK_a = 4.74$, $C_T = 1 \times 10^{-2}$ M.

Hasselbach equation such that

$$1 = \log \frac{[A^-]}{[HA]} \tag{2.20}$$

and taking the antilogarithm of both sides yields

$$10 = \frac{[A^-]}{[HA]} \tag{2.21}$$

Assume that the only species present are HA and A^- such that

$$[HA] + [A^-] = 1 \tag{2.22}$$

Rearranging equation (2.22) to solve for [HA] and substituting into equation (2.21) gives

$$10 = \frac{[A^-]}{1 - [A^-]} \tag{2.23}$$

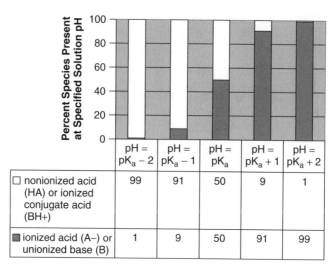

Figure 2.11. Percent of ionogenic (ionizable) species present for weak acids and bases when solution pH is 2 units above or below the acid dissociation constant.

and therefore $[A^-] = 0.909$. In an analogous manner, it is possible to calculate that the fraction of $[A^-]$ present in a solution in which the pH is 2 units above the pK_a (i.e., pH = $pK_a + 2$) is 0.990. According to the Henderson–Hasselbach equation, 50% of each species is present when the pH is equal to the pK_a. Therefore, depending on whether the compound is an acid or a base (Figure 2.11), an analyte is either 99% nonionized or ionized when the pH value is 2 units above or below the pK_a.

The purpose of applying master variable diagrams, distribution diagrams, and the Henderson–Hasselbach equation to ionizable organic chemicals is to better understand the species present at any solution pH. Organic compounds can be extracted from liquids in either the ionized or nonionized form. Generally, however, for ionizable compounds, it is best to adjust the solution pH to force the compound to exist in the ionized state or in the nonionized state as completely as possible. Less than optimal results may be obtained if the ionizable compound is extracted within the window of the $pK_a \pm 2$ log units. When the pH is equal to the pK_a, half of the compound is ionized and half of the compound is nonionized. Mixed modes of extraction are required to transfer the compound completely from one phase to another. The "2 units" rule of thumb is very important for an analyst to understand and apply when developing extraction protocol for acidic or basic compounds. More information concerning graphical methods for solving acid–base equilibrium problems can be found in Bard [1], Snoeyink and Jenkins [35], and Langmuir [36].

2.1.4. Distribution of Hydrophobic Ionogenic Organic Compounds

Some highly hydrophobic weak acids and bases exhibit substantial hydrophobicity even in the ionized state. For highly hydrophobic ionogenic organic compounds, not only is transfer of the neutral species between the aqueous phase and the immiscible phase important, but the transfer of the hydrophobic, ionized, organic species as free ions or ion pairs may also be significant [37]. Mathematically, this is described by refining the n-octanol/water partition coefficient, as defined in equation (2.7), to reflect the pH-dependent distribution between water (W) and n-octanol (O) of chemical X in both the ionized and nonionized forms. If chemical X is a weak acid, HA, the distribution ratio is

$$D_{OW}(HA, A^-) = \frac{[HA]_{O,\,total}}{[HA]_W + [A^-]_W} \qquad (2.24)$$

where $[HA]_{O,\,total}$ is the sum of all neutral species, free ions, and ions paired with inorganic counterions that transfer to octanol [8,37].

For example, the ratio of the n-octanol/water distribution coefficient of the nondissociated species to that of the ionic species is nearly 10,000 for 3-methyl-2-nitrophenol, but only about 1000 for pentachlorophenol because of the greater significance of the hydrophobicity of the ionized form of pentachlorophenol. The logarithm of the n-octanol/water distribution coefficient of pentachlorophenol as the phenolate is about 2 (determined at pH 12, and 0.1 M KCl), which indicates significant distribution of the ionized form into the n-octanol phase [8,37]. Extraction of such highly hydrophobic ionogenic organic compounds can result from mixed-mode mechanisms that incorporate both the hydrophobic and ionic character of the compound.

2.2. LIQUID–LIQUID EXTRACTION

In liquid–liquid extraction (LLE), phases A and B are both liquids. The two liquid phases must be immiscible. For that reason, LLE has also been referred to as *immiscible solvent extraction*. In practice, one phase is usually aqueous while the other phase is an organic solvent. An extraction can be accomplished if the analyte has favorable solubility in the organic solvent. Chemists have used organic solvents for extracting substances from water since the early nineteenth century [38].

Miscibility

Solvent manufacturer Honeywell Burdick & Jackson [39] defines solvents as *miscible* if the two components can be mixed together in all proportions without forming two separate phases. A solvent miscibility chart (Figure 2.12) is a useful aid for determining which solvent pairs are immiscible and would therefore be potential candidates for use in LLE. More solvent combinations are miscible than immiscible, and more solvents are immiscible with water than with any other solvent. Solvents miscible with water in all proportions include acetone, acetonitrile, dimethyl acetamide, *N,N*-dimethylformamide, dimethyl sulfoxide, 1,4-dioxane, ethyl alcohol, glyme, isopropyl alcohol, methanol, 2-methoxyethanol, *N*-methylpyrrolidone, *n*-propyl alcohol, pyridine, tetrahydrofuran, and trifluoroacetic acid [40].

Density

Another consideration when selecting an extraction solvent is its density [41]. Solvents that are more dense than water will form the lower layer of the pair when mixed together, while solvents that are less dense than water will form the upper layer or "float" on water. For example, ethyl ether has a density of 0.7133 g/mL at 20°C and would constitute the upper phase when combined with water, which has a density of 0.9982 g/mL at that temperature. On the other hand, the density of chloroform is 1.4892 at 20°C. Therefore, water would form the top layer in a water–chloroform solvent pair.

Solubility

Although solvents may form two visibly distinct phases when mixed together, they are often somewhat soluble in each other and will, in fact, become mutually saturated when mixed with each other. Data on the solubility of various solvents in water (Table 2.2) and on the solubility of water in other solvents (Table 2.3) should be consulted when selecting an extraction solvent pair. For example, 1.6% of the solvent dichloromethane (or methylene chloride) is soluble in water. Conversely, water is 0.24% soluble in dichloromethane. According to Table 2.3, when the phases are separated for recovery of the extracted analyte, the organic solvent layer will contain water. Similarly, according to Table 2.2, after extraction the depleted aqueous phase will be saturated with organic solvent and may pose a disposal problem. (*Author's note:* I previously recounted [43] my LLE experience with disposal of extracted aqueous samples that were cleaned of pesticide residues but saturated with diethyl ether. Diethyl ether is 6.89% soluble in water at 20°C.)

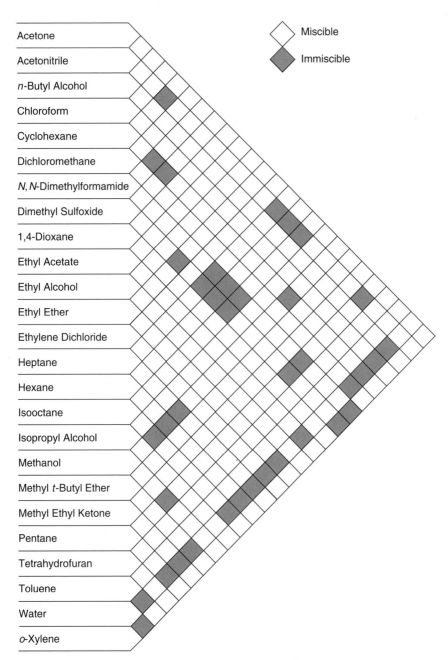

Figure 2.12. Solvent miscibility chart. (Reprinted with permission from Ref. 39. Copyright ©
2002 Honeywell Burdick & Jackson.) Available online at
http://www.bandj.com/BJProduct/SolProperties/Miscibility.html

Table 2.2. Solubility in Water

Solvent	Solubility (%)[a]
Isooctane	0.0002 (25°C)
Heptane	0.0003 (25°C)
1,2,4-Trichlorobenzene	0.0025
Cyclohexane	0.006 (25°C)
Cyclopentane	0.01
Hexane	0.014
o-Dichlorobenzene	0.016 (25°C)
1,1,2-Trichlorotrifluoroethane	0.017 (25°C)
o-Xylene	0.018 (25°C)
Pentane	0.04
Chlorobenzene	0.05
Toluene	0.052 (25°C)
n-Butyl chloride	0.11
Methyl isoamyl ketone	0.54
n-Butyl acetate	0.68
Ethylene dichloride	0.81
Chloroform	0.815
Dichloromethane	1.60
Methyl isobutyl ketone	1.7
Methyl t-butyl ether	4.8
Triethylamine	5.5
Methyl n-propyl ketone	5.95
Ethyl ether	6.89
n-Butyl alcohol	7.81
Isobutyl alcohol	8.5
Ethyl acetate	8.7
Propylene carbonate	17.5 (25°C)
Methyl ethyl ketone	24.0

Source: Reprinted with permission from Ref. 40. Copyright © (2002) Honeywell Burdick & Jackson.

[a]Solvents are arranged in order of increasing solubility in water, the maximum weight percent (w/w) of each solvent that can be dissolved in water (at 20°C unless otherwise indicated).

2.2.1. Recovery

As defined earlier,

$$K_D = \frac{[X]_B}{[X]_A} \tag{2.2}$$

Table 2.3. Solubility of Water in Each Solvent

Solvent	Solubility (%)[a]
Isooctane	0.006
Pentane	0.009
Cyclohexane	0.01
Cyclopentane	0.01
Heptane	0.01 (25°C)
Hexane	0.01
1,1,2-Trichlorotrifluoroethane	0.011 (25°C)
1,2,4-Trichlorobenzene	0.020
Toluene	0.033 (25°C)
Chlorobenzene	0.04
Chloroform	0.056
n-Butyl chloride	0.08
Ethylene dichloride	0.15
Dichloromethane	0.24
o-Dichlorobenzene	0.31 (25°C)
n-Butyl acetate	1.2
Ethyl ether	1.26
Methyl isoamyl ketone	1.3
Methyl t-butyl ether	1.5
Methyl isobutyl ketone	1.9 (25°C)
Ethyl acetate	3.3
Methyl n-propyl ketone	3.3
Triethylamine	4.6
Propylene carbonate	8.3 (25°C)
Methyl ethyl ketone	10.0
Isobutyl alcohol	16.4
n-Butyl alcohol	20.07

Source: Reprinted with permission from Ref. 42. Copyright ©
(2002) Honeywell Burdick & Jackson.

[a]Solvents are arranged in order of increasing solubility of water
in each solvent, the maximum weight percent (w/w) of water
that can be dissolved in the solvent (at 20°C unless otherwise
indicated).

Analytes distribute themselves between aqueous and organic layers according to the Nernst distribution law, where the distribution coefficient, K_D, is
equal to the analyte ratio in each phase at equilibrium.

The analyte distributes itself between the two immiscible liquids according to the relative solubility in each solvent [1,38,44,45]. To determine the
effect of the distribution coefficient on an extraction, consider the following
example.

Example

A 1-L aqueous sample containing 100 parts per billion (ppb) of a compound having a molecular weight of 250 g/mol is extracted once with 150 mL of organic extracting solvent. Assume that the K_D value is 5. Given this information, the molarity of the original sample is 4.0×10^{-10} M. Calculate the percent of the analyte extracted into the organic extracting solvent at equilibrium.

Step 1. Calculate the moles of analyte in the original sample.

$$\text{moles in original sample} = \text{molarity of sample (in mol/L)}$$
$$\times \text{volume extracted (in L)}$$

Therefore,

$$\text{moles in original sample} = 4.0 \times 10^{-10} \ M \times 1 \ L = 4.0 \times 10^{-10} \ \text{mol} \quad (2.25)$$

Step 2. Calculate the moles of analyte left in the aqueous phase after extraction.

$$K_D = \frac{(\text{moles in original sample} - \text{moles left in water after extraction})/\text{extraction solvent volume (in L)}}{\text{moles left in water after extraction}/\text{volume of original sample (in L)}}$$

$$(2.26)$$

Therefore,

$$\frac{\text{moles left in water}}{\text{after extraction}} = \frac{\text{moles in original sample}}{\{[K_D \times \text{extraction solvent volume (in L)}]/\text{volume of original sample (in L)}\} + 1}$$

such that,

$$\text{moles left in water after extraction}$$
$$= \frac{4.0 \times 10^{-10} \ \text{mol}}{[(5 \times 0.150 \ L)/1 \ L] + 1} = 2.2857 \times 10^{-10} \ \text{mol}$$

Step 3. Calculate the moles of analyte extracted into layer B (i.e., the extracting solvent) at equilibrium.

moles of analyte extracted into organic solvent

$$= \text{moles of analyte in original sample} - \text{moles left in water after extraction}$$

$$= 4.0 \times 10^{-10} \ \text{mol} - 2.2857 \times 10^{-10} \ \text{mol} = 1.7143 \times 10^{-10} \ \text{mol} \quad (2.27)$$

Step 4. Calculate the percent of analyte extracted into the organic solvent at equilibrium. The recovery factor, R_X, is the fraction of the analyte extracted divided by the total concentration of the analyte, multiplied by 100 to give the percentage recovery:

$\% \ R_X$ = percent of analyte extracted into organic solvent

$$= \frac{\text{moles of analyte extracted into organic solvent}}{\text{moles of analyte in original sample}} \times 100$$

$$= \frac{1.7143 \times 10^{-10} \text{ mol}}{4.0 \times 10^{-10} \text{ mol}} \times 100 = 42.857\% \tag{2.28}$$

If the problem is reworked such that the volume of the extracting solvent is 50 mL instead of 150 mL, the percent of analyte extracted into the organic solvent, calculated by repeating steps 1 through 4, is determined to be only 20% (Table 2.4) as compared to 42.857% if an extracting solvent of 150 mL is used. If after separating the phases, the aqueous sample is extracted with a second sequential extraction volume of 50 mL, again 20% of what remained available for extraction will be removed. However, that represents only 16% additional recovery, or a cumulative extraction of 36% after two sequential extractions (i.e., 2×50 mL). If after separating the phases, the aqueous sample is extracted with a third sequential extraction volume of 50 mL, again 20% of what remained available for extraction will be removed. That represents only 12.8% of additional recovery or a cumulative extraction of 48.8% after three sequential extractions (i.e., 3×50 mL). Analogous to a hapless frog that jumps halfway out of a well each time it jumps, never to escape the well, LLE recovery is an equilibrium procedure in which exhaustive extraction is driven by the principle of repeated extractions.

The percent recovery obtained with a single extraction of 150 mL of organic solvent is compared to that for three sequential extractions of 50 mL each for K_D values of 500, 250, 100, 50, and 5 (Table 2.4). In sequential extractions, the same percent recovery is extracted each time (i.e., the frog jumps the same percentage of the distance out of the well each time). That is, at a K_D value of 500, 96.154% is extracted from the original sample using an organic solvent volume of 50 mL; 96.154% of the analyte remaining in solution after the first extraction is removed during the second sequential extraction by 50 mL; and 96.154% of the analyte remaining in solution after the second extraction is removed during the third sequential extraction by 50 mL.

When K_D is equal to 500, the first extraction using 50 mL recovers 96.154% of the original analyte; the second sequential extraction produces

Table 2.4. Distribution Coefficient Effects on Single and Repeated Extractions

K_d	Single Extraction 1 × 150 mL Percent Extracted	Single Extraction 1 × 50 mL Percent Extracted	Second Sequential Extraction			Third Sequential Extraction		
			1 × 50 mL Repeat Percent Extracted	1 × 50 mL Additional Recovery	2 × 50 mL Cumulative Extraction	1 × 50 mL Repeat Percent Extracted	1 × 50 mL Additional Recovery	3 × 50 mL Cumulative Extraction
500	98.684	96.154	96.154	3.697	99.851	96.154	0.142	99.993
250	97.403	92.593	92.593	6.859	99.451	92.593	0.508	99.959
100	93.750	83.333	83.333	13.890	97.223	83.333	2.315	99.538
50	88.235	71.429	71.429	20.411	91.839	71.429	5.832	97.671
5	42.857	20.000	20.000	16.000	36.000	20.000	12.800	48.800

additional recovery of 3.697% of the original analyte; and the third sequential extraction produces further recovery of 0.142% of the original analyte, for a cumulative recovery after three sequential extractions (3 × 50 mL) of 99.993%. The cumulative recovery after three extractions of 50 mL each is greater than that calculated for recovery from a single extraction of 150 mL of organic solvent (i.e., 98.684%).

The effect of concentration on recovery by single or repeated extractions can be examined. Instead of assuming a concentration of 4.0×10^{-10} M for the aqueous sample to be extracted as stated in the original problem, the values in Table 2.4 can be recalculated after substitution with a concentration of 0.01 M. If the same four steps outlined previously are followed, it can be demonstrated that the recovery values in Table 2.4 are identical regardless of concentration. The most desirable analytical protocols are independent of sample concentration in the range of samples to be analyzed.

The operation conducted in steps 1 through 4 above can be summarized by the following equation such that the recovery factor of analyte X, expressed as a percent, is

$$\% R_X = \frac{100K_D}{K_D + (V_O/V_E)} \tag{2.29}$$

where V_O is the volume of the original sample and V_E is the extraction solvent volume. (Note that the recovery factor is independent of sample concentration.) The recovery factor can also be expressed in the equivalent form

$$\% R_X = 100 \left[\frac{K_D(V_E/V_O)}{1 + K_D(V_E/V_O)} \right] = 100 \left[\frac{K_D(V)}{1 + K_D(V)} \right] \tag{2.30}$$

where $V = V_E/V_O$ is known as the *phase ratio*.

Therefore, applying equation (2.29) to the previous example in which a 1-L aqueous sample containing 100 ppb of a compound having a molecular weight of 250 g/mol is extracted once with 150 mL of organic extracting solvent, and assuming that K_D is 5, substitution yields.

$$R_X = \frac{100 \times 5}{5 + (1.0 \text{ L}/0.150 \text{ L})} = 42.857\%$$

If the analyte is partially dissociated in solution and exists as the neutral species, free ions, and ions paired with counterions, the distribution ratio, D, analogous to equation (2.24), would be

$$D = \frac{\text{concentration of X in all chemical forms in the organic phase}}{\text{concentration of X in all chemical forms in the aqueous phase}} \tag{2.31}$$

In this instance, the value for D would be substituted for K_D in equation (2.29).

The formula for expressing repeated extractions is

$$\% \, R_X = \left\{ 1 - \left[\frac{1}{1 + K_D(V_E/V_O)} \right]^n \right\} \times 100 \qquad (2.32)$$

Applying equation (2.32) to the previous calculation having three successive multiple extractions where $K_D = 5$, $V_E = 50$ mL, $V_O = 1$ L, and $n = 3$, the cumulative recovery is calculated to be 48.8% (Table 2.4).

Repeated extractions may be required to recover the analyte sufficiently from the aqueous phase. Neutral compounds can have substantial values of K_D. However, organic compounds that form hydrogen bonds with water, are partially soluble in water, or are ionogenic (weak acid or bases) may have lower distribution coefficients and/or pH-dependent distribution coefficients. Additionally, the sample matrix itself (i.e., blood, urine, or wastewater) may contain impurities that shift the value of the distribution coefficient relative to that observed in purified water.

Investigation of the principle of repeated extractions demonstrates that:

- The net amount of analyte extracted depends on the value of the distribution coefficient.
- The net amount of analyte extracted depends on the ratio of the volumes of the two phases used.
- More analyte is extracted with multiple portions of extracting solvent than with a single portion of an equivalent volume of the extracting phase.
- Recovery is independent of the concentration of the original aqueous sample.

2.2.2. Methodology

The LLE process can be accomplished by shaking the aqueous and organic phases together in a separatory funnel (Figure 2.13a). Following mixing, the layers are allowed to separate. Flow from the bottom of the separatory funnel is controlled by a glass or Teflon stopcock and the top of the separatory funnel is sealed with a stopper. The stopper and stopcock must fit tightly and be leakproof. Commonly, separatory funnels are globe, pear, or cylindrically shaped. They may be shaken mechanically, but are often shaken manually.

With the stopcock closed, both phases are added to the separatory funnel. The stopper is added, and the funnel is inverted without shaking. The stop-

(a)   (b)

Figure 2.13. Liquid–liquid extraction apparatus: (a) separatory funnel and (b) evaporative Kuderna–Danish sample concentrator. (Reprinted with permission from Ref. 46. Copyright © 2002 Kimble/Kontes.)

cock is opened immediately to relieve excess pressure. When the funnel is inverted, the stem should be pointed away from yourself and others. The funnel should be held securely with the bulb of the separatory funnel in the palm of one hand, while the index finger of the same hand is placed over the stopper to prevent it from being blown from the funnel by pressure buildup during shaking. The other hand should be positioned to hold the stopcock end of the separatory funnel, and for opening and closing the stopcock.

The separatory funnel should be gently shaken for a few seconds, and frequently inverted and vented through the stopcock. When pressure builds up less rapidly in the separatory funnel, the solvents should be shaken more vigorously for a longer period of time while venting the stopcock occasionally. The separatory funnel should be supported in an upright position in an iron ring padded with tubing to protect against breakage.

When the layers are completely separated (facilitated by removing the stopper), the lower layer should be drawn off through the stopcock, and the upper layer should be removed through the top of the separatory funnel. The relative position of each layer depends on the relative densities of the two immiscible phases. During an extraction process, all layers should be saved until the desired analyte is isolated. A given solvent layer can easily be determined to be aqueous or organic by testing the solubility of a few drops in water.

Once the analyte has been extracted into phase B, it is usually desirable to reduce the volume of the extracting solvent. This can be accomplished with specialized glassware such as a Kuderna–Danish sample concentrator (Figure 2.13b), which is widely used for concentrating semivolatile compounds dissolved in volatile solvents. The concentrator consists of three primary components held together by hooks and/or clamps: a central flask with sufficient capacity to hold the extracting solvent, a tapered receiving vessel to contain the concentrated extract, and a distilling–condensing column that allows the solvent vapor to pass while retaining the analyte. The apparatus should be placed over a vigorously boiling water bath to bathe the central flask in steam. The solvent should then be allowed to escape into a hood or recovered via an additional solvent recovery system. Alternatively, a mechanical rotary evaporator may be used to evaporate excess extracting solvent, or other evaporating units that evaporate solvent with an inert gas should be used.

Performing LLE of analytes from drinking water is relatively straightforward. However, if your "aqueous" sample is blood, urine, or wastewater, the extraction process can become more tedious. Quite often in such samples, a scum forms at the layer interface, due to the presence of nonsoluble debris and the formation of emulsions. Analysts overcome this difficulty using techniques such as adding salts, chilling the sample, or centrifugation. Applying a continuous LLE technique can be useful also.

Continuous LLE is a variant of the extraction process that is particularly applicable when the distribution coefficient of the analyte between phases A and B is low. Additionally, the apparatus for conducting continuous LLE (Figures 2.14 and 2.15) automates the process somewhat. The analyst is freed from manually shaking the phases in a separatory funnel to effect a separation allowing multiple extractions to be performed simultaneously. Since the phases are not shaken to mix them, this procedure also helps avoid the formation of emulsions. The apparatus can be assembled to perform extraction alone (Figure 2.14), or extraction and concentration (Figure 2.15). The extractor performs ·on the principle that organic solvent cycles continuously through the aqueous phase, due to constant vaporization and condensation of the extracting solvent. Continuous LLE apparatus designed for heavier-than-water or lighter-than-water extracting solvents is available.

2.2.3. Procedures

A general extraction scheme (Figure 2.16) can be devised to extract semivolatile organics from aqueous solution such that important categories of organic compounds (i.e., bases, weak acids, strong acids, and neutrals) are fractionated from each other and isolated in an organic solvent. Many

Figure 2.14. Continuous liquid–liquid extraction apparatus designed for samples where the extracting solvent is heavier than water. (Reprinted with permission from Ref. 46. Copyright © 2002 Kimble/Kontes.)

pharmaceuticals and pesticides are ionogenic or neutral compounds, and could be recovered by this procedure. Such a scheme is based on pH control of the aqueous sample. The K_D value of a base in acidic conditions is low as is the K_D value of an acid in basic conditions, because in each instance the compound would be ionized. In these situations, the ionized base or acid would therefore tend to remain in the aqueous solution when mixed with an organic extracting solvent. Neutral compounds tend to transfer to the organic extracting phase regardless of solution pH.

If an aqueous sample hypothetically containing inorganics and organics, including bases, strong acids, weak acids, and neutrals, is adjusted to pH 2 and extracted with an organic solvent (Figure 2.16, step 1), a separation in which the inorganics and bases will remain in the aqueous phase is effected. The inorganics prefer the aqueous phase, due to charge separation in ionic bonds, and at pH 2, the ionogenic organic bases will be positively charged and thereby prefer the aqueous phase. The neutral, strongly acidic, and weakly acidic organic compounds will have higher K_D values under these conditions and will prefer to transfer to the organic phase from the aqueous phase.

To isolate the organic bases from inorganic compounds and to recover the organic bases in an organic solvent, the acidified aqueous solution from

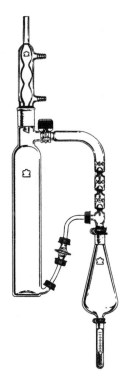

Figure 2.15. Continuous liquid–liquid extraction apparatus designed for samples where the extracting solvent is heavier than water in which both extraction and concentration are performed with the same apparatus. (Reprinted with permission from Ref. 46. Copyright © 2002 Kimble/Kontes.)

which the neutral and acidic compounds were removed is adjusted to pH 10 and extracted with an organic solvent (Figure 2.16, step 2). At pH 10, the K_D values of nonionized organic bases should be favorable for extraction into an organic solvent, while inorganic compounds preferentially remain in the aqueous solution.

To separate strongly acidic organic compounds from weakly acidic and neutral compounds, the organic phase containing all three components is mixed with a sodium bicarbonate (pH 8.5) solution (Figure 2.16, step 3). This seeming reversal of the process, that is, extracting compounds back into an aqueous phase from the organic phase, is called *washing, back-extraction,* or *retro-extraction*. Under these pH conditions, the organic phase retains the nonionized weakly acidic and neutral compounds, while ionized strong acids transfer into the aqueous washing solution.

The organic solvent phase containing only weakly acidic and neutral compounds is sequentially back-extracted with an aqueous (pH 10) solution of sodium hydroxide (Figure 2.16, step 4). Neutral compounds remain in the organic solvent phase, while weak organic acids, ionized at this pH, will be extracted into the aqueous phase.

Step 1: Adjust aqueous sample to pH2. Extract with organic solvent.

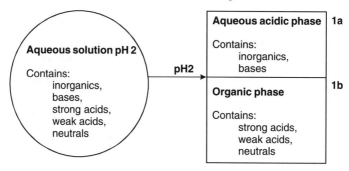

Step 2: Adjust aqueous acidic phase, 1a, to pH 10. Extract with organic solvent.

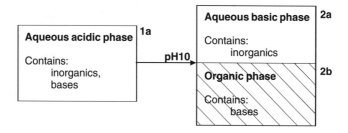

Step 3: Extract organic phase, 1b, with bicarbonate solution (pH 8.5).

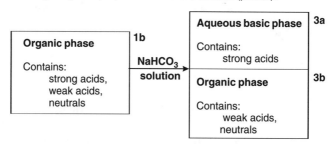

Figure 2.16. General extraction scheme. Hatched boxes represent isolation of organic compound categories in an organic phase.

The aqueous basic phase containing strong acids (Figure 2.16, step 5) and the aqueous basic phase containing weak acids (Figure 2.16, step 6) are each separately adjusted to pH 2 and extracted with organic solvent. Two organic solutions result: one containing recovered strong organic acids and the other containing weak organic acids.

Step 4: Extract organic phase, 3b, with hydroxide solution (pH 10).

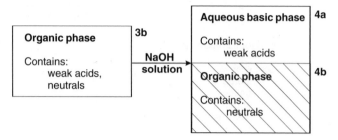

Step 5: Adjust aqueous basic phase, 3a, to pH 2. Extract with organic solvent.

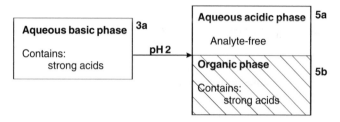

Step 6: Adjust aqueous basic phase, 4a, to pH 2. Extract with organic solvent.

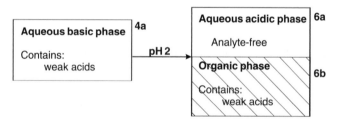

Figure 2.16. *(Continued)*

2.2.4. Recent Advances in Techniques

Historically, analysts performing LLE have experienced difficulties such as exposure to large volumes of organic solvents, formation of emulsions, and generation of mountains of dirty, expensive glassware. To address these problems, other sample preparation techniques, such as solid-phase extraction (SPE) and solid-phase microextraction (SPME), have experienced increased development and implementation during the previous two decades. However, advances in microfluidics amenable to automation are fueling a resurgence of LLE applications while overcoming some of the inherent difficulties associated with them.

Fujiwara et al. [47] devised instrumentation for online, continuous ion-pair formation and solvent extraction, phase separation, and detection. The procedure was applied to the determination of atropine in synthetic urine, and of atropine and scopolamine in standard pharmaceuticals. Aqueous sample solution was pumped at a flow rate of 5 mL/min. The organic extracting solvent, dichloromethane, was pumped at a flow rate of 2 mL/min and mixed with the aqueous sample stream to produce an aqueous-to-organic volume ratio of 2.5. The mixture was passed through an extraction coil composed of a 3-m PTFE tube [0.5 mm inside diameter (ID)] where associated ion pairs were transferred from the aqueous into the organic phase. The phases were separated using a Teflon membrane. The organic phase transversed the phase-separating membrane and passed onward in the stream to the detector while the aqueous stream was wasted.

Tokeshi et al. [48] performed an ion-pair solvent extraction successfully on a microchannel-fabricated quartz glass chip. An aqueous Fe complex (Fe–4,7-diphenyl-1,10-phenanthrolinedisulfonic acid) and a chloroform solution of capriquat (tri-n-octylmethylammonium chloride) were introduced separately into a microchannel (250 μm) to form a parallel two-phase laminar flow producing a liquid–liquid aqueous–organic interface (Figure 2.17). The authors noted that in the microchannel, the aqueous–organic interface did not attain the upper–lower arrangement produced by differences in specific gravity normally observed in LLE. In the microchannel environment, surface tension and frictional forces are stronger than specific gravity, resulting in an interface that is side by side and parallel to the sidewalls of the microchannel. The ion-pair product extracted from aqueous solution into

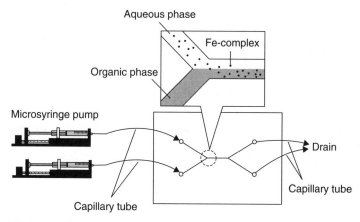

Figure 2.17. Schematic diagram of microextraction system on a glass chip. (Reprinted with permission from Ref. 48. Copyright © 2000 American Chemical Society.)

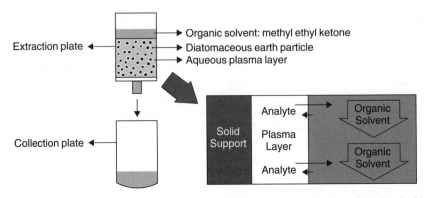

Figure 2.18. Schematic representation of automated liquid–liquid extraction. (Reprinted with permission from Ref. 50. Copyright © 2001 American Chemical Society.)

chloroform within 45 seconds when the flow was very slow or stopped, corresponding with molecular diffusion time. The extraction system required no mechanical stirring, mixing, or shaking.

Solid-supported LLE is a new approach reported by Peng et al. [49,50]. They exploited the efficiency of 96-channel, programmable, robotic liquid-handling workstation technology to automate methodology for this LLE variation. A LLE plate was prepared by adding inert diatomaceous earth particles to a 96-well plate with hydrophobic GF/C glass fiber bottom filters. Samples and solvents were added to the plate sequentially. LLE occurred in the interface between the two liquid phases and on the surface of individual particles in each well (Figure 2.18). The organic phase extracts were eluted under gentle vacuum into a 96-well collection plate. The approach was used for initial purification of combinatorial library samples and for quantitative analysis of carboxylic acid–based matrix metalloprotease inhibitor compounds in rat plasma.

2.3. LIQUID–SOLID EXTRACTION

When a liquid is extracted by a solid, phase A of the Nernst distribution law [equation (2.2)] refers to the liquid sample, and phase B, the extracting phase, represents the solid (or solid-supported liquid) phase:

$$K_D = \frac{[X]_B}{[X]_A} \qquad (2.2)$$

Classically, batch-mode liquid–solid extractions (LSEs), were used to concentrate semivolatile organic compounds from liquids into the solid phase. The liquid sample was placed in contact with the flowable, bulk solid extracting phase, an equilibrium between the two phases was allowed to occur, followed by physical separation (by decanting or filtering) of the solid and liquid phases. During the past quarter century, different approaches to solid-phase extractions of semivolatile organic compounds have emerged, including three described here: solid-phase extraction (SPE), solid-phase microextraction (SPME), and stir bar sorptive extraction (SBSE). Like LLE, SPE is designed to be a total, or exhaustive, extraction procedure for extracting the analyte completely from the entire sample volume via the sorbent. Unlike LLE, SPE is a nonequilibrium or pseudoequilibrium procedure. Unlike SPE, SPME is an equilibrium procedure that is not intended to be an exhaustive extraction procedure. SPME is an analytical technique in its own right that is inherently different from SPE or LLE. SBSE is physically a scaled-up version of SPME, but in principle it is more closely related to LLE (as it has been applied to date), in that it is an equilibrium partitioning procedure that unlike SPME more easily presents the opportunity to achieve exhaustive extraction. Each variation on the theme of liquid–solid extraction is an important addition to the analyst's arsenal of procedures for recovering semivolatile organics from liquids.

2.3.1. Sorption

To understand any of the solid-phase extraction techniques discussed in this chapter, it is first necessary to understand the physical–chemical processes of sorption. Schwarzenbach et al. [8] make the distinction between *absorption* (with a "b") meaning *into* a three-dimensional matrix, like water uptake in a sponge, and *adsorption* (with a "d") as meaning *onto* a two-dimensional surface (Figure 2.19). Absorption, also referred to as *partitioning*, occurs when analytes pass into the bulk of the extracting phase and are retained. Adsorption is the attraction of an analyte to a solid that results in accumulation of the analyte's concentration at porous surfaces of the solid. Absorption results from weaker interactive forces than adsorption. Because adsorption and/or absorption processes are sometimes difficult to distinguish experimentally [52] and often occur simultaneously, the general term *sorption* will be used here when referring to these processes. The term *sorbent* will refer to the solid extracting phase, including certain solid-supported liquid phases. To predict and optimize extraction, it is important for the analyst to be aware of the nature of the sorbent used.

Although different processes may dominate in different situations, it can be assumed that multiple steps occur during sorption of an organic compound from liquids "into" or "onto" a solid phase. Any of the steps may

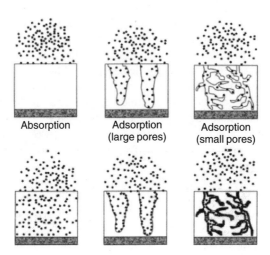

Figure 2.19. Schematic representation of absorptive versus adsorptive extraction and adsorption in small versus large pores. (Reprinted with permission from Ref. 51. Copyright © 2000 Elsevier Science.)

become a rate-limiting process in controlling sorption of an analyte. The analyte may interact with a solid-phase sorbent in at least four ways:

1. Through absorption, the analyte may interact with the sorbent by penetrating its three-dimensional structure, similar to water being absorbed by a sponge. Three-dimensional penetration into the sorbent is a particularly dominating process for solid-supported liquid phases. In the absorption process, analytes do not compete for sites; therefore, absorbents can have a high capacity for the analyte.

2. The analyte may interact two-dimensionally with the sorbent surface through adsorption due to intermolecular forces such as van der Waals or dipole–dipole interactions [53]. Surface interactions may result in displacement of water or other solvent molecules by the analyte. In the adsorption process, analytes may compete for sites; therefore, adsorbents have limited capacity. Three steps occur during the adsorption process on porous sorbents: *film diffusion* (when the analyte passes through a surface film to the solid-phase surface), *pore diffusion* (when the analyte passes through the pores of the solid-phase), and *adsorptive reaction* (when the analyte *binds, associates,* or *interacts* with the sorbent surface) [54].

3. If the compound is ionogenic (or ionizable) in aqueous solution (as discussed earlier), there may be an electrostatic attraction between the

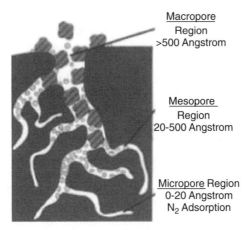

Macropore
Region
>500 Angstrom

Mesopore
Region
20-500 Angstrom

Micropore Region
0-20 Angstrom
N₂ Adsorption

Figure 2.20. Micro-, macro-, and mesopores in a porous sorbent. (Reprinted with permission from Ref. 56. Copyright © 1996 Barnebey Sutcliffe Corporation.)

analyte and charged sites on the sorbent surface. Sorbents specifically designed to exploit these types of ionic interactions are referred to as *ion-exchange* (either anion- or cation-exchange) *sorbents*.

4. Finally, it is possible that the analyte and the sorbent may be chemically reactive toward each other such that the analyte becomes covalently bonded to the solid-phase sorbent. This type of sorption is generally detrimental to analytical recovery and may lead to slow or reduced recovery, also termed *biphasic desorption*. All of these interactions have the potential of operating simultaneously during sorption [8,54,55].

For porous sorbents, most of the surface area is not on the outside of the particle but on the inside pores of the sorbent (Figure 2.20) in complex, interconnected networks of *micropores* (diameters smaller than 2 nm), *mesopores* (2 to 50 nm), also known as *transitional pores*, and *macropores* (greater than 50 nm) [57]. Most of the surface area is derived from the small-diameter micropores and the medium-diameter transitional pores [56]. Porous sorbents vary in pore size, shape, and tortuosity [58] and are characterized by properties such as particle diameter, pore diameter, pore volume, surface areas, and particle-size distribution.

Sorption tendency is dependent on the characters of the sorbent, the liquid sample (i.e., solvent) matrix, and the analyte. Much of the driving force for extracting semivolatile organics from liquids onto a solid sorbent results from the favorable energy gains achieved when transferring between phases.

For some of the sorbents discussed in this section on liquid–solid extraction, the solid-supported liquid sorbent phase performing the extraction may appear to the naked eye to be a solid when it is actually a liquid. The chromatographic method of employing two immiscible liquid phases, one of which is supported on a solid phase, was introduced by Martin and Synge in 1941 [59]. The liquid sorbent phase was mechanically added to the solid support material, which can lead to problems with *bleeding*, or *stripping*, of the liquid phase from the supporting solid material. Therefore, in the 1960s, covalently bonded phases were developed that overcame some of these problems by actually anchoring the liquid phase to the solid support. When the liquid extracting phase merely coats a solid support instead of bonding to the surface, it continues to behave primarily like a liquid; that is, the solid-supported liquid phase still has three-dimensional freedom of motion and the sorptive behavior observed is dominated by absorption processes. When the liquid extracting phase is covalently bonded to the surface, it no longer acts primarily like a bulk liquid, since there is freedom of movement in two dimensions only; translational and rotational movement are restricted; and retention on this type of phase can no longer be described solely by absorption processes. Retention on a liquid phase covalently bonded to a porous solid support does not result from a pure absorption or a pure adsorption mechanism.

Is analyte recovery using a solid-supported liquid phase classified as LLE or LSE? In Section 2.2.4, a process described as solid-supported LLE [49,50] was discussed in which the liquid sorbent phase was distributed on the surfaces of individual particles (Figure 2.18). The solid-supported phases in the LSE section have been arbitrarily distinguished as liquids mechanically supported on solid devices, such as the liquid-coated fused silica fibers used for SPME or the liquid-coated glass sheath of a stirring bar in used SBSE, rather than liquids supported on finely divided solid particles.

2.4. SOLID-PHASE EXTRACTION

The historical development of solid-phase extraction (SPE) has been traced by various authors [60,61]. After a long latency period (from biblical times to 1977) when the theoretical "science" of SPE was known but not frequently practiced, technological breakthroughs in sorbents and devices fueled the growth of SPE use that continues today. The modern era of SPE, which resulted in today's exponential growth in applications of this technique, began in 1977 when the Waters Corporation introduced commercially available, prepackaged disposable cartridges/columns containing bonded silica sorbents. The term *solid-phase extraction* was coined in 1982 by employees of the J.T. Baker Chemical Company [62–65].

The most commonly cited benefits of SPE that led to early advances relative to LLE are reduced analysis time, reduced cost, and reduced labor (because SPE is faster and requires less manipulation); reduced organic solvent consumption and disposal [66–68], which results in reduced analyst exposure to organic solvents; and reduced potential for formation of emulsions [43]. The potential for automation of SPE increased productivity because multiple simultaneous extractions can be accomplished [43]. SPE provides higher concentration factors (i.e., K_D) than LLE [68] and can be used to store analytes in a sorbed state or as a vehicle for chemical derivatization [69]. SPE is a multistaged separation technique providing greater opportunity for selective isolation than LLE [66,68,70,71], such as fractionation of the sample into different compounds or groups of compounds [69]. The use of SPE for all of these objectives is being exploited by today's SPE researchers.

Solid-phase extraction refers to the nonequilibrium, exhaustive removal of chemical constituents from a flowing liquid sample via retention on a contained solid sorbent and subsequent recovery of selected constituents by elution from the sorbent [72]. The introduction of sorbents exhibiting a very strong affinity for accumulating semivolatile organic compounds from water was the primary advance in the 1970s that propelled the technique into widespread use. The affinity, which was strong enough to be analytically useful from sorbents that were inexpensive enough to be economically feasible, was useful in both pharmaceutical and environmental applications. Mathematically, a *strong affinity* equates to a large K_D value in equation (2.2) because the concentration in the sorbent extracting phase, $[X]_B$, is large relative to the sample extracted. For this reason, SPE is sometimes referred to as *digital chromatography*, indicating the all-or-nothing extremes in the sorptive nature of these sorbents, caused by the strong attraction for the analyte by the sorbent. SPE drives liquid chromatographic mechanisms to their extreme, such that K_D approaches infinity, representing total accumulation of the analyte during retention, and K_D approaches zero during subsequent elution or release of the analyte.

Some analysts mistakenly refer to SPE sorbents as "filters" and the SPE process as "filtration" because of the porous character of many of the sorbents used for SPE. The molecules of the analyte that exist in true homogeneous solution in the sample are not filtered; they become associated with the solid phase through sorption. However, sorbent particles do act as *depth filters* toward particulate matter that is not in true homogeneous solution in the sample. Particulate matter can become lodged in the interstitial spaces between the sorbent particles or in the intraparticulate void volume, or pore space, within sorbent particles. The filtering of particulate matter is generally detrimental to the analysis and can lead to *plugging* of the extraction sorbent or channeling the flow through the sorbent. Fritz [73] summarizes that the severity of a plugging problem in SPE depends on (1) the concentration, type, and size of the particulates in the sample; (2) the pore size of the sorbent; and (3) the surface area of the sorbent bed.

While particulate matter can cause plugging and channeling of the sorbent in SPE as described above, analysts performing SPE extraction and other analytical procedures must also be concerned with the potential for the analyte's association with particulate and colloidal matter contamination in the sample. Complex equilibria govern partitioning of organic analytes among the solution phase, colloidal material, and suspended particulate matter. Depending on the chemical nature of the analyte and the contamination, some of the analyte molecules can become sorbed to the contaminating particulate and/or colloidal matter in the sample [74]. Analytes can adhere to biological particulates such as cellular debris or bind to colloidal proteins. Similarly, analytes can adhere to environmental particulates or associate with colloidal humic substances. If the sample is not filtered, particulates can partially or entirely elute from the sorbent, leading to both a dissolved and particulate result when the sample is analyzed [75]. In addition to concern about the potential for suspended solids in the water sample plugging the SPE sorbent and analytes of interest adsorbing onto particulates, loss of the analyte may occur if small particulates pass through the pores of the sorbent bed [73].

To avoid these problems and ensure consistent results, sample particulate matter should be removed by filtration prior to SPE analysis [43]. If measuring the degree to which the analyte is bound to contaminants in the solution or, conversely, the degree to which the analyte is unassociated, or in true solution is important, the sample should be filtered prior to analysis by SPE or LLE. Glass-fiber filters, which have no organic binders, should be inert toward the analyte of interest while trapping particulate matter [43]. Particles with a diameter of 1 μm or greater tend to settle out of solution by gravity. Nominal filter sizes of 0.7, 0.45, or 0.22 μm are commonly reported in literature in conjunction with preparation of a sample for SPE. An appropriate level of filtration should be determined for the particular sample matrix being analyzed and used consistently prior to SPE analysis. The material retained on the filter may be analyzed separately to determine the level of bound analyte. The analyst must carefully assess whether rinsing the filter with water or an organic solvent and recombining the rinsings with the filtered sample meet the objectives sought and are appropriate for the given analysis.

Prefiltering samples prior to SPE in a standardized manner using glass-fiber filters having no organic binders and testing the analytes of interest to establish that they are not adsorbed on the filter selected is recommended [43]. Alternatively, Simpson and Wynne [76] present the counter viewpoint that sample filtration is not always appropriate when the analyte adheres to biological or environmental particulates. They suggest that SPE devices more tolerent to the buildup of matrix solids, such as in-line filters, high-flow frits, or large-particle-size beds, should be tested. The analyst must be knowledgeable about the particulate/colloidal matter present in the sample matrix in order to consider these technical decisions about sample processing.

2.4.1. Sorbents in SPE

Appropriate SPE sorbent selection is critical to obtaining efficient SPE recovery of semivolatile organics from liquids. Henry [58] notes that an SPE sorbent "must be able to sorb rapidly and reproducibly, defined quantities of sample components of interest." Fritz [73] states that "successful SPE has two major requirements: (1) a high, reproducible percentage of the analytical solutes must be taken up by the solid extractant; and (2) the solutes must then be easily and completely eluted from the solid particles." The sorption process must be reversible. In addition to reversible sorption, SPE sorbents should be porous with large surface areas, be free of leachable impurities, exhibit stability toward the sample matrix and the elution solvents, and have good surface contact with the sample solution [68,73].

   Obviously, knowledge of the chemistry and character of commonly used SPE sorbents is important to achieving successful extractions. Liska [60] describes developments from the late 1960s until the early 1980s as the "age of searching" for a universal SPE sorbent that culminated in the introduction of polymeric materials and bonded silicas. These sorbents have proven useful for a wide variety of applications. However, the realization that no single optimal sorbent for all purposes exists prompts current efforts to optimize a sorbent for a particular application [60], that is, for a specific analyte in a specific matrix. Poole et al. [77] categorize the SPE sorbents available today as either general purpose, class specific, or compound specific. This discussion covers polar, polymeric, bonded silica, and graphitized carbon sorbents of general applicability as well as functionalized polymeric resins, ion-exchange sorbents, controlled-access sorbents, immunoaffinity sorbents, and molecularly imprinted polymers designed for more specific purposes.

Polar Sorbents

The earliest applications of *chromatography*, a term coined by Tswett in 1906, used polar sorbents to separate analytes dissolved in nonpolar solvents. Using light petroleum as the nonpolar mobile phase, Tswett separated a colored extract from leaves using column chromatography on a polar calcium carbonate column [78,79]. The alternate system, in which the sorbent is nonpolar while a polar solvent is used, was not used in chromatography until the late 1940s to early 1950s [80–83]. Howard and Martin [83] introduced the term *reversed-phase* to describe separation of fatty acids using solid-supported liquid paraffin or *n*-octane as nonpolar stationary phases that were eluted with polar aqueous solvents. At that time, these systems appeared to be "reversed" to the "normal" arrangement of polar stationary phases used with less polar eluents. Although reversed-phase applications outnumber normal-phase chromatographic applications today, the nomenclature still applies.

The most common polar sorbents used for normal-phase SPE are silica $(SiO_2)_x$, alumina (Al_2O_3), magnesium silicate $(MgSiO_3$ or Florisil$)$, and the bonded silica sorbents in which silica is reacted with highly polar functional groups to produce aminopropyl $[(SiO_2)_x-(CH_2)_3NH_2]$-, cyanopropyl $[(SiO_2)_x-(CH_2)_3CN]$-, and diol $[(SiO_2)_x-(CH_2)_3OCH_2CH(OH)CH_2(OH)]$- modified silica sorbents (Figure 2.21). Polar SPE sorbents are often used to

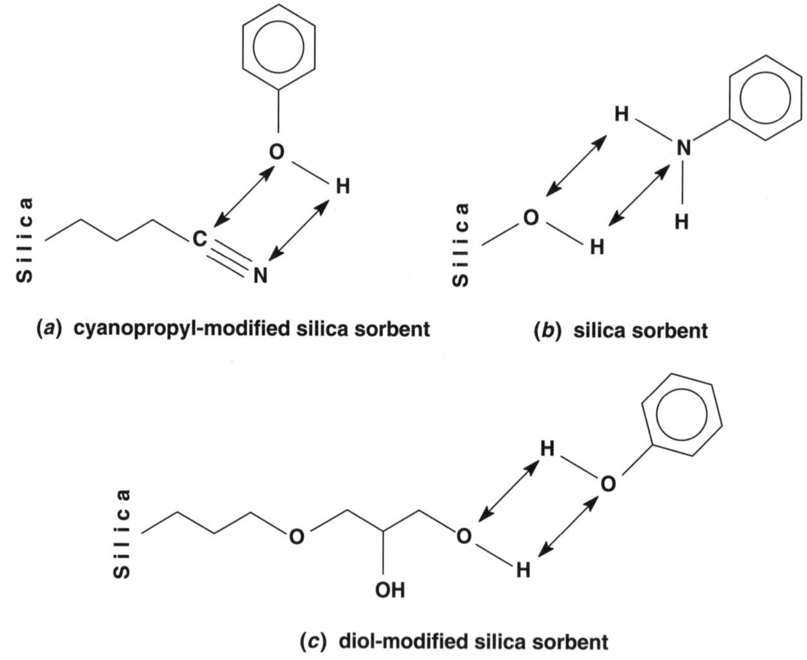

(a) cyanopropyl-modified silica sorbent   (b) silica sorbent

(c) diol-modified silica sorbent

Figure 2.21. Interactions between analytes and polar sorbents via dipolar attraction or hydrogen bonding.

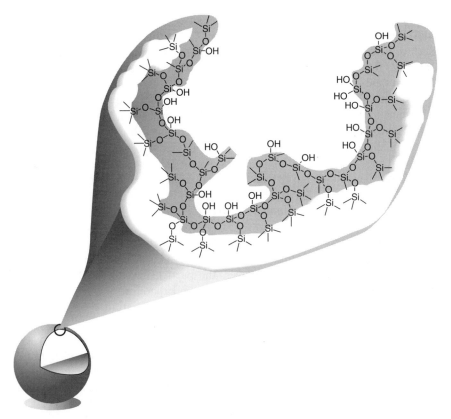

Figure 2.22. Representation of an unbonded silica particle. (Reprinted with permission from Ref. 84. Copyright © 2002 Waters Corporation.)

remove matrix interferences from organic extracts of plant and animal tissue [73]. The hydrophilic matrix components are retained by the polar sorbent while the analyte of interest is eluted from the sorbent. The interactions between solute and sorbent are controlled by strong polar forces including hydrogen bonding, dipole–dipole interactions, π–π interactions, and induced dipole–dipole interactions [75].

Porous silica (Figure 2.22) is an inorganic polymer $(SiO_2)_x$ used directly as a sorbent itself and for the preparation of an important family of sorbents known as chemically bonded silicas that are discussed later. Silica consists of siloxane backbone bridges, –Si–O–Si–, and silanol groups, –Si–OH. Colin and Guiochon [85] proposed that there are five main types of silanol group sites on the surface of a silica particle, depending on the method of preparation and pretreatment of the silica, including free silanol, silanol with physically adsorbed water, dehydrated oxide, geminal silanol, and bound and reactive silanol. Porous silica consists of a directly accessible external surface and internal pores accessible only to molecules approximately less than 12,000 Da [86]. Pesek and Matyska [87] have reviewed the chemical and physical properties of silica.

Silica particles used for SPE sorbents are typically irregularly shaped, 40 to 60 μm in diameter. Silica particles used for sorbents in high-performance liquid chromatographic (HPLC) columns are generally spherical and 3 to 5 μm in diameter. Due to the differences in size and shape, SPE sorbents are less expensive than HPLC sorbents. Much greater pressures are required to pump solvents through the smaller particle sizes used in HPLC.

Apolar Polymeric Resins

Synthetic styrene–divinylbenzene and other polymers, particularly the trade-marked XAD resins developed by Rohm & Haas, were used for SPE in the late 1960s and early 1970s. However, the particle size of the XAD resins is too large for efficient SPE applications, and therefore the resins require additional grinding and sizing. Also, intensive purification procedures are needed for XAD resins [73,75].

In the latter half of the 1990s, porous, highly cross-linked polystyrene–divinylbenzene (PS-DVB) resins with smaller, spherical particle sizes more suitable for SPE uses became available (Figure 2.23). The new generation of apolar polymeric resins is produced in more purified form, reducing the level of impurities extracted from the sorbent. Polymeric resins are discussed in more detail by Huck and Bonn [69], Fritz [73], Thurman and Mills [75], and Pesek and Matyska [87].

The enhanced performance of PS-DVB resins is due to their highly hydrophobic character and greater surface area as compared to the bonded silica sorbents, which are discussed in the following section. The strong sorption properties of PS-DVB resins may arise from the aromatic, poly-

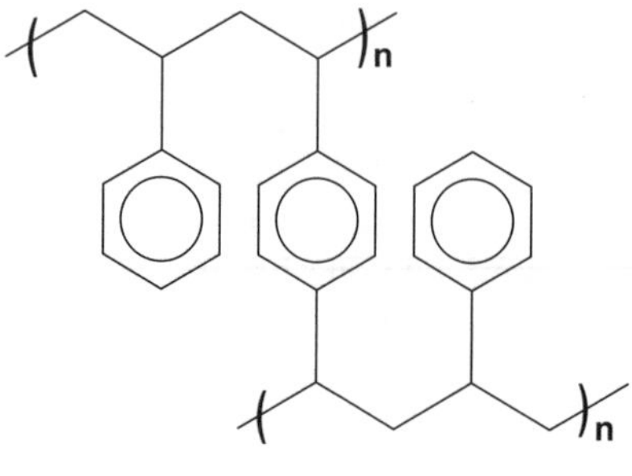

Figure 2.23. Cross-linked styrene–divinylbenzene copolymer.

meric structure that can interact with aromatic analytes via π–π interactions. However, because PS-DVB sorbents are highly hydrophobic, they are less selective. Also, PS-DVB sorbents exhibit low retention of polar analytes.

Polymeric organic sorbents can reportedly be used at virtually any pH, 2 to 12 [75] or 0 to 14 [73,88], increasing the potential to analyze simultaneously multiresidue samples containing acidic, basic, and neutral compounds. Polymeric sorbents contain no silanol groups and thereby avoid the problems caused by residual silanol groups when bonded silica sorbents are used [73,75].

The PS-DVB sorbents can be more retentive than the bonded silica sorbents. Polymeric sorbents have been shown to be capable of retaining chemicals in their ionized form even at neutral pH. Pichon et al. [88] reported SPE recovery of selected acidic herbicides using a styrene–divinylbenzene sorbent so retentive that no adjustment of the pH of the solution was necessary to achieve retention from water samples at pH 7. At pH 7 the analytes were ionized and thereby retained in their ionic form. To effect retention of acidic compounds in their nonionized form using bonded silica sorbents, it is necessary to lower the pH of the sample to approximately 2. Analysis at neutral pH can be preferable to reduced pH because at lower pHs undesirable matrix contaminants, such as humic substances in environmental samples, can be coextracted and coeluted with the analytes of interest and subsequently may interfere with chromatographic analyses.

Bonded Silica Sorbents

The first class of sorbents used for modern-era SPE were bonded-phase silicas. In the early 1970s, bonded silica sorbents found popularity as a stationary phase for HPLC. HPLC was not commonly used until the early 1970s, nor SPE until the late 1970s, until the application of silanized, or bonded silica sorbents, was realized. May et al. [89] and Little and Fallick [90] are credited with the first reports of applying bonded phases to accumulate organic compunds from water [60]. The first article about SPE on commercially available bonded-phase silica (an octadecyl, C_{18}, phase) was published by Subden et al. [91] and described the cleanup of histamines from wines.

Chemically bonded silica sorbents are currently the most commonly used solid phase for SPE. Bonded stationary phases are prepared by "grafting" organic nonpolar, polar, or ionic ligands (denoted R) to a silica particle via covalent reaction with the silanol groups on its surface. The importance of this advancement to chromatography in general and particularly to solid-phase extraction was the ability to produce highly hydrophobic phases that were more attractive to organic solutes in aqueous solution than any other

PRINCIPLES OF EXTRACTION

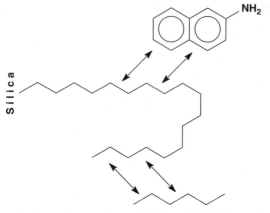

reversed-phase octadecyl (C$_{18}$) modified silica sorbent

Figure 2.24. Interactions between analytes and nonpolar bonded silica sorbents via van der Waals forces.

sorbents available at the time. Reversed-phase bonded silica sorbents having alkyl groups covalently bonded to the silica gel backbone interact primarily with analytes via van der Waals forces (Figure 2.24).

Bonded-phase sorbents are stable to aqueous solvents over a pH range of 1 to 8.5, above which the silica backbone itself begins to dissolve and below which the Si–C bond is attacked. Manufacturers have continued to extend these ranges through improved products, and researchers have stretched the limits of these restrictions. The development of bonded silica sorbents led to a proliferation of pharmaceutical and environmental applications for extracting semivolatile organics from aqueous solution.

The bonded phases produced by manufacturers vary according to the nature of the silica used to prepare the bonded phase and in the reactants and reaction conditions used. The variations are closely guarded, proprietary manufacturing processes. However, it is generally known that the most common commercially manufactured bonded-phase sorbents are based on chemical reaction between silica and organosilanes via the silanol groups on the silica surface to produce chemically stable Si–O–Si–C covalent linkages to the silica backbone [75,87]. Nonpolar, polar, or ionic bonded phases can be prepared by varying the nature of the organic moiety bonded to the silica surface.

Bonded phases can be obtained as monomeric or polymeric coverage of an organic ligand group, R, on the silica surface depending on whether a monofunctional (R$_3$SiX) or a trifunctional (RSiX$_3$) reactant is used, respec-

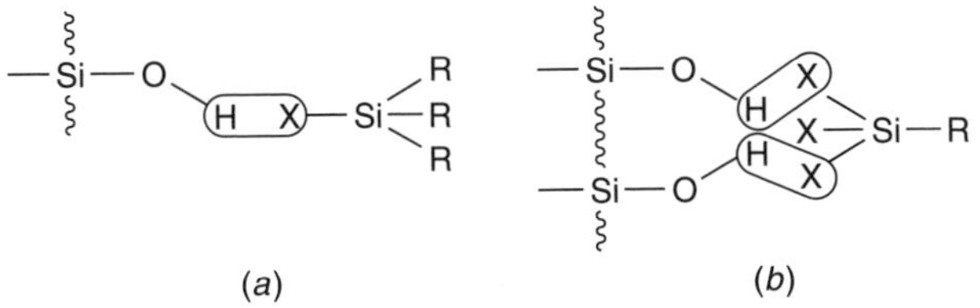

Figure 2.25. Reaction of a (a) monofunctional or (b) trifunctional organosilane with silanol groups on the silica surface.

tively (Figure 2.25). The organosilane contains a reactive group, X, that will interact chemically with the silanol groups on the silica surface. Typically, the reactant is an organochloro- or organoalkoxysilane in which the moiety, X, is chloro, methoxy, or ethoxy.

One or two Si–X groups can remain unreacted per bonded functional group because of the stoichiometry observed when trifunctional reactant modifiers are used. Hydrolysis of the Si–X group occurs in the workup procedure and results in the re-formation of new silanol groups (Figure 2.26), thereby reducing the hydrophobic character of the sorbent surface. The reactions result in the formation of a cross-linked polymeric network and/or a multilayer adsorbent. The monomeric types of bonded sorbents are obtained by using monofunctional organosilanes such as alkyldimethylmonochlorosilane to preclude the possibility of re-forming unreacted silanol groups.

A polymeric surface structure can result in slower mass transfer of the analyte in the polymer coating compared with the more "brush- or bristle-like" bonding of monomeric phases and thereby lead to higher efficiencies with monomeric phases. However, Thurman and Mills [75] note that the trifunctional reagent yields a phase that is more stable to acid because the

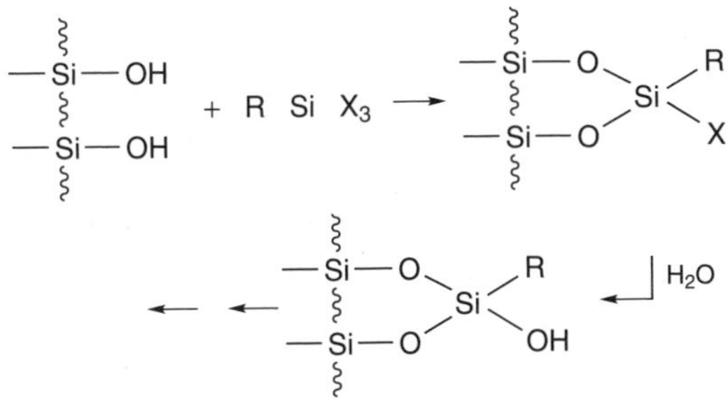

Figure 2.26. Reformation of additional silanol groups during processing when trifunctional modifiers are used.

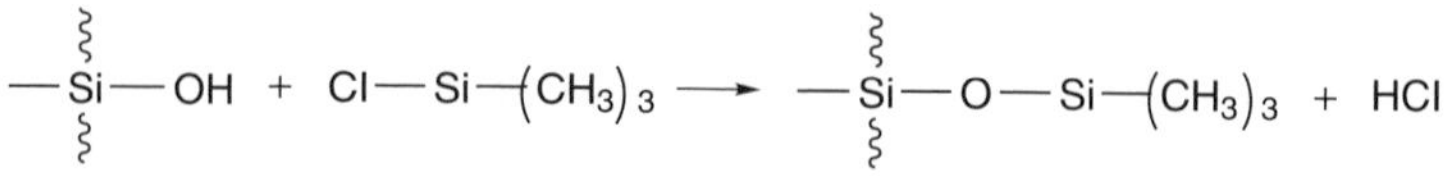

PRINCIPLES OF EXTRACTION

$$—\underset{\overset{|}{\underset{|}{}}}{\overset{|}{Si}}—OH \;+\; Cl—Si—(CH_3)_3 \;\longrightarrow\; —\underset{\overset{|}{\underset{|}{}}}{\overset{|}{Si}}—O—Si—(CH_3)_3 \;+\; HCl$$

Figure 2.27. Accessible silanol groups are endcapped by reaction with trimethylchlorosilane.

organosilane is attached to the silica surface by multiple linkages to the silica backbone.

Silanol groups can be left unreacted on the silica surface, due to reaction conditions or steric inhibition, or generated during subsequent processing of polymeric bonded phases. In either case, they can have an effect on the sorption of the target analyte. Hennion [92] notes that silanol groups are uncharged at pH 2 and become increasingly dissociated above pH 2. Experimentally observable effects due to negatively charged silanols are evident above pH 4. The presence of unmasked silanol groups may have a positive, negative, or little effect, depending on the specific analyte of interest [93]. A positively charged competing base, such as triethylamine or tetrabutylammonium hydrogen sulfate, can be added to the sample to mask residual silanols.

To reduce the number of accessible silanol groups remaining on the sorbent, a technique known as *capping* or *endcapping* is sometimes used. With this technique, a small silane molecule such as trimethylchlorosilane is allowed to react with the bonded silica (Figure 2.27) to produce a more hydrophobic surface.

When using bonded silica SPE sorbents (or HPLC columns), a monomeric or polymeric phase may be best for a given analyte–matrix situation. Similarly, an endcapped or unendcapped product may be best. The preceding discussion should be helpful to analysts when consulting with manufacturers regarding the nature of the bonded surface of the sorbents produced. Hennion [92] recently published a table listing characteristics of some common, commercially available bonded silicas, including data on porosity, mean particle diameter, functionality of the silane used for bonding (i.e., mono- or trifunctional), endcapping, and percent carbon content.

Bonded silica sorbents are commercially available with many variations in the organic ligand group, R. Common bonded phases produced for reversed-phase applications include hydrophobic, aliphatic alkyl groups, such as octadecyl (C_{18}), octyl (C_8), ethyl (C_2), or cyclohexyl, covalently bonded to the silica gel backbone. Aromatic phenyl groups can also be attached. The R ligand can contain cyanopropyl or diol hydrophilic functional groups that result in polar sorbents used in normal-phase applications. Ionic functional groups, including carboxylic acid, sulfonic acid, aminopropyl, or quaternary amines, can also be bonded to the silica sorbent to produce ion-exchange sorbents.

The primary disadvantages of the bonded silica sorbents are their limited pH stability and the ubiquitous presence of residual silanol groups. Despite these difficulties, the bonded silicas have been the workhorse sorbents of SPE applications for the last two decades and are still the most commonly used SPE sorbents.

Graphitized Carbon Sorbents

Graphitized carbon sorbents are earning a reputation for the successful extraction of very polar, extremely water soluble organic compounds from aqueous samples. The retention behavior of the graphitized carbon sorbents is different than that of the apolar polymeric resins or the hydrophobic bonded silica sorbents. Two types of graphitized carbon sorbents, graphitized carbon blacks (GCBs) and porous graphitic carbons (PGCs), are commercially available for SPE applications.

GCBs do not have micropores and are composed of a nearly homogeneous surface array of graphitelike carbon atoms. Polar adsorption sites on GCBs arise from surface oxygen complexes that are few in number but interact strongly with polar compounds. Therefore, GCBs behave both as a nonspecific sorbent via van der Waals interactions and as an anion-exchange sorbent via electrostatic interactions [92,94,95]. GCBs have the potential for simultaneous extraction of neutral, basic, and acidic compounds. In some cases no pH adjustment of the sample is necessary. Desorption can be difficult because GCB is very retentive.

PGC sorbents have even more highly homogeneous hydrophobic surfaces than GCB sorbents. PGCs are macroporous materials composed of flat, two-dimensional layers of carbon atoms arranged in graphitic structure. The flat, homogeneous surface of PGC arranged in layers of carbons with delocalized π electrons makes it uniquely capable of selective fractionation between planar and nonplanar analytes such as the polychlorinated biphenyls [92,94,95].

Functionalized Polymeric Resins

Adding polar functional groups to cross-linked, apolar polymeric resins by covalent chemical modification has developed particularly for generation of SPE sorbents suitable for recovery of polar compounds. Hydrophilic functional groups such as acetyl, benzoyl, o-carboxybenzoyl, 2-carboxy-3/4-nitrobenzoyl, 2,4-dicarboxybenzoyl, hydroxymethyl, sulfonate, trimethyl-ammonium, and tetrakis(p-carboxyphenyl)porphyrin have been chemically introduced into the structural backbone of PS-DVB copolymers [96]. Generation of a macroporous copolymer consisting of two monomer components, divinylbenzene (lipophilic) and N-vinylpyrrolidone (hydrophilic), produced a hydrophilically–lipophilically balanced SPE sorbent [69]. Chemically modifying apolar polymeric sorbents in this way improves wettability, surface contact between the aqueous sample and the sorbent surface, and mass transfer by making the surface of the sorbent less hydrophobic (i.e., more hydrophilic [73,75,96,97]). The sulfonate and trimethylammonium derivatives are used as ion-exchange sorbents, a type of sorbent that is considered in a later section.

Higher breakthrough volumes (i.e., indicating greater attraction of the sorbent for the analyte) for selected polar analytes have been observed when the hydrophilic functionalized polymeric resins are used as compared to classical hydrophobic bonded silicas or nonfunctionalized, apolar polymeric resins. In addition to having a greater capacity for polar compounds, functionalized polymeric resins provide better surface contact with aqueous samples. The bonded silica sorbents and the polymeric resins (discussed in earlier sections) have hydrophobic surfaces and require pretreatment, or conditioning, with a hydrophilic solvent to activate the surface to sorb analytes. Using covalent bonding to incorporate hydrophilic character permanently in the sorbent ensures that it will not be leached from the sorbent as are the common hydrophilic solvents (e.g., methanol, acetonitrile, or acetone) used to condition bonded silica sorbents or polymeric resins [69,73,96].

Ion-Exchange Sorbents

SPE sorbents for ion exchange are available based on either apolar polymeric resins or bonded silica sorbents. Ion-exchange sorbents contain ionized functional groups such as quaternary amines or sulfonic acids, or ionizable functional groups such as primary/secondary amines or carboxylic acids. The charged functional group on the sorbent associates with the oppositely charged counterion through an electrostatic, or ionic, bond (Figure 2.28).

The functional group on the sorbent can be positively or negatively charged. When the sorbent contains a positively charged functional group and the exchangeable counterion on the analyte in the liquid sample matrix is negatively charged, the accumulation process is called *anion exchange.* Conversely, if the functional group on the sorbent surface is negatively charged and the exchangeable counterion on the analyte in the liquid sample matrix is positively charged, the accumulation process is called *cation exchange.*

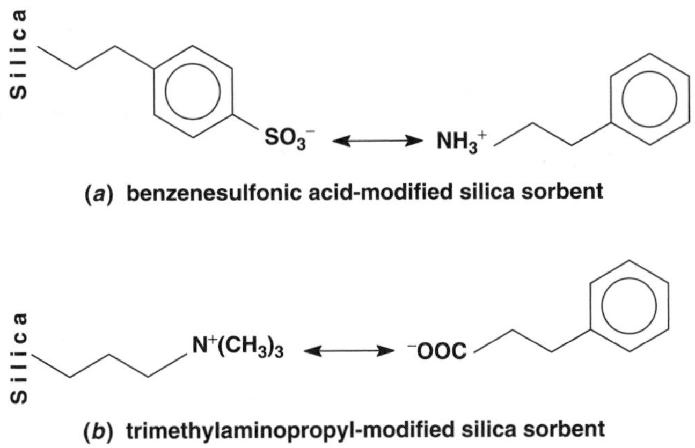
Silica
SO3−
NH3+
(a) benzenesulfonic acid-modified silica sorbent
Silica
N+(CH3)3
−OOC
(b) trimethylaminopropyl-modified silica sorbent

Figure 2.28. Interactions between analytes and ion-exchange sorbents: (*a*) strong cation-exchange sorbent and (*b*) strong anion-exchange sorbent.

The theoretical principles of acid–base equilibria discussed earlier in this chapter apply to the sorbent, the analyte, and the sample in ion-exchange processes. The pH of the sample matrix must be adjusted in consideration of the pK_a of the sorbent (Table 2.5) and the pK_a of the analyte such that the sorbent and the analyte are oppositely charged under sample loading conditions.

Anion-exchange sorbents for SPE contain weakly basic functional groups such as primary or secondary amines which are charged under low-pH conditions or strongly basic quaternary ammonium groups which are charged at all pHs. Cation-exchange sorbents for SPE contain weakly acidic functional

Table 2.5. Ionization Constants of Ion-Exchange Sorbents

Ion-Exchange Sorbents	Sorbent pK_a
Cation exchange	
$-CH_2CH_2COOH$	4.8
$-CH_2CH_2CH_2SO_3H$	<1.0
$-CH_2CH_2\phi SO_3H$	≪1.0
Anion exchange	
$-CH_2CH_2CH_2NHCH_2CH_2NH_2$	10.1 and 10.9
$-CH_2CH_2CH_2N(CH_2CH_3)_2$	10.7
$-CH_2CH_2CH_2N^+(CH_3)_3Cl^-$	Always charged

Source: Data from Ref. 98.

groups such as carboxylic acids, which are charged under high-pH conditions, or strongly acidic aromatic or aliphatic sulfonic acid groups, which are charged at all pH levels. "Weakly" acidic or basic ion-exchange sorbents are pH dependent because they dissociate incompletely, while "strongly" acidic or basic ion-exchange sorbents are pH independent because they dissociate completely.

In SPE, the ionic interaction between an ion-exchange sorbent and an analyte is a stronger attraction than the hydrophobic interactions achievable with apolar polymeric resins or with aliphatic/aromatic bonded silica sorbents. In ion exchange, the distribution coefficient, K_D [equation (2.2)], generally increases with the charge and bulkiness of the exchanging ion [73]. The kinetics of the ion-exchange process is slower than with nonpolar or polar interaction mechanisms. Simpson [99] discusses the kinetic effects on SPE by ion-exchange extraction.

The counterion associated with the sorbent when it is manufactured is replaced by another ion of like charge existing on the analyte to achieve retention. However, analyte retention is affected by the ionic strength of the sample matrix because other ions present will compete with the analyte of interest for retention by ion-exchange mechanisms [75].

Controlled-Access Sorbents

Controlled-access sorbents are intended to be either "inclusive" or "exclusive" of large molecules and macromolecules. Wide-pore, or large-pore, sorbents are designed intentionally to allow accessibility of macromolecules to the internal pore structure of the sorbent such that they will be retained. Conventional SPE sorbents commonly have pores of 60 Å, whereas wide-pore SPE sorbents have pores of 275 to 300 Å [75].

Conversely, restricted access materials or restricted access media (RAM) retain small molecules while excluding macromolecules such as biological proteins in their presence (Figure 2.29). Small molecules are retained by sorption processes in the pores of the sorbent while the large molecules are excluded and elute at the interstitial volume of the sorbent. This separation leads to size-selective disposal of interfering macromolecular matrix constituents.

Unlike conventional steric exclusion sorbents, RAM sorbents exhibit *bifunctional* or *dual-zone* character, in that the inner and outer surfaces are different. The outer surface is designed to exclude macromolecules physically and is rendered chemically hydrophilic to discourage retention of biomolecules. Small molecules penetrate to an inner surface, where they are retained by any of the various other sorptive surface chemistries already discussed [92].

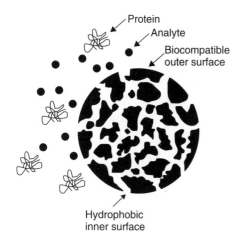

Protein
Analyte
Biocompatible
outer surface

Hydrophobic
inner surface

Figure 2.29. Schematic representation of a sorbent particle for restricted-access media chromatography. This medium allows proteins and macromolecules to be excluded and elute in the solvent front, while small analyte molecules enter the pores and are retained. (Reprinted with permission from Ref. 100. Copyright © 2000 Elsevier Science.)

Immunoaffinity or Immunosorbents

The driving force behind development of more selective sorbents is minimizing the problem of coextracting matrix interferences that are usually present in much greater concentration than the trace levels of the analyte of interest. More selective sorbents also permit extraction of larger sample volumes, thereby reducing the level of detection of the analyte of interest.

A recent approach to producing highly selective sorbents for SPE is based on molecular recognition technology and utilizes antibodies immobilized by covalent reaction onto solid supports such as silica (Figure 2.30). Preparation of immunoaffinity sorbents for SPE was reviewed by Stevenson [101] and Stevenson et al. [102]. Using immunosorbents, efficient cleanup is achieved from complex biological and environmental samples.

Antibodies can cross-react with closely related analytes within a chemical family. This disadvantage has been used to advantage in SPE. Therefore, immunosorbents have been designed for a single analyte, a single analyte and its metabolites, or a class of structurally related analytes [92]. The approach is therefore useful for chemical class-specific screening of compounds, such as triazines, phenylureas, or polyaromatic hydrocarbons. The specificity of the antibody is used for extraction by chemical class. Following SPE, analytical chromatographic techniques such as HPLC and GC separate structurally similar analytes for quantification.

Molecularly Imprinted Polymeric Sorbents

Another approach to selective SPE based on molecular recognition is the development of molecularly imprinted polymers (MIPs), which are said to

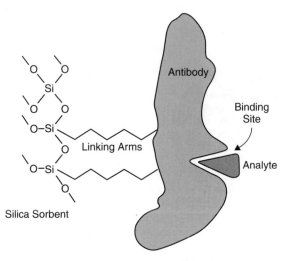

Figure 2.30. Diagrammatic representation of an immunoaffinity SPE binding an analyte. (Reprinted with permission from Ref. 75. Copyright © 1998 John Wiley & Sons, Inc.)

be an attempt to synthesize antibody mimics [92,101]. Produced by chemical synthesis, MIPs are less expensive and more easily and reproducibly prepared than immunosorbents that are prepared from biologically derived antibodies [102].

SPE sorbents that are very selective for a specific analyte are produced by preparing (MIPs) in which the target analyte is present as a molecular template when the polymer is formed. Sellergren [103] is credited with first reporting of the use of MIP sorbents for SPE. Subsequently, MIP-SPE has been applied to several biological and environmental samples [92,104–106].

MIP sorbents are prepared by combining the template molecule with a monomer and a cross-linking agent that causes a rigid polymer to form around the template (Figure 2.31). When the template is removed, the polymer has *cavities* or *imprints* designed to retain the analyte selectively. Retention of the analyte on these sorbents is due to shape recognition, but other physicochemical properties, including hydrogen bonding, ionic interactions, and hydrophobic interactions, are important to retention as well [92,104,107].

MIP-SPE sorbents are stable in both aqueous and organic solvents and are very selective for the analyte of interest. Increased selectivity relative to other sorbents produces increased sensitivity because larger sample volumes can be extracted. Also, increased selectivity results in efficient sample cleanup of the analyte in the presence of complex biological or environmental matrix

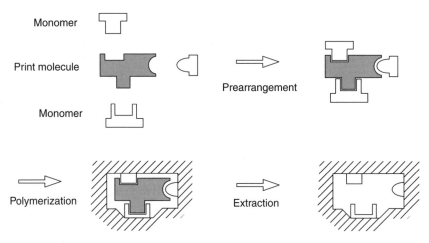

Monomer

Print molecule

Prearrangement

Monomer

Polymerization

Extraction

Figure 2.31. Schematic depiction of the preparation of molecular imprints. (Reprinted with permission from Ref. 105. Copyright © 2000 Elsevier Science.)

interferences. However, desorption is usually more difficult if any sorbent has increased affinity for the analyte.

One problem noted in MIP-SPE is incomplete removal of the template molecule from the polymer, resulting in leaching of the analyte during subsequent trace analyses. Stringent cleaning of the sorbent and analytical confirmation of the lack of interfering compound can reduce this problem. Alternatively, another approach has been to use a structural analog of the target analyte as the template used to create the MIP sorbent [105,106]. This approach is successful if the structural analog creates an imprint that is selective for the target analyte and if the structural analog and the target analyte can be separated chromatographically for quantitation after extraction.

Mixed-Mode Sorbents and Multiple-Mode Approaches

Each of the types of SPE sorbents discussed retains analytes through a primary mechanism, such as by van der Waals interactions, polar dipole–dipole forces, hydrogen bonding, or electrostatic forces. However, sorbents often exhibit retention by a secondary mechanism as well. Bonded silica ion-exchange sorbents primarily exhibit electrostatic interactions, but the analyte also experiences nonpolar interaction with the bonded ligand. Nonpolar bonded silicas primarily retain analytes by hydrophobic interactions but exhibit a dual-retention mechanism, due to the silica backbone and the presence of unreacted surface silanol groups [72]. Recognition that a dual-

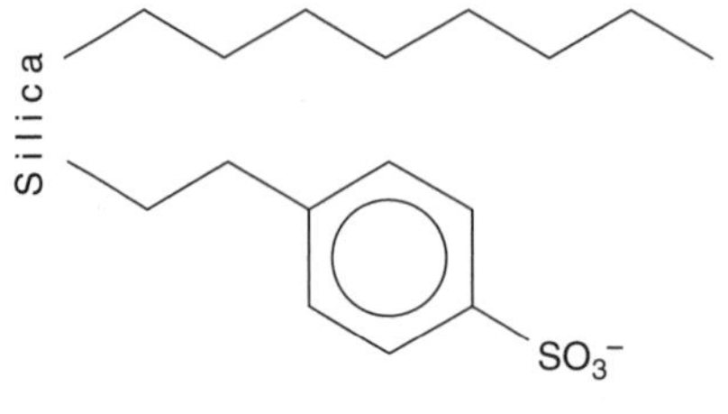

Figure 2.32. Example of a mixed-mode sorbent consisting of silica modified with octyl (C_8) alkyl chains and strong cation-exchange sites bonded on the same sorbent particle.

retention mechanism is not always detrimental to an analysis [93] has led to the production of mixed-mode sorbents by design. The development of mixed-mode sorbents and multiple-mode approaches to capitalize on multiple retention mechanisms has evolved as a logical extension of the observation of secondary interactions [108].

A mixed-mode sorbent is designed chemically to have multiple retentive sites on an individual particle (Figure 2.32). These sites exploit different retention mechanisms by chemically incorporating different ligands on the same sorbent. For example, sorbents have been manufactured that contain hydrophobic alkyl chains and cation-exchange sites on the same sorbent particle [92]. Mixed-mode sorbents exploit interaction with different functional groups on a single analyte or different functional groups on multiple analytes. Mixed-mode SPE sorbents are particularly useful for the extraction of analytes from bodily fluids [68].

Alternatively, there are several different mechanical approaches to achieving multiple-mode retention (Figure 2.33). Sorbent particles of different types (i.e., a hydrophobic sorbent and an ion-exchange sorbent) that exhibit separate mechanisms of retention can be homogeneously *admixed*, or *blended*, in the same column, or they can be *layered* into the same column by packing one phase over another [97]. Additionally, multiple phases can be *stacked* by arranging in tandem series sorbents of different retention mechanisms contained in separate columns. The technique of stacking or sequencing sorbents in tandem columns, termed *chromatographic mode sequencing* (CMS), can produce very selective isolation of analytes [109].

2.4.2. Sorbent Selection

Thurman and Mills [75] point out that knowing the analyte structure is the clue to effective isolation by SPE. A sorbent selection chart (Figure 2.34) is a useful guide for matching the analyte with the appropriate sorbent. Most manufacturers of SPE sorbents provide such guidelines either in printed product literature or on the Internet. To use a sorbent selection scheme, the analyst must be prepared to answer the following questions:

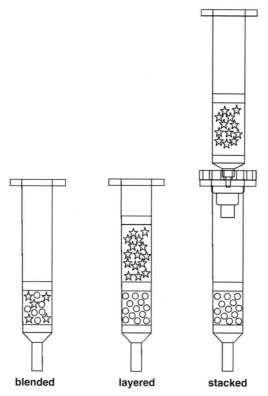

blended   layered   stacked

Figure 2.33. SPE multiple-mode approaches.

- Is the sample matrix miscible primarily with water or organic solvents?
- If the sample matrix is water soluble, is the analyte ionized or non-ionized?
- If ionized, is the analyte permanently ionized (pH independent) or ionizable (pH dependent); is the analyte anionic or cationic?
- If the analyte is nonionized or ionization can be controlled (by pH suppression or ion pairing), is it nonpolar (hydrophobic), moderately polar, or polar (hydrophilic)?
- If the sample matrix is organic solvent miscible, is it miscible only in nonpolar organic solvents such as hexane, or is it also miscible in polar organic solvents such as methanol?
- Is the analyte nonpolar (hydrophobic), moderately polar, or polar (hydrophilic)?

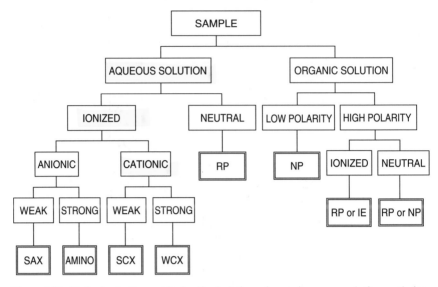

Figure 2.34. Method selection guide for the isolation of organic compounds from solution. SAX, strong anion exchanger; SCX, strong cation exchanger; WCX, weak cation exchanger; RP, reversed-phase sampling conditions; NP, normal-phase sampling conditions; IE, ion-exchange sampling conditions. (Reprinted with permission from Ref. 77. Copyright © 2000 Elsevier Science.)

Various types of sorbents used for SPE can be grouped (Table 2.6) according to the primary mechanism by which the sorbent and the analyte interact [32,72]. Reversed-phase bonded silica sorbents having alkyl groups such as octadecyl (C_{18}, C18), octyl (C_8, C8), or ethyl (C_2, C2) covalently bonded to the silica gel backbone or cyclohexyl (CH) or phenyl groups and sorbents composed of polymeric resins such as polystyrene–divinylbenzene

Table 2.6. SPE Sorbent–Analyte Interaction Mechanisms

Primary Interaction Mechanism	Sorbents	Energy of Interaction[a] (kcal/mol)
Van der Waals	Octadecyl, octyl, ethyl, phenyl, cyclohexyl, styrene–divinylbenzene, graphitized carbon	1–10
Polar/dipole–dipole	Cyano, silica, alumina Florisil	1–10
Hydrogen bonding	Amino, diol	5–10
Electrostatic	Cation exchange, anion exchange	50–200

[a] Data from Ref. 97.

interact primarily with analytes via van der Waals forces. Nonionic water-soluble compounds can be retained by reversed-phase sorbents but may not be as well retained as analytes that are soluble in methanol or methanol–water miscible mixtures. Normal-phase polar sorbents, such as silica, alumina, and Florisil, and cyano (CN) bonded phases interact by polar-dipole/dipole forces between polar functional groups in the analyte and the polar surface of the sorbent. Amino (NH_2) and diol sorbents interact with analytes by hydrogen bonding. Hexane-soluble analytes are best retained by normal-phase sorbents such as silica or Florisil or polar functionally substituted bonded phases such as amino or diol. Strong cation-exchange (SCX) and strong anion-exchange (SAX) sorbents interact primarily through electrostatic attractions between the sorbent and the analyte. Graphitized carbon sorbents exhibit both nonspecific van der Waals interactions and anion-exchange, or electrostatic, attraction for analytes.

2.4.3. Recovery

Recovery from spiked samples is calculated by measuring the amount of analyte eluted from the sorbent and comparing the original concentration to the concentration remaining after SPE. Retention and elution are two separate phases of the SPE method. However, the value measured is the overall recovery, which depends on both the sorption and elution efficiencies. Therefore, protocol development is confounded by the interdependence of sorption and desorption processes:

$$\text{recovery} = \text{sorption efficiency} \times \text{desorption efficiency} \qquad (2.33)$$

If sorption is 50% efficient but desorption is 100% efficient, the recovery measured is 50% and it is impossible to know whether sorption or desorption was inefficient or if reduced recovery was produced by a combination of both. Therefore, method development requires either optimizing sorption while controlling desorption, or vice versa using an iterative approach [67,72]. Alternatively, a statistical factorial design can be used to determine and optimize quickly variables important to SPE [110]. Using either approach, it is important to consider the major factors influencing retention, including sample pH, sample volume, and sorbent mass.

Dependence of Sorption on Sample pH

If a compound is ionizable, the extraction will be pH dependent. Data collected by Suzuki et al. [111] are graphically represented for selected data in Figure 2.35 to illustrate the influence of pH on SPE recovery. The effects

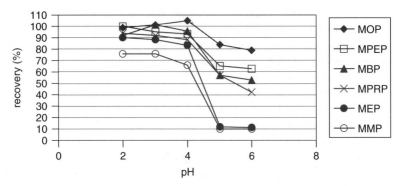

Figure 2.35. Dependence of SPE sorption on sample pH. Graphic based on selected data from Ref. 111.

of sample pH on SPE recovery of phthalic acid monoesters were evaluated using a styrene–divinylbenzene apolar polymeric phase. The effect of pH on the recovery of the free acid form of the monomethyl (MMP), mono-ethyl (MEP), mono-n-propyl (MPRP), mono-n-butyl (MBP), mono-n-pentyl (MPEP), and mono-n-octyl (MOP) phthalates was determined. The data clearly illustrate the principles discussed in Section 2.1.4.

Phathalic acid monoesters are weakly acidic compounds, due to the presence of a carboxyl group. At pH 2, the SPE recovery ranges from 76% for monomethyl phthalate to 99% for mono-n-octyl phthalate and 100% for mono-n-pentyl phthalate. As the pH increases, recovery gradually decreases but declines rapidly between pH 3 and 5. Recovery levels off between pH 5 and 6. The appearance of the data leads to the conclusion that the pK_a of the phthalic acid monoesters is between 3 and 5. The pK_a of this family of compounds appears to be approximately the same for each member of the series; that is, the electronic character of the carboxylic acid group is relatively unaffected by changes in the chain length of the alkyl group. At pH 2, these compounds are therefore nonionized, and at pH 6 they exist substantially in the ionized state. However, even at pH 6, recovery ranges from 10% for monomethyl phthalate to 79% for mono-n-octyl phthalate. This illustrates two principles discussed earlier in the chapter. First, even in the ionized state, these compounds retain a substantial degree of hydrophobicity. Second, the styrene–divinylbenzene sorbent is highly retentive, as illustrated by the degree of retention of the phthalic acid monoesters in the ionized state.

The order of recovery in the data at pH 2 and 6 is correlated approximately with the increase in the number of carbons in the alkyl chain, which in turn is roughly correlated with an increase in hydrophobicity. This example is a good illustration of the difficulty in recovering all analytes effectively from a single extraction when they range from hydrophilic to hydrophobic extremes [43]. Potential ways to increase the recovery of the least hydrophobic compound in this series, that is, the monomethyl phthalate, might include increasing the mass of the sorbent, decreasing the volume of the sample, or adding salt to the sample for a salting-out effect. However, using these approaches to improve recovery of the monomethyl phthalate may indeed reduce recovery of the most hydrophobic components in this family of compounds.

If, in this example, the best recovery were observed for the monomethyl phthalate and the least recovery observed for the mono-*n*-octyl phthalate (i.e., the order in recovery at pH 2 were reversed), an inadequate volume or eluotropic strength of the elution solvent might be the cause of reduced recovery for the more hydrophobic analytes.

Dependence of Sorption on Sample Volume

Breakthrough volume is the maximum sample volume from which 100% recovery can be achieved [112]. Since that value is somewhat difficult to predict or derive experimentally (as are peaks in the stock market), it is helpful to use Poole and Poole's [113] definition, which arbitrarily defines breakthrough volume as the point at which 1% of the sample concentration at the entrance of the sorbent bed is detected at the outlet of the sorbent bed. The type and quantity of sorbent, hydrophobicity and ionizability of the analytes, and sample volume and pH interactively determine the breakthrough volume. The breakthrough volume for a specific mass of sorbent can be established by either loading variable-volume samples of constant concentration or variable-volume samples of variable concentration, in which case the latter comprises a constant molar amount loaded [112]. Alternatively, methods exist for predicting the breakthrough volume [113].

Selected data published by Patsias and Papadopoulou-Mourkidou [114] illustrate sorption's dependence on sample volume (Figure 2.36). Their research pursues development of an automated online SPE-HPLC methodology for analysis of substituted anilines and phenols. Recovery (%) was measured for numerous compounds on various polymeric sorbents, but the only data presented here are those in which a styrene–divinylbenzene polymeric sorbent was used for analysis of aniline, phenol, 4-nitroaniline, and 4-nitrophenol. Aqueous sample volumes of 5, 10, 25, 50, 75, 100, 125, and 150 mL were acidified to pH 3 before SPE.

Recovery for 4-nitroaniline and 4-nitrophenol begins to decrease when the analytes break through from the sorbent between the sample volumes of 10 and 25 mL. Breakthrough volumes for phenol and aniline are less

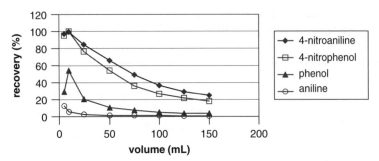

Figure 2.36. Dependence of SPE sorption on sample volume. Graphic based on selected data from Ref. 114.

than 5 mL under these conditions. The difference in the dependence of sorption upon sample loading volume between the parent molecules aniline and phenol and the nitro-substituted derivative compounds is a function of the characteristic hydrophobicity of the analytes involved as influenced by the acid dissociation constant of the analyte and the pH of the solution. The hydrophobic substituent parameter values, π_x [equation (2.8)], for para-substituted nitroaniline relative to aniline and for para-substituted nitro-phenol relative to phenol (see Table 2.1) are positive, indicating that the nitro-substituted compounds are more hydrophobic than the parent compounds. The relative differences in hydrophobicity are reflected in the degree of recovery illustrated for these compounds in Figure 2.36. At each sample volume tested, the recovery is greater for the nitro-substituted compounds than for phenol and aniline.

Using a styrene–divinylbenzene sorbent, as in this example, the primary interaction mechanism is via van der Waals forces; therefore, the more hydrophobic the compound, the larger the breakthrough volume will be and the larger the sample size from which quantitative recovery can be expected. This observation can be generalized to other sorbents by stating that regardless of the primary interaction mechanism between the analyte and the sorbent (see Table 2.6), it holds true that the stronger the interaction, the larger the breakthrough volume will be.

Dependence of Sorption on Sorbent Mass

Increasing the amount of sorbent will increase the sample volume that can be passed through the sorbent before breakthrough. The dependence of sorption on sorbent mass is illustrated (Figure 2.37) for SPE recovery from a 50-mL sample volume (72 ppb) in which two C_8 columns, each contain-

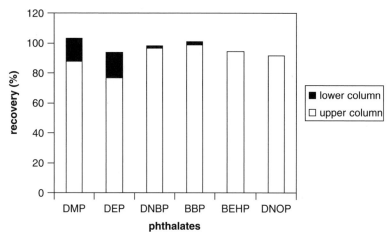

Figure 2.37. Dependence of SPE sorption on sorbent mass. Graphic based on data from Ref. 115.

ing 1.0 g of sorbent, were arranged in tandem [115]. Selected phthalates, including dimethyl phthalate (DMP), diethyl phthalate (DEP), di-*n*-butyl phthalate (DNBP), butyl benzyl phthalate (BBP), bis(2-ethylhexyl) phthalate (BEHP), and di-*n*-octyl phthalate (DNOP), were monitored. After sample loading was complete, the two columns were separated and eluted separately with 10 mL of hexane to establish the recovery for each separate mass of sorbent. The analytes in Figure 2.37 are arranged on the *x*-axis from left to right in order of increasing hydrophobicity. The results demonstrate that 1.0 g of C_8 sorbent (the upper column in the two-column tandem arrangement) is enough to sorb BEHP and DNOP but is not enough to sorb DMP, DEP, DNBP, and BBP completely. The latter compounds are less hydrophobic than the former, and the breakthrough volumes are therefore smaller. Approximately 16% recovery for DMP and DEP was detected in the bottom column of the tandem stack. A small amount of DNBP and BBP (about 2%) was also recovered from the bottom column. BEHP and DNOP were retained completely on the upper column. BEHP and DNOP are highly hydrophobic, and the breakthrough volumes are larger. BEHP and DNOP require a smaller amount of sorbent to achieve optimized recoveries. For DMP, DEP, DNBP, and BBP, the van der Waals interactions with the sorbent are less, so more sorbent mass is needed for sorption. For BEHP and DNOP, van der Waals forces are strong, so less sorbent mass is required for sorption. Optimum recovery of all six compounds from this sample volume requires 2.0 g of C_8 sorbent.

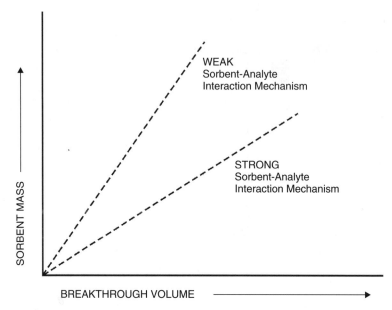

Figure 2.38. SPE interaction between sorbent mass and breakthrough volume.

Analyte sorption is dependent on both sample volume and sorbent mass (Figure 2.38). For a given amount of sorbent, the breakthrough volume is smaller for an analyte that interacts less strongly with the sorbent. For any given sample volume up to and including the breakthrough volume, the analyte that interacts more strongly with the sorbent will require a smaller amount of sorbent to achieve quantitative recovery.

Dependence of Sorption on Sample Concentration

Concentration-dependent recovery is an analytical chemist's nightmare. If an SPE method is to be useful, the analyst must demonstrate that sorption is not dependent on sample concentration in the expected concentration range of samples to be analyzed.

Dependence of Desorption on Eluting Solvent Strength

Relative elution solvent strength (or eluotropic strength) is depicted in solvent polarity charts (Figure 2.39). The relative elution strength for a solvent on a polar, normal-phase sorbent such as silica or alumina increases in reverse order to that measured on a nonpolar, reversed-phase sorbent. Ac-

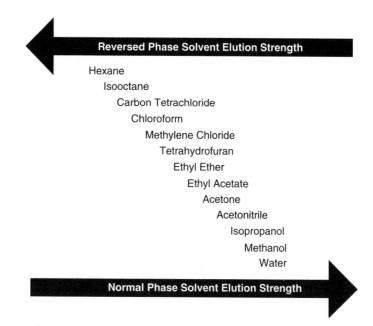

Figure 2.39. Solvent polarity chart indicates relative elution strength. (Reprinted with permission from Ref. 116. Copyright © 2002 Alltech Associates.)

cording to this chart, water is considered to be a weak solvent and hexane a strong solvent on reversed-phase sorbents. The eluting power increases as the solvent polarity decreases. Mixtures of miscible solvents can provide elution solvents of intermediate eluotropic strength.

When selecting a desorption solvent, the effect of the solvent on recovery of sample matrix contaminants should be considered. If available, a control sample matrix should be screened against potential elution solvents to assess which solvents can be used to maximize recovery of the analyte of interest and minimize the elution of sample contaminants.

Suzuki et al. [111] screened three solvents—methylene chloride, diethyl ether, and benzene—to determine their ability to produce optimum elution of phthalic acid monoesters sorbed on a styrene–divinylbenzene polymer (Figure 2.40). The effect of elution solvent strength on the recovery of the free acid form of the monomethyl (MMP), ethyl (MEP), *n*-propyl (MPRP), *n*-butyl (MBP), *n*-pentyl (MPEP), and *n*-octyl (MOP) phthalates is compared. The phthalic acid monoesters are arranged in Figure 2.40 in the order of increasing number of carbons in the alkyl chain, which in turn is roughly correlated with an increase in hydrophobicity.

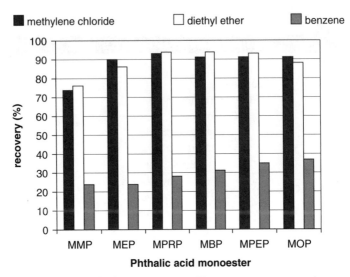

Figure 2.40. Dependence of SPE desorption on elution solvent eluotropic strength. Graphic based on selected data from Ref. 111.

When using benzene, recovery of the analytes upon elution increased with increasing hydrophobicity of the analyte but ranged from a low of 24% for the monomethyl phthalate to a high of 37% for the mono-*n*-octyl phthalate. Although benzene is expected to be a strong eluent on an apolar polymeric sorbent, it was not in this instance. Benzene may be incapable of wetting the sorbent in the presence of absorbed/adsorbed water because of its nonpolar nature. The layer of sorbed water on a sorbent phase is difficult to remove completely, even after drying with vacuum, and may be the cause of the inadequate recovery observed in this data when benzene is used. Similar results have been observed in other instances when hexane was used as an eluting solvent [112,117].

Recovery using methylene chloride or diethyl ether as eluting solvents was 86% or more for the monoesters depicted in Figure 2.40, except for the monomethyl phthalate. Relative to benzene, the polar character of methylene chloride and diethyl ether improves the wetability of the apolar sorbent having polar water molecules sorbed to the surface. The reduced recovery of mono-methyl-phthalate using methylene chloride or diethyl ether is probably due to incomplete sorption (i.e., the breakthrough volume may have been exceeded) rather than to incomplete desorption, because the more hydrophobic components were more completely desorbed.

Dependence of Desorption on Eluting Solvent Volume

Using SPE, the initial sample volume (V_i) divided by the final, or eluting, solvent volume (V_f) indicates the degree of concentration expected on 100% recovery (e.g., an optimized method for a 1000-mL sample loading volume recovered with a 10-mL eluting solvent volume is expected to produce a 100-fold increase in concentration). Therefore, the smallest amount of solvent that produces efficient recovery is generally used to produce the greatest degree of sample concentration. However, desorbing the sample using a larger volume of a solvent of lower eluting strength rather than a smaller volume of a solvent of stronger eluting strength can leave strongly retained contaminants on the sorbent as the analyte of interest is recovered.

Selected phthalates were extracted from a 50-mL sample volume (25 ppb) by SPE using 1.0 g of C_8 sorbent [115]. Extraction of dimethyl phthalate (DMP), diethyl phthalate (DEP), di-*n*-butyl phthalate (DNBP), butyl benzyl phthalate (BBP), bis(2-ethylhexyl) phthalate (BEHP), and di-*n*-octyl phthalate (DNOP) illustrates (Figure 2.41) the dependence of elution, and therefore recovery, upon solvent volume. The recovery of all analytes in this example increased with increasing elution volume from 5 mL to 10 mL of hexane. In this graph, the analytes are arranged within each elution volume and compared in order of increasing hydrophobicity. The least hydrophobic members, DMP and DEP, of this group are probably retained incompletely by 1.0 g of C_8 sorbent. Among all the members of this group of analytes, the

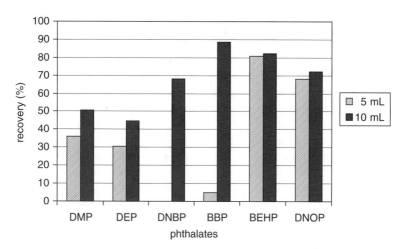

Figure 2.41. Dependence of SPE desorption on eluting solvent volume. Graphic based on data from Ref. 115.

extremely hydrophobic BEHP and DNOP compounds are eluted best when using a 5-mL elution volume of hexane. Perhaps the extreme hydrophobicity of BEHP and DNOP or their extended chain length relative to the other compounds makes it possible for hexane to better interact with these analytes than those with shorter chain lengths that are more intimately associated with the layer of water sorbed on the sorbent surface. Figure 2.41 clearly illustrates the importance of examining the dependence of desorption on sample volume.

2.4.4. Methodology

Generally, SPE consists of four steps (Figure 2.42): column preparation, or prewash, sample loading (retention or sorption), column postwash, and sample desorption (elution or desorption), although some of the recent advances in sorbent technology reduce or eliminate column preparation procedures. The prewash step is used to condition the stationary phase if necessary, and the optional column postwash is used to remove undesirable contaminants. Usually, the compounds of interest are retained on the sorbent while interferences are washed away. Analytes are recovered via an elution solvent.

SPE is not a single type of chromatography. SPE is a nonequilibrium procedure combining nonlinear modes of chromatography (Figure 2.43): the sample loading, or retention step, involves frontal chromatography and the sample desorption, or elution, step involves stepwise, or gradient, desorption, or displacement development [43,119]. In contrast, HPLC is a form of linear, or elution, chromatography that leads to dilution of the analyte as opposed to concentration of the analyte that is achieved with SPE.

In HPLC, the sample is introduced via elution development (Figure 2.44a) in which "the mixture is applied as a small quantity at the head of the column ... and the individual components are separated by being transported along the stationary phase by the continuous addition and movement of the mobile phase" [120]. Sample introduction in SPE is conducted as frontal chromatography (Figure 2.44b) in which there is "the continuous addition of the dissolved mixture to the column, with the result that the least sorbed compound is obtained in a pure state" [120]. Linear chromatography is distinguished from nonlinear chromatography by the different way in which the sample is fed into the sorbent. Therefore, SPE results in greater concentration of the analyte in the final elution volume than in the original sample, while HPLC, for example, dilutes the sample in the effluent relative to the original sample.

SPE sorbents are commercially available in three formats: contained within cartridges, in columns fashioned like syringe barrels, or in disks

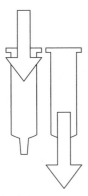

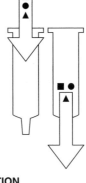

CONDITIONING
Conditioning the sorbent prior to sample application ensures reproducible retention of the compound of interest (the isolate).

RETENTION
■ Adsorbed isolate
● Undesired matrix constituents
▲ Other undesired matrix components

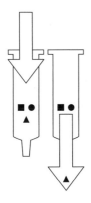

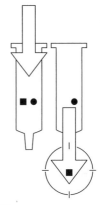

RINSE
▲ Rinse the columns to remove undesired matrix components

ELUTION
● Undesired components remain
■ Purified and concentrated isolate ready for analysis

Figure 2.42. Four basic steps for solid-phase extraction. (Reprinted with permission from Ref. 118. Copyright © 2002 Varian, Inc.)

(Figure 2.45). Bulk sorbent phases can also be purchased. Typical column housings are manufactured of polypropylene or glass, and the sorbent is contained in the column by using porous frits made of polyethylene, stainless steel, or Teflon. Pesek and Matyska [87] describe three types of disk construction: (1) the sorbent is contained between porous disks, which are inert with respect to the solvent extraction process; (2) the sorbent is en-

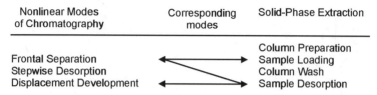

Nonlinear Modes of Chromatography	Corresponding modes	Solid-Phase Extraction
		Column Preparation
Frontal Separation		Sample Loading
Stepwise Desorption		Column Wash
Displacement Development		Sample Desorption

Figure 2.43. Nonlinear modes of chromatography. (Reprinted with permission from Ref. 43. Copyright © 2000 Marcel Dekker, Inc.)

meshed into a web of Teflon or other inert polymer; and (3) the sorbent is trapped in a glass fiber or paper filter. The commercial availability of SPE sorbents in 96-well formats (i.e., 96 individual columns contained in a single molded block) has made parallel processing with robotic automated workstations possible. Solvents can be passed through SPE sorbents by positive pressure, or hand pumping, or can be pulled through by vacuum.

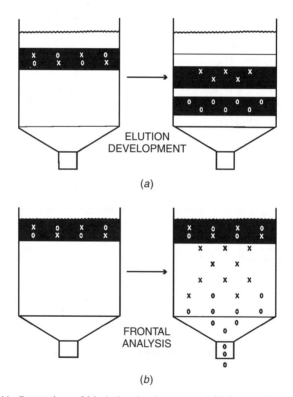

Figure 2.44. Comparison of (*a*) elution development and (*b*) frontal chromatography.

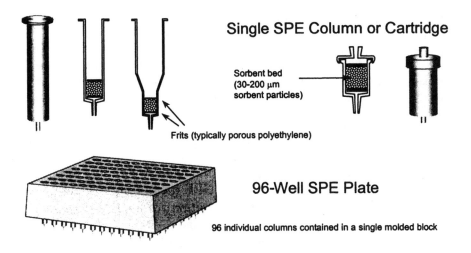

Single SPE Column or Cartridge

Sorbent bed
(30-200 μm
sorbent particles)

Frits (typically porous polyethylene)

96-Well SPE Plate

96 individual columns contained in a single molded block

High-pressure (online) SPE cartridge

PVDF body (I.D. 1-3 mm), containing 20-40 mg of sorbent between two steel frits

Figure 2.45. SPE formats. (Reprinted with permission from Ref. 87. Copyright © 2000 Marcel Dekker, Inc.)

2.4.5. Procedures

Ionized Analytes

Ionic water-soluble compounds can be retained by ion-exchange sorbents or by reversed-phase (RP) sorbents if ionization is controlled by ion suppression (i.e., by pH control that produces the nonionized form). In ion-exchange SPE, retention occurs at a sample pH at which the analyte is in its ionic form, whereas the analyte is desorbed in its neutral form; if the analytes are ionic over the entire pH range, desorption occurs by using a solution of appropriate ionic strength [92].

Alternatively, ionic compounds can be recovered from solution on hydrophobic sorbents using ion-pair SPE (IP-SPE). Carson [121] notes that advantages of IP-SPE over ion-suppression RP-SPE or ion-exchange SPE include selectivity, compatibility with aqueous samples and rapid evaporative concentration of eluents, and potential application to multiclass multiresidue analysis. IP reagents (e.g., 1-dodecanesulfonic acid for pairing with basic analytes or tetrabutylammonium hydrogen sulfate for pairing with acidic analytes) are molecules typically composed of a long-chain aliphatic hydrocarbon and a polar acidic or basic functional group. IP reagents improve the sorption of analytes on hydrophobic sorbents in two ways: (1) the reagent and analyte form a neutral complex pair, and (2) the IP reagent usually contains a hydrophobic and/or bulky portion of the molecule that increases the overall hydrophobicity of the complex relative to the unpaired analyte.

Multistage SPE

Basic SPE procedures consist of four steps, as illustrated earlier (Figure 2.42). However, using multiple processing steps such as selective sorption, selective desorption and multiple mode processes such as chromatographic mode sequencing (Figure 2.33) are possible and can lead to increased selectivity [66,70,71]. The number of theoretical plates of an SPE column is roughly two orders of magnitude less than for HPLC columns. However, SPE columns have considerable capacity for chemical class separations and can be used to isolate compounds selectively from multicomponent samples. Multistage procedures exploit differences in analyte hydrophobicity, polarity, and ionogenicity. Multistage processes lead to multiple extracts or fractions that separate components and lead to improvement in the subsequent analytical results. Selective sorption in SPE can be accomplished by controlling the sample matrix or the sorbent. Selective desorption is accomplished by utilizing differences in the eluotropic strength, ionic strength, pH, or volume of the eluting solvent to produce multistep serial elution of the sorbent. Chromatographic mode sequencing (CMS) is the serial use of different chromatographic sorbents for SPE [109].

Automation

During the past decade, SPE process automation has become a reality. High-throughput 96-well workstations and extraction plates are commercially available and allow numerous samples to be processed simultaneously [122]. Among the advantages of automated SPE, Rossi and Zhang [100] list timesaving; high throughput with serial sample processing (25 to 50 samples per hour) and even greater throughput using parallel processing systems (up to 400 samples per hour); improved precision and accuracy; reduced analyst exposure to pathogenic or hazardous samples; reduced tedium; and the possibility of automated method development. The advantages of automated systems outweigh the limitations, but the disadvantages should be considered and include the potential for carryover, systematic errors that can occur undetected and decrease precision, and sample stability issues.

2.4.6. Recent Advances in SPE

Microfluidics and miniaturization hold great promise in terms of sample throughput advantages [100]. Miniaturization of analytical processes into microchip platforms designed for micro total analytical systems (μ-TASs) is a new and rapidly developing field. For SPE, Yu et al. [123] developed a microfabricated analytical microchip device that uses a porous monolith sorbent with two different surface chemistries. The monolithic porous polymer was prepared by in situ photoinitiated polymerization within the channels of the microfluidic device and used for on-chip SPE. The sorbent was prepared to have both hydrophobic and ionizable surface chemistries. Use of the device for sorption and desorption of various analytes was demonstrated [123].

As analytical capabilities improve, multiple procedures are linked together in series to effect analyses. Procedures combined in this manner are called *hyphenated techniques*. Ferrer and Furlong [124] combined multiple techniques—accelerated solvent extraction (ASE) followed by online SPE coupled to ion trap HPLC/MS/MS—to determine benzalkonium chlorides in sediment samples. Online SPE, especially coupled to HPLC, is being used more routinely. This approach allowed online cleanup of the ASE extract prior to introduction to the analytical column.

2.5. SOLID-PHASE MICROEXTRACTION

Solid-phase microextraction (SPME) was introduced by Arthur and Pawliszyn [125]. The original concept of miniaturizing extractions (*microextraction*) using solid-phase sorbents has evolved (Figure 2.46) into a family of different approaches that strain the ability of the term *SPME* to adequately describe all techniques. According to Lord and Pawliszyn [51], one problem in the terminology applied today is that the extracting phases are not always solids. However, changing the term to *stationary-phase microextraction* or *supported-phase microextraction* in reference to the extraction phase being stationary during extraction or supported on a solid framework would not be all-inclusive either; although usually true (Figure 2.46*a,b,c,e,f*), it is not always true that the sorbent phase is stationary or supported (Figure 2.46*d*). For this discussion, all of the configurations depicted in Figure 2.46 will be considered as variations on the basic SPME theme. Most SPME applications published to date use sorption via exposure of the sample to a thin layer of sorbent coated on the outer surface of fibers (Figure 2.46*a*) or on the internal surface of a capillary tube (Figure 2.46*b*). One application of in-tube, suspended-particle SPME (which appears to this author to be a

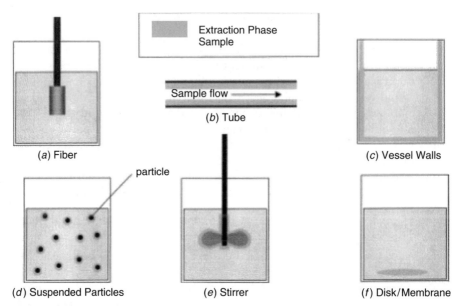

Figure 2.46. Configurations of solid-phase microextraction: (*a*) fiber, (*b*) tube, (*c*) vessel walls, (*d*) suspended particles, (*e*) stirrer, and (*f*) disk/membrane. (Reprinted with permission from Ref. 51. Copyright © 2000 Elsevier Science.)

miniaturized version of classical batch LSE and a hybrid of Figure 2.46*b* and *d*) has been published [126] and is discussed further in Section 2.5.4. The "stirrer" variation of SPME (Figure 2.46*e*) is rapidly evolving into a term and acronym in its own right [i.e., stir bar sorptive extraction (SBSE)] and is discussed later in this chapter.

Understanding analytical nomenclature is important, but it is more important to understand the underlying common extraction mechanism that leads to grouping all the approaches depicted in Figure 2.46. Exhaustive extraction of analyte from the sample matrix is not achieved by SPME, nor is it meant to occur (although SBSE techniques approach exhaustive extraction and therefore probably do deserve their own acronym). By SPME, samples are analyzed after equilibrium is reached or at a specified time prior to achieving equilibrium. Therefore, SPME operationally encompasses non-exhaustive, equilibrium and preequilibrium, batch and flow-through micro-extraction techniques. Thus defined, SPME is distinctly different from SPE because SPE techniques, including semimicro SPE (SM-SPE) and mini-aturized SPE (M-SPE) [73], are exhaustive extraction procedures.

The distribution constant, K_{fs}, between the coated fiber SPME sorbent and the aqueous sample matrix is given by

$$K_D = \frac{[X]_B}{[X]_A} = K_{fs} = \frac{C_f}{C_s} \qquad (2.34)$$

where C_f is the concentration of analyte in the fiber sorbent and C_s is the concentration of analyte in the aqueous sample phase. As with the other extraction techniques discussed, if the value of K_{fs} is larger, the degree of concentration of the target analytes in the sorbent phase is greater, and the analytical procedure is more sensitive [127].

When equilibrium conditions are reached, the number of moles, n, of analyte extracted by the fiber coating is independent of increases in extraction time, such that

$$n = \frac{K_{fs} V_f V_s C_0}{K_{fs} V_f + V_s} \qquad (2.35)$$

where V_f is the fiber coating volume, V_s the sample volume, and C_0 the initial concentration of a given analyte in the sample [51,128–130]. K_{fs} values are influenced by temperature, salt, pH, and organic solvents [130].

Examination of equation (2.35) leads to the conclusion that when the sample volume is very large (i.e., $K_{fs} V_f \ll V_s$), the amount of extracted analyte is independent of the volume of the sample, such that

$$n = K_{fs} V_f C_0 \qquad (2.36)$$

If the amount of extracted analyte is independent of sample volume, the concentration extracted will correspond directly to the matrix concentration [51,128]. Therefore, SPME is directly applicable for field applications in air and water sampling.

However, it is not necessary to continue an extraction by SPME until equilibrium is reached. A quantitative result may be achieved by careful control of time and temperature. Ulrich [130] notes that important kinetic considerations of the relationship between analyte concentration and time by SPME include:

- The time of extraction is independent of the concentration of analyte in the sample.
- The relative number of molecules extracted at a distinct time is independent of analyte concentration.
- The absolute number of molecules extracted at a distinct time is linearly proportional to the concentration of analyte.

One of the major advantages of SPME is that it is a solventless sample preparation procedure, so solvent disposal is eliminated [68,131]. SPME is a relatively simple, straightforward procedure involving only sorption and desorption [132]. SPME is compatible with chromatographic analytical systems, and the process is easily automated [131,133]. SPME sampling devices are portable, thereby enabling their use in field monitoring.

SPME has the advantages of high concentrating ability and selectivity. Conventional SPE exhaustively extracts most of the analyte (>90%) from a sample, but only 1 to 2% of the sample is injected into the analytical instrument. SPME nonexhaustively extracts only a small portion of the analyte (2 to 20%), whereas all of the sample is injected [68,73,75]. Furthermore, SPME facilitates unique investigations, such as extraction from very small samples (i.e., single cells). SPME has the potential for analyses in living systems with minimal disturbance of chemical equilibria because it is a nonexhaustive extraction system [51].

Despite the advantages of an equilibrium, nonexhaustive extraction procedure, there are also disadvantages. Matrix effects can be a major disadvantage of a sample preparation method that is based on equilibration rather than exhaustive extraction [134]. Changes in the sample matrix may affect quantitative results, due to alteration of the value of the distribution constant relative to that obtained in a pure aqueous sample [68,134].

SPME can be used to extract semivolatile organics from environmental waters and biological matrices as long as the sample is relatively clean. Extraction of semivolatile organic compounds by SPME from dirty matrices is more difficult [134]. One strategy for analyzing semivolatiles from dirty matrices is to heat the sample to drive the compound into the sample headspace for SPME sampling; another approach is to rinse the fiber to remove nonvolatile compounds before analysis [134].

2.5.1. Sorbents

For structural integrity, SPME sorbents are most commonly immobilized by coating onto the outside of fused silica fibers or on the internal surface of a capillary tube. The phases are not bonded to the silica fiber core except when the polydimethylsiloxane coating is 7 μm thick. Other coatings are cross-linked to improve stability in organic solvents [135]. De Fatima Alpendurada [136] has reviewed SPME sorbents.

Apolar, Single-Component Absorbent Phase

Polydimethylsiloxane (PDMS) is a single-component, nonpolar liquid absorbent phase coated on fused silica commercially available in film thicknesses of 7, 30, and 100 μm [137]. The PDMS phases can be used in conjunction with analysis by GC or HPLC. The thickest coating, 100 μm, used for volatile compounds by headspace procedures is not discussed in this chapter. The intermediate coating level, 30 μm, is appropriate for use with nonpolar semivolatile organic compounds, while the smallest-diameter coating, 7 μm, is used when analyzing nonpolar, high-molecular-weight compounds. The use of PDMS fibers is restricted to a sample pH between 4 and 10 [136].

Polar, Single-Component Absorbent Phase

Polyacrylate (PA) is a single-component polar absorbent coating commercially available in a film thickness of 85 μm [137]. The sorbent is used with GC or HPLC analyses and is suitable for the extraction of polar semivolatile compounds.

Porous, Adsorbent, Blended Particle Phases

Multiple-component phases were developed to exploit adsorbent processes for SPME. Adsorbent blended phases commercially available for SPME contain either divinylbenzene (DVB) and/or Carboxen particles suspended in either PDMS, a nonpolar phase, or Carbowax (CW), a moderately polar phase [55]. The solid particle is suspended in a liquid phase to coat it onto the fiber.

PDMS-DVB is a multiple-component bipolar sorbent coating. PDMS-DVB is commercially available in a film thickness of 65 μm for SPME of volatile, amine, or nitroaromatic analytes for GC analyses or in a film thickness of 60 μm for SPME of amines and polar compounds for final determination by HPLC [137]. DVB is suspended in the PDMS phase [135].

CW-DVB is a multiple-component, polar sorbent manufactured in 65- or 70-μm film thicknesses for GC analyses. SPME using CW-DVB is appropriate for the extraction of alcohols and polar compounds [137]. DVB is suspended in the Carbowax phase [135].

Carboxen/PDMS is a multiple-component bipolar sorbent (75 or 85 μm thickness) used for SPME of gases and low-molecular-weight compounds with GC analyses [137]. Carboxen is suspended in the PDMS phase [135]. Carboxen is a trademark for porous synthetic carbons; Carboxen 1006 used in SPME has an even distribution of micro-, meso-, and macropores. Carboxens uniquely have pores that travel through the entire length of the particle, thus promoting rapid desorption [135]. Among the SPME fibers currently available, the 85-μm Carboxen/PDMS sorbent is the best choice for extracting analytes having molecular weights of less than 90, regardless of functional groups present with the exception of isopropylamine [138]. The Carboxen particles extract analytes by adsorption.

DVB/Carboxen-PDMS is a multiple-component bipolar phase that contains a combination of DVB-PDMS (50 μm) layered over Carboxen-PDMS (30 μm) [55,137]. This arrangement expands the analyte molecular weight range, because larger analytes are retained in the meso- and macropores of the outer DVB layer, while the micropores in the inner layer of Carboxen retain smaller analytes [55]. The dual-layered phase is used for extraction of odor compounds and volatile and semivolatile flavor compounds with GC analysis. DVB sorbents have a high affinity for small amines; consequently, the combination coating of DVB over Carboxen is the best sorbent choice for extracting isopropylamine [138].

CW/templated resin (TPR), 50 μm, is used for analysis of surfactants by HPLC. The templated resin in CW/TPR is a hollow, spherical DVB formed by coating DVB over a silica template. When the silica is dissolved, the hollow, spherical DVB particle formed has no micro- or mesopores [135].

2.5.2. Sorbent Selection

Analyte size, concentration levels, and detection limits must all be taken into consideration when selecting SPME sorbents [55]. Physical characteristics, including molecular weight, boiling point, vapor pressure, polarity, and presence of functional groups, of the analytes of interest must be considered [135]. Analyte size is important because it is related to the diffusion coefficient of the analyte in the sample matrix and in the sorbent.

When selecting an SPME sorbent (Table 2.7), the polarity of the sorbent coating should match the polarity of the analyte of interest, and the coating should be resistant to high-temperature conditions and extremes in pH, salts, and other additives [130]. In addition to selecting sorbents having a high affinity for the analyte of interest, it is important to select sorbents with a lack of affinity for interfering compounds [134].

Recovery

Extraction recovery can be optimized by changing sample conditions such as pH, salt concentration, sample volume, temperature, and extraction time [130,132,133,136]. Currently, all commercially available SPME sorbents are neutral, such that the sample pH should be adjusted to ensure that the analyte of interest is also neutral [131].

The detection limits for SPME headspace sampling are equivalent to SPME liquid sampling for volatile compounds. However, semivolatile organic compounds diffuse slowly into the headspace so that SPME headspace sampling is not appropriate for semivolatile compounds [134].

Table 2.7. SPME Fiber Selection Guide

Analyte Class	Fiber Type	Linear Range
Acids (C2–C8)	Carboxen-PDMS	10 ppb–1 ppm
Acids (C2–C15)	CW-DVB	50 ppb–50 ppm
Alcohols (C1–C8)	Carboxen-PDMS	10 ppb–1 ppm
Alcohols (C1–C18)	CW-DVB	50 ppb–75 ppm
	Polyacrylate	100 ppb–100 ppm
Aldehydes (C2–C8)	Carboxen-PDMS	1 ppb–500 ppb
Aldehydes (C3–C14)	100 μm PDMS	50 ppb–50 ppm
Amines	PDMS-DVB	50 ppb–50 ppm
Amphetamines	100 μm PDMS	100 ppb–100 ppm
	PDMS-DVB	50 ppb–50 ppm
Aromatic amines	PDMS-DVB	5 ppb–1 ppm
Barbiturates	PDMS-DVB	500 ppb–100 ppm
Benzidines	CW-DVB	5 ppb–500 ppb
Benzodiazepines	PDMS-DVB	100 ppb–50 ppm
Esters (C3–C15)	100 μm PDMS	5 ppb–10 ppm
Esters (C6–C18)	30 μm PDMS	5 ppb–1 ppm
Esters (C12–C30)	7 μm PDMS	5 ppb–1 ppm
Ethers (C4–C12)	Carboxen-PDMS	1 ppb–500 ppb
Explosives (nitroaromatics)	PDMS-DVB	1 ppb–1 ppm
Hydrocarbons (C2–C10)	Carboxen-PDMS	10 ppb–10 ppm
Hydrocarbons (C5–C20)	100 μm PDMS	500 ppt–1 ppb
Hydrocarbons (C10–C30)	30 μm PDMS	100 ppt–500 ppb
Hydrocarbons (C20–C40+)	7 μm PDMS	5 ppb–500 ppb
Ketones (C3–C9)	Carboxen-PDMS	5 ppb–1 ppm
Ketones (C5–C12)	100 μm PDMS	5 ppb–10 ppm
Nitrosamines	PDMS-DVB	1 ppb–200 ppb
PAHs	100 μm PDMS	500 ppt–1 ppm
	30 μm PDMS	100 ppt–500 ppb
	7 μm PDMS	500 ppt–500 ppb

Source: Reprinted from Ref. 135. Copyright © (1999) Marcel Dekker, Inc.

Thicker phase coatings extract a greater mass of analyte, but the extraction time is longer than for a thinner coating [135]. Because the coated fiber sorbents are reused multiple times, ease and completeness of desorption of the fiber is an issue in order to reduce sample carryover [134].

2.5.3. Methodology

Although various ways to implement SPME are proposed and are being developed (Figure 2.46), there are two primary approaches to conducting SPME (Figure 2.47): with the sorbent coated on the outer surface of fibers

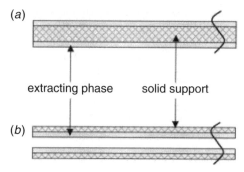

Figure 2.47. Two different implementations of the SPME technique: (*a*) polymer coated on outer surface of fiber; (*b*) polymer coated on internal surface of capillary tube. (Reprinted with permission from Ref. 51. Copyright © 2000 Elsevier Science.)

or with the sorbent coated on the internal surface of a capillary tube [51]. The fiber design can be interfaced with either GC or HPLC. However, the in-tube design has developed as an easier approach for interfacing SPME with HPLC.

In the fiber design, a fused silica core fiber is coated with a thin film (7 to 100 μm) of liquid polymer or a solid sorbent in combination with a liquid polymer (Figure 2.47*a*). Fiber lengths are generally 1 cm, although different-sized fibers can be prepared. In addition to standard fused silica fibers, silica fibers coated in a thin layer of plastic are also available. The plastic coating makes the fiber more flexible, and the sorbent phase coating bonds to the plastic layer better than the bare fused silica [55]. The in-tube design for SPME uses 0.25-mm-ID capillary tubes with about 0.1 μL of coating of the sorbent on the internal surface of the tube [51].

The theoretical calculations of the phase volume of the sorbent are facilitated by considering the fiber to be a right cylinder. The dimensions of the fused silica fiber are accurately known so that the volume of the fused silica core can be subtracted from the total volume of the fiber to yield the phase volume of the sorbent.

SPME (Figure 2.48) can be conducted as a direct extraction in which the coated fiber is immersed in the aqueous sample; in a headspace configuration for sampling air or the volatiles from the headspace above an aqueous sample in a vial (headspace SPME analyses are discussed elsewhere); or by a membrane protection approach, which protects the fiber coating, for analyses of analytes in very polluted samples [136]. The SPME process consists of two steps (Figure 2.49): (a) the sorbent, either an externally coated fiber or an internally coated tube, is exposed to the sample for a specified period of time; (b) the sorbent is transferred to a device that interfaces with an ana-

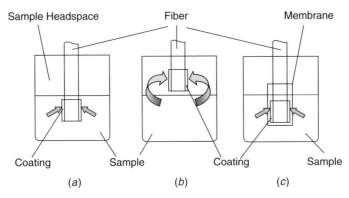

Figure 2.48. Modes of SPME operation: (*a*) direct extraction; (*b*) headspace SPME; (*c*) membrane-protected SPME. (Reprinted with permission from Ref. 51. Copyright © 2000 Elsevier Science.)

lytical instrument for thermal desorption using GC or for solvent desorption when using HPLC.

In the fiber mode, the sorbent coated fiber is housed in a microsyringe-like protective holder. With the fiber retracted inside the syringe needle, the needle is used to pierce the septum of the sample vial. The plunger is depressed to expose the sorbent-coated fiber to the sample. After equilibrium is reached or at a specified time prior to reaching equilibrium, the fiber is retracted into the protection of the microsyringe needle and the needle is withdrawn from the sample. The sorbent is then interfaced with an analytical instrument where the analyte is desorbed thermally for GC or by solvents for HPLC or capillary electrophoresis. For the in-tube mode, a sample aliquot is repeatedly aspirated and dispensed into an internally coated capillary. An organic solvent desorbs the analyte and sweeps it into the injector [68,130,133]. An SPME autosampler has been introduced by Varian, Inc., that automates the entire process for GC analyses.

Procedures

Determination of the optimum time for which the SPME sorbent will be in direct contact with the sample is made by constructing an extraction-time profile of each analyte(s) of interest. The sorption and desorption times are greater for semivolatile compounds than for volatile compounds. To prepare the extraction-time profile, samples composed of a pure matrix spiked with the analyte(s) of interest are extracted for progressively longer times. Constant temperature and sample convection must be controlled. Stirring the

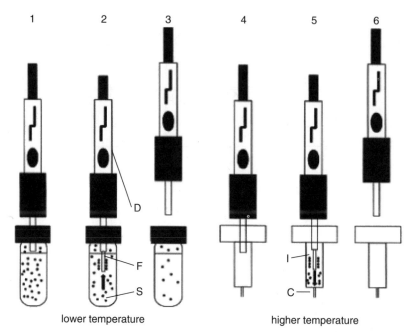

Figure 2.49. Principle of SPME: 1, introduction of syringe needle of the SPME device (D) into the sample vial and close to the sample (S), 2, moving the fiber (F) into the position outside the syringe and into the sample (extraction), 3, moving the fiber back into the syringe needle and subsequent transfer of the device to the GC injector port (1) and capillary head (C), 4, penetration of the septum with syringe needle, 5, moving the fiber into the position outside the syringe (desorption), 6, moving the fiber back into the syringe needle and withdrawing the needle. (Reprinted with permission from Ref. 130. Copyright © 2000 Elsevier Science.)

sample during sorption is necessary to reduce the diffusion layer at the sample matrix/sorbent interface and reach equilibrium faster [132]. A graph is prepared of time plotted on the x-axis and the detector response, or amount of analyte extracted, plotted on the y-axis (Figure 2.50). The extraction-time profile enables the analyst to select a reasonable extraction time while taking into consideration the detection limit of the analyte [134,136].

The SPME extraction-time profile prepared in this manner is typically composed of three distinct stages: the initial period of greatest amount of analyte extracted per time in which the graph rises sharply and has the greatest slope (however, small errors in the time measurement can lead to large errors in estimating the amount of analyte extracted); second, the profile enters an intermediate stage in which the slope of the plot is positive but smaller in magnitude relative to the initial stages of the plot; and finally, under ideal conditions equilibrium is reached such that the plot is a plateau

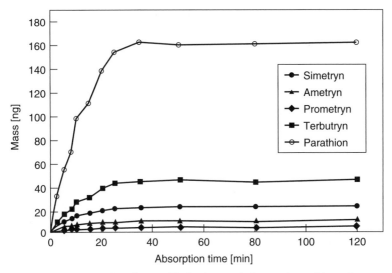

Figure 2.50. SPME absorption–time profile for four *s*-triazines and parathion using magnetic stirring. (Reprinted with permission from Ref. 139. Copyright © 1997 Elsevier Science.)

where the slope is equal to zero and there is no further increase in analyte extracted regardless of increases in contact time (Figure 2.51). Under equilibrium conditions, small errors in the time measurement produce small errors in estimating the amount of analyte extracted. Essentially, it is appropriate to conduct SPME under either the intermediate or equilibrium conditions in order to minimize the standard deviation of the analytical

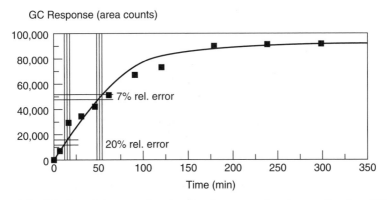

Figure 2.51. Selection of the extraction time based on extraction time profile of *p,p'*-DDT. (Reprinted with permission from Ref. 128. Copyright © 1997 John Wiley & Sons, Inc.)

measurements. In the first stage of the extraction-time profile, contact times are short, which shortens the overall analytical time, but the degree of error in the measurement is large. To reach true equilibrium, contact times may be long, but the degree of error in the measurement is small. Choosing a contact time within the intermediate region of the extraction-time profile strikes a balance between the contact time required for measurement and the anticipated degree of error. When intermediate contact times are used that do not reach equilibrium, the longest reasonable extraction time should be selected for quantitation in order to maximize the limit of detection and minimize the relative error of determination.

Quantitation of extraction under nonequilibrium conditions is based on the proportional relationship between the sorbed analyte and initial concentration [68]. Calibration of the SPME technique can be based on internal calibration using isotopically labeled standards or standard addition if recovery is matrix dependent. External calibration can be used if the standard matrix and the sample matrix are closely similar or identical [128,132,134].

2.5.4. Recent Advances in Techniques

Mullett et al. [126] recently published an automated application of a variation on the in-tube SPME approach for the analysis of propranolol and other β-blocker class drugs. The analytes were extracted from serum samples using a molecularly imprinted polymeric (MIP) adsorbent phase. MIP phases were discussed earlier as an emerging type of sorbent being used for SPE analyses. MIP phases are polymeric sorbents prepared in the presence of a target analyte that performs as a molecular template. When the template is removed, cavities that are selective recognition sites for the target analyte remain in the sorbent. In this approach, the MIP sorbent based on propranolol was passed through a 50-μm sieve and the fines removed by sedimentation in methanol. A slurry of the sorbent in methanol was placed into an 80-mm length of polyether ether ketone (PEEK) tubing of 0.76 mm ID such that the particles were not packed but suspended in the tube to allow easy flow through of the sample (Figure 2.46d). The MIP SPME capillary column was placed between the injection loop and the injection needle of an HPLC autosampler. The extraction process utilized the autosampler to aspirate and dispense the sample repeatedly across the extraction sorbent in the capillary column. In this technique, the sorbent is a "solid-phase" and the procedure is a "microscale extraction." The technique is not SPE because the particles are loosely packed and the sample passes back and forth through the column. However, the surface contact area between the sorbent and the sample is much greater than in the coated fiber or coated inner surface tubing SPME procedures described earlier. To this author, the extraction phase of the SPME procedural variation reported in this paper is more closely related to classical batch LSE, with a miniaturization of scale, than it is to classical SPME. Regardless of terminology, the approach taken in this paper is analytically elegant, and along with other examples discussed in this chapter, well illustrates the fact that the lines between strict definitions of LLE and LSE procedures and among LSE procedures are becoming blurred as analysts derive new procedures. The techniques available represent a continuum array of extraction approaches for today's analyst.

Koster et al. [140] conducted on-fiber derivatization for SPME to increase the detectability and extractability of drugs in biological samples. Amphetamine was used as a model compound. The extraction was performed by direct immersion of a 100-µm polydimethylsiloxane-coated fiber into buffered human urine. On-fiber derivatization was performed with pentafluorobenzoyl chloride either after or simultaneously with extraction.

2.6. STIR BAR SORPTIVE EXTRACTION

Stir bar sorptive extraction (SBSE), an approach theoretically similar to SPME, was recently introduced [141] for the trace enrichment of organic compounds from aqueous food, biological, and environmental samples. A stir bar is coated with a sorbent and immersed in the sample to extract the analyte from solution. To date, reported SBSE procedures were not usually operated as exhaustive extraction procedures; however, SBSE has a greater capacity for quantitative extraction than SPME. The sample is typically stirred with the coated stir bar for a specified time, usually for less than 60 minutes, depending on the sample volume and the stirring speed, to approach equilibrium. SBSE improves on the low concentration capability of in-sample solid-phase microextraction (IS-SPME).

The stir bar technique has been applied to headspace sorptive extraction (HSSE) [142–144]. However, headspace techniques are discussed elsewhere, as they are more applicable to volatile organic compounds than to the semivolatile organic compounds that comprise the focus of this chapter.

2.6.1. Sorbent and Analyte Recovery

To date, the only sorbent used reportedly for coating the stir bar is polydimethylsiloxane (PDMS), although the use of stir bars coated with polar sorbents is predicted for the future [141]. Using this sorbent, the primary mechanism of interaction with organic solutes is via absorption or partitioning into the PDMS coating such that the distribution constant [equation (2.37)] between PDMS and water ($K_{\text{PDMS/W}}$) is proposed to be proportional to the octanol–water partition coefficient (K_{OW}) [141]:

$$K_D = \frac{[X]_B}{[X]_A} = K_{PDMS/W} \approx K_{OW} \qquad (2.37)$$

According to the theoretical development for this technique given in Baltussen et al. [141],

$$K_{OW} \approx K_{PDMS/W} = \frac{[X]_{PDMS}}{[X]_W} = \frac{m_{PDMS}}{m_W} \times \frac{V_W}{V_{PDMS}} \qquad (2.38)$$

where $[X]_{PDMS}$ and $[X]_W$, and m_{PDMS} and m_W, are the analyte concentration and the analyte mass in the PDMS and water phase, respectively, while V_{PDMS} and V_W represent the volume of the PDMS sorbent and water phase, respectively. Therefore, the parameters determining the mass of an analyte recovered by SBSE using the PDMS sorbent are the partition coefficient of the analyte (K_{OW}) and the phase ratio (V_W/V_{PDMS}) of the volume of the water phase to the volume of the PDMS coating on the stir bar.

Baltussen et al. [141] theoretically compared recovery by SBSE using a stir bar assumed to be coated with a 100-μL volume of PDMS to recovery by IS-SPME having an assumed coating volume of 0.5 μL of PDMS. For the extraction of a 10-mL sample of water, it was demonstrated (Figure 2.52) that with SBSE, a more favorable extraction of analytes having lower K_{OW} values should be possible than with SPME. The small volume of the PDMS sorbent used in SPME results in a large phase ratio that implies [equation (2.38)] that a high octanol–water partition coefficient is required for efficient extraction. For SPME using PDMS, the analyte K_{OW} value is estimated (Figure 2.52) to be 20,000 (log K_{OW} = 4.3) or greater for high recovery efficiency from a 10-mL sample volume [141,145], whereas, using SBSE with PDMS, analytes with a K_{OW} value of 500 (log K_{OW} = 2.7) or greater can be extracted more quantitatively [141] due to the higher volume of PDMS coating for SBSE devices relative to SPME fibers. However, since larger volumes of PDMS are used in SBSE than in SPME, more time is required to reach equilibrium because more analyte mass will be transferred to the PDMS sorbent phase [145].

In comparing the same compounds while using PDMS sorbent, recovery from aqueous solution by SBSE was demonstrated [141] to be greater than recovery by SPME. Tredoux et al. [146] noted enrichment factors for benzoic acid in beverages to be approximately 100 times higher for SBSE relative to SPME, and Hoffmann et al. [147] reported sensitivities 100 to 1000 times higher by SBSE than by SPME for the extraction of analytes in orange juice and wine.

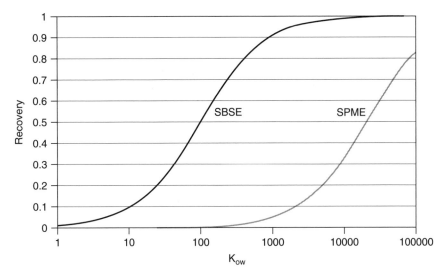

Figure 2.52. Theoretical recovery of analytes in SBSE and SPME from a 10-mL water sample as a function of their octanol–water partitioning constant. Volume of PDMS on SPME fiber: 0.5 μL; volume of PDMS on SBSE stir bar: 100 μL. (Reprinted with permission from Ref. 141. Copyright © 1999 John Wiley & Sons, Inc.)

2.6.2. Methodology

The stir bar consists of a stainless steel rod encased in a glass sheath (Figure 2.53). The glass is coated with PDMS sorbent. The length of the stir bar is typically 10 to 40 mm. The PDMS coating varies from 0.3 to 1 mm, resulting in PDMS phase volumes of 55 to 220 μL [145]. With a larger stir bar, more PDMS coating is deposited, and consequently, a larger sample volume can be extracted.

A thermodesorption unit that will accept the PDMS-coated stir bar is used to transfer the analytes into a gas chromatograph (Figure 2.54). The analyte is desorbed from the stir bar and cryofocused on a precolumn. Subsequent flash heating transfers analytes into the gas chromatograph. After desorption, the stir bar can be reused.

Procedures

Extraction of aqueous samples occurs during stirring at a specified speed for a predefined time. After a given stirring time, the bar is removed from the sample and is usually thermally desorbed into a gas chromatograph.

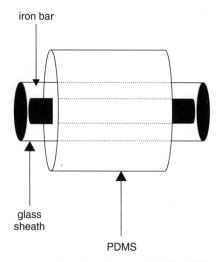

Figure 2.53. Schematic representation of a stir bar applied for SBSE. (Reprinted with permission from Ref. 145. Copyright © 2001 American Chemical Society.)

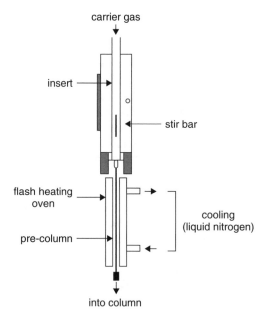

Figure 2.54. Schematic representation of the desorption unit. (Reprinted with permission from Ref. 145. Copyright © 2001 American Chemical Society.)

However, Popp et al. [148] desorbed extracted polycyclic aromatic hydrocarbons by ultrasonic treatment of the stir bar in acetonitrile or acetonitrile–water mixtures in order to perform liquid chromatographic analyses of the extract.

Although the development of this technique is still in its infancy, SBSE should have many useful analytical applications. Extraction remains a balancing act between sorbent mass and sample volume, and it appears that the primary advantage of SBSE using the PDMS sorbent (i.e., greater concentration capability than SPME) will also be its greatest disadvantage. The nonselective sorptive capability of the PDMS sorbent co-concentrates undesirable matrix components from solution. SBSE produces analyte accumulation in the sorbent but not sample cleanup. Sandra et al. [149] reported that for SBSE of fungicides in wine, standard addition methods were necessary for quantification due to matrix effects of the wine on recovery, and Ochiai et al. [150] added surrogate internal standards to compensate for sample matrix effects and coextracted analytes. Benijts et al. [151] also reported matrix suppression when SBSE on PDMS was applied to the enrichment of polychlorinated biphenyls (PCBs) from human sperm. The lipophilic medium lowered recoveries from the sperm matrix proportionally with PCB polarity.

Nevertheless, SBSE is attractive because it is a solventless enrichment technique. That coupled with the rapidity and ease of use of this procedure will make it a desirable approach for analysts. The introduction of more selective sorbents will overcome problems with matrix effects.

2.6.3. Recent Advances in Techniques

SBSE appears to be particularly useful for the extraction of a variety of components from beverages and sauces. Applications have included coffee [144], soft drinks [150], orange juice [147], lemon-flavored beverages [146], wine [147,149,150], balsamic vinegar [150], and soy sauce [150].

SBSE was recently applied [152] to the analysis of off-flavor compounds, including 2-methylisoborneol (2-MIB) and geosmin, in drinking water. These organic compounds cause taste and odor problems at very low concentrations and are notoriously difficult to extract. Detection limits by SBSE ranged from 0.022 to 0.16 ng/L. The recoveries ranged from 89 to 109% with relative standard deviations of 0.80 to 3.7%.

Vercauteren et al. [145] used SBSE to determine traces of organotin compounds in environmental samples at part per quadrillion (ppq) levels. The limits of detection reported using SBSE are the lowest ever determined for these compounds.

PRINCIPLES OF EXTRACTION

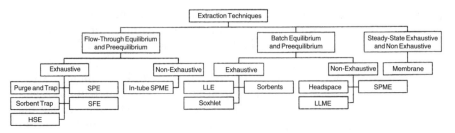

Figure 2.55. Classification of sample preparation techniques. (Reprinted with permission from Ref. 155. Copyright © 2001 NRC Research Press.)

2.7. METHOD COMPARISON

LLE, SPE, SPME, and SBSE applications for the extraction of semivolatile organics from liquids were discussed. Others [134,153,154] have compared sample preparation techniques. When examined collectively for perspective, the sample processing techniques can be perceived as variations on a single theme as practiced by today's analysts (Figure 2.55).

Two fundamentals drive extraction procedures: (1) determining the value of K_D for a given analyte–sample matrix–sorbent combination, which will indicate if the process is an equilibrium procedure (in nonequilibrium procedures, K_D approaches infinity during sorption), and (2) determining if the majority of the analyte ($>90\%$) is recovered from the sample (Table 2.8), which will indicate if the process used is exhaustive. K_D is the continuum that relates the procedures discussed here and those to be developed in the future. As commonly implemented, K_D values for the studied procedures decrease in the order $K_{D(SPE)} > K_{D(LLE)} \simeq K_{D(SBSE)} > K_{D(SPME)}$. As commonly practiced, SPE and SPME exist at opposite ends of the continuum in method fundamentals. LLE is an equilibrium procedure, but through application of repeated extractions, nearly quantitative, or exhaustive, recovery of analytes can be achieved. SBSE is a recently emerging procedure that appears to lie on the extraction continuum between LLE and SPME. The capacity of SBSE for exhaustive extraction is greater than SPME but less

Table 2.8. Extraction Method Fundamentals

SPE	Nonequilibrium	Exhaustive
LLE	Equilibrium	Exhaustive
SBSE	Equilibrium	Nonexhaustive
SPME	Equilibrium	Nonexhaustive

than LLE. The capacity for quantitative, or exhaustive, transfer is related to the K_D value and the total mass of sorbent utilized. More sorbent mass is typically present in SBSE than in SPME; therefore, more analyte is transferred to the sorbent in SBSE.

Compared to nonequilibrium methods, equilibrium methods tend to be simpler, less expensive, more selective, therefore require less cleanup, require determination of preequilibrium/equilibrium status, are time, temperature, and matrix dependent, and require internal standards for calibration [43,75,128,156].

Extraction approaches differ, but the choice of methodology depends on the analyst's objectives and resources and the client's expectations. In practice, an analyst may prefer equilibrium or nonequilibrium procedures. However, no stigma should be placed on whether an extraction method is exhaustive or nonexhaustive or equilibrium or nonequilibrium.

AKNOWLEDGMENTS

The author wishes to acknowledge the editorial and graphical assistance of Ms. Amy Knox, Ms. Sandra Pigg, and Ms. Binney Stumpf.

REFERENCES

1. A. J. Bard, *Chemical Equilibrium*, Harper & Row, New York, 1966, pp. 107, 138.
2. D. Mackay and T. K. Yuen, *Water Pollut. Res. J. Can.*, **15**(2), 83 (1980).
3. D. Mackay, W. Y. Shiu, and K. C. Ma, Henry's law constant, in R. S. Boethling and D. Mackay, eds., *Handbook of Property Estimation Methods for Chemicals: Environmental and Health Sciences*, CRC Press, Boca Raton, FL, 2000, p. 69.
4. R. G. Thomas, Volatilization from water, in W. J. Lyman, W. F. Reehl, and D. H. Rosenblatt, eds., *Handbook of Chemical Property Estimation Methods: Environmental Behavior of Organic Compounds*, McGraw-Hill, New York, 1982, p. 15-1.
5. C. F. Grain, Vapor pressure, in W. J. Lyman, W. F. Reehl, and D. H. Rosenblatt, eds., *Handbook of Chemical Property Estimation Methods: Environmental Behavior of Organic Compounds*, McGraw-Hill, New York, 1982, p. 14-1.
6. D. Mackay, W. Y. Shiu, and K. C. Ma, *Illustrated Handbook of Physical-Chemical Properties and Environmental Fate for Organic Chemicals*, Vol. II, *Polynuclear Aromatic Hydrocarbons, Polychlorinated Dioxins, and Dibenzofurans*, Lewis Publishers, Chelsea, MI, 1992, pp. 3, 250–252.

7. M. L. Sage and G. W. Sage, Vapor pressure, in R. S. Boethling and D. Mackay, eds., *Handbook of Property Estimation Methods for Chemicals: Environmental and Health Sciences*, CRC Press, Boca Raton, FL, 2000, p. 53.

8. R. P. Schwarzenbach, P. M. Gschwend, and D. M. Imboden, *Environmental Organic Chemistry*, Wiley, New York, 1993, pp. 56, 57, 77, 109, 111, 124, 131, 163, 178–181, 255–341.

9. K. Verschueren, *Handbook of Environmental Data on Organic Chemicals*, 3rd ed., Van Nostrand Reinhold, New York, 1996, pp. 4–6, 20, 22.

10. W. J. Lyman, Solubility in water, in W. J. Lyman, W. F. Reehl, and D. H. Rosenblatt, eds., *Handbook of Chemical Property Estimation Methods: Environmental Behavior of Organic Compounds*, McGraw-Hill, New York, 1982, p. 2-1.

11. W. J. Lyman, Solubility in various solvents, in W. J. Lyman, W. F. Reehl, and D. H. Rosenblatt, eds., *Handbook of Chemical Property Estimation Methods: Environmental Behavior of Organic Compounds*, McGraw-Hill, New York, 1982, p. 3-1.

12. D. Mackay, Solubility in water, in R. S. Boethling and D. Mackay, eds., *Handbook of Property Estimation Methods for Chemicals: Environmental and Health Sciences*, CRC Press, Boca Raton, FL, 2000, p. 125.

13. R. E. Ney, *Where Did That Chemical Go? A Practical Guide to Chemical Fate and Transport in the Environment*, Van Nostrand Reinhold, New York, 1990, pp. 10, 13, 18, 32.

14. J. Traube, *Annalen*, **265**, 27 (1891).

15. C. Tanford, *The Hydrophobic Effect: Formation of Micelles and Biological Membranes*, Wiley, New York, 1973, pp. 2–4, 10–11, 19, 20, 34.

16. E. Tomlinson, *J. Chromatogr.*, **113**, 1 (1975).

17. J. W. McBain, *Colloid Science*, D.C. Heath, Boston, 1950.

18. H. S. Scheraga, *Acc. Chem. Res.*, **12**, 7 (1979).

19. M. J. M. Wells, C. R. Clark, and R. M. Patterson, *J. Chromatogr.*, **235**, 43 (1982).

20. H. S. Frank and M. W. Evans, *J. Chem. Phys.*, **13**, 507 (1945).

21. C. Hansch and A. Leo, *Substituent Constants for Correlation Analysis in Chemistry and Biology*, Wiley, New York, 1979, pp. 13–17.

22. A. Leo, Octanol/water partition coefficients, in R. S. Boethling and D. Mackay, eds., *Handbook of Property Estimation Methods for Chemicals: Environmental and Health Sciences*, CRC Press, Boca Raton, FL, 2000, p. 89.

23. I. E. Bush, *The Chromatography of Steroids*, Macmillan, New York, 1961, pp. 11, 18.

24. C. Hansch and A. Leo, *Exploring QSAR*, Vol. I, *Fundamentals and Applications in Chemistry and Biology*, American Chemical Society, Washington, DC, 1995, p. 97.

25. H. Meyer, *Arch. Exp. Pathol. Pharmakol.*, **42**, 110 (1899).

26. E. Overton, *Studien uber die Narkose*, Fischer, Jena, Germany, 1901.

27. R. Collander, *Physiol. Plant.*, **7**, 420 (1954).

28. W. J. Lyman, Octanol/water partition coefficient, in W. J. Lyman, W. F. Reehl, and D. H. Rosenblatt, eds., *Handbook of Chemical Property Estimation Methods: Environmental Behavior of Organic Compounds*, McGraw-Hill, New York, 1982, p. 1-1.

29. T. Fujita, J. Iwasa, and C. Hansch, *J. Am. Chem. Soc.*, **86**, 5175 (1964).

30. M. J. M. Wells and C. R. Clark, *Anal. Chem.*, **64**, 1660 (1992).

31. M. Nakamura, M. Nakamura, and S. Yamada, *Analyst*, **121**, 469 (1996).

32. M. J. M. Wells and L. Z. Yu, *J. Chromatogr. A*, **885**, 237 (2000).

33. D. Mackay, W. Y. Shiu, and K. C. Ma, *Illustrated Handbook of Physical-Chemical Properties and Environmental Fate for Organic Chemicals*, Vol. I, *Monoaromatic Hydrocarbons, Chlorobenzenes, and PCBs*, Lewis Publishers, Chelsea, MI, 1992, p. 141.

34. EnviroLand, version 2.50, 2002.
    www.hartwick.edu/geology/enviroland

35. V. L. Snoeyink and D. Jenkins, *Water Chemistry*, Wiley, New York, 1980.

36. D. Langmuir, *Aqueous Environmental Geochemistry*, Prentice Hall, Upper Saddle River, NJ, 1997.

37. C. T. Jafvert, J. C. Westall, E. Grieder, and R. P. Schwarzenbach, *Environ. Sci. Technol.*, **24**, 1795 (1990).

38. I. M. Kolthoff, E. B. Sandell, E. J. Meehan, and S. Bruckenstein, *Quantitative Chemical Analysis*, 4th ed., Macmillan, New York, 1969, pp. 335–375.

39. Honeywell Burdick & Jackson, Miscibility, 2002.
    www.bandj.com/BJProduct/SolProperties/Miscibility.html

40. Honeywell Burdick & Jackson, Solubility in water, 2002.
    www.bandj.com/BJProduct/SolProperties/SolubilityWater.html

41. Honeywell Burdick & Jackson, Density, 2002.
    www.bandj.com/BJProduct/SolProperties/Density.html

42. Honeywell Burdick & Jackson, Solubility of water in each solvent, 2002.
    www.bandj.com/BJProduct/SolProperties/SolWaterEach.html

43. M. J. M. Wells, Handling large volume samples: applications of SPE to environmental matrices, in N. J. K. Simpson, ed., *Solid-Phase Extraction: Principles, Techniques, and Applications*, Marcel Dekker, New York, 2000, pp. 97–123.

44. J. R. Dean, Classical approaches for the extraction of analytes from aqueous samples, in *Extraction Methods for Environmental Analysis*, Wiley, Chichester, West Sussex, England, 1998, pp. 23–33.

45. A. J. Holden, Solvent and membrane extraction in organic analysis, in A. J. Handley, ed., *Extraction Methods in Organic Analysis*, Sheffield Academic Press, Sheffield, Yorkshire, England, 1999, pp. 5–53.

46. Kimble/Kontes, Inc., product literature, 2002.

47. T. Fujiwara, I. U. Mohammadzai, K. Murayama, and T. Kumamaru, *Anal. Chem.*, **72**, 1715 (2000).

48. M. Tokeshi, T. Minagawa, and T. Kitamori, *Anal. Chem.*, **72**, 1711 (2000).

49. S. X. Peng, C. Henson, M. J. Strojnowski, A. Golebiowski, and S. R. Klopfenstein, *Anal. Chem.*, **72**, 261 (2000).

50. S. X. Peng, T. M. Branch, and S. L. King, *Anal. Chem.*, **73**, 708 (2001).

51. H. Lord and J. Pawliszyn, *J. Chromatogr. A*, **885**, 153 (2000).

52. K. S. W. Sing, Historical perspectives of physical adsorption, in J. Fraissard and C. W. Conner, eds., *Physical Adsorption: Experiment, Theory and Applications*, Kluwer Academic, Dordrecht, The Netherlands, 1997, pp. 3–8.

53. K.-U. Goss and R. P. Schwarzenbach, *Environ. Sci. Technol.*, **35**(1), 1 (2001).

54. W. W. Eckenfelder, Jr., Granular carbon adsorption of toxics, in P. W. Lankford and W. W. Eckenfelder, eds., *Toxicity Reduction in Industrial Effluents*, Wiley, New York, 1990, pp. 203–208.

55. R. E. Shirey and R. F. Mindrup, *SPME-Adsorption versus Absorption: Which Fiber Is Best for Your Application?* product literature, T400011, Sigma-Aldrich Co., 1999.
www.supelco.com

56. Barnebey & Sutcliffe Corporation, Activated carbon technologies, in *Introduction to Activated Carbons*, 1996.

57. W. J. Thomas and B. Crittenden, *Adsorption Technology and Design*, Butterworth-Heinemann, Woburn, MA, 1998, pp. 8, 9, 31, 70.

58. M. Henry, SPE technology: principles and practical consequences, in N. J. K. Simpson, ed., *Solid-Phase Extraction: Principles, Techniques, and Applications*, Marcel Dekker, New York, 2000, pp. 125–182.

59. A. J. P. Martin and R. L. M. Synge, *Biochem. J.*, **35**, 1358 (1941).

60. I. Liska, *J. Chromatogr. A*, **885**, 3 (2000).

61. N. J. K. Simpson and M. J. M. Wells, Introduction to solid-phase extraction, in N. J. K. Simpson, ed., *Solid-Phase Extraction: Principles, Techniques, and Applications*, Marcel Dekker, New York, 2000, pp. 1–17.

62. M. Zief, L. J. Crane, and J. Horvath, *Am. Lab.*, **14**(5), 120, 122, 125–126, 128, 130 (1982).

63. M. Zief, L. J. Crane, and J. Horvath, *Int. Lab.*, **12**(5), 102, 104–109, 111 (1982).

64. G. D. Wachob, *LC, Liq. Chromatogr. HPLC Mag.*, **1**(2), 110–112 (1983).

65. G. D. Wachob, *LC, Liq. Chromatogr. HPLC Mag.*, **1**(7), 428–430 (1983).

66. M. J. M. Wells, Off-line multistage extraction chromatography for ultraselective herbicide residue isolation, in *Proceedings of the 3rd Annual International Symposium on Sample Preparation and Isolation Using Bonded Silicas*, Analytichem International, Harbor City, CA, 1986, pp. 117–135.

67. M. J. M. Wells and J. L. Michael, *J. Chromatogr. Sci.*, **25**, 345 (1987).

68. J. S. Fritz and M. Macka, *J. Chromatogr. A*, **902**, 137 (2000).

69. C. W. Huck and G. K. Bonn, *J. Chromatogr. A*, **885**, 51 (2000).

70. M. J. M. Wells and J. L. Michael, *Anal. Chem.*, **59**, 1739 (1987).

71. M. J. M. Wells and G. K. Stearman, Coordinating supercritical fluid and solid-phase extraction with chromatographic and immunoassay analysis of herbicides, in M. T. Meyer and E. M. Thurman, eds., *Herbicide Metabolites in Surface Water and Groundwater*, ACS Symposium Series 630, American Chemical Society, Washington, DC, 1996, pp. 18–33.

72. M. J. M. Wells, Essential guides to method development in solid-phase extraction, in I. D. Wilson, E. R. Adlard, M. Cooke, and C. F. Poole, eds., *Encyclopedia of Separation Science*, Vol. 10, Academic Press, London, 2000, pp. 4636–4643.

73. J. S. Fritz, *Analytical Solid-Phase Extraction*, Wiley-VCH, New York, 1999, 264 pp.

74. M. J. M. Wells and V. D. Adams, Determination of anthropogenic organic compounds associated with fixed or suspended solids/sediments: an overview, in R. A. Baker, ed., *Organic Substances and Sediments in Water: Processes and Analytical*, Vol. 2, Lewis Publishers, Chelsea, MI, 1991, pp. 409–479.

75. E. M. Thurman and M. S. Mills, *Solid-Phase Extraction: Principles and Practice*, Wiley, New York, 1998, 344 pp. (Vol. 147 in Chemical Analysis: A Series of Monographs on Analytical Chemistry and Its Applications).

76. N. J. K. Simpson and P. M. Wynne, The sample matrix and its influence on method development, in N. J. K. Simpson, ed., *Solid-Phase Extraction: Principles, Techniques, and Applications*, Marcel Dekker, New York, 2000, pp. 19–38.

77. C. F. Poole, A. D. Gunatilleka, and R. Sethuraman, *J. Chromatogr. A*, **885**, 17 (2000).

78. M. S. Tswett, *Proc. Warsaw Soc. Nat. Sci. Biol. Sec.*, **14**(6) (1903).

79. M. S. Tswett, *Ber. Dtsch. Bot. Ges.*, **24**, 234, 316, 384 (1906).

80. R. J. Boscott, *Nature*, **159**, 342 (1947).

81. J. Boldingh, *Experientia*, **4**, 270 (1948).

82. A. J. P. Martin, *Biochem. Soc. Symp.*, **3**, 12 (1949).

83. G. A. Howard and A. J. P. Martin, *Biochem. J.*, **46**, 532 (1950).

84. Waters Corporation, product literature, 2002.

85. H. Colin and G. Guiochon, *J. Chromatogr.*, **141**, 289 (1977).

86. Regis Technologies, Inc., product literature, 2002.

87. J. J. Pesek and M. T. Matyska, SPE sorbents and formats, in N. J. K. Simpson, ed., *Solid-Phase Extraction: Principles, Techniques, and Applications*, Marcel Dekker, New York, 2000, pp. 19–38.

88. V. Pichon, C. Cau Dit Coumes, L. Chen, S. Guenu, and M.-C. Hennion, *J. Chromatogr. A*, **737**, 25 (1996).

89. W. E. May, S. N. Chesler, S. P. Cram, B. H. Gump, H. S. Hertz, D. P. Enagonio, and S. M. Dyszel, *J. Chromatogr. Sci.*, **13**, 535 (1975).

90. J. N. Little and G. J. Fallick, *J. Chromatogr.*, **112**, 389 (1975).

91. R. E. Subden, R. G. Brown, and A. C. Noble, *J. Chromatogr.*, **166**, 310 (1978).

92. M.-C. Hennion, *J. Chromatogr. A*, **856**, 3 (1999).

93. M. J. M. Wells, *J. Liq. Chromatogr.*, **5**, 2293 (1982).

94. E. Matisova and S. Skrabakova, *J. Chromatogr. A*, **707**, 145 (1995).

95. M.-C. Hennion, *J. Chromatogr. A*, **885**, 73 (2000).

96. M. D. Leon-Gonzalez and L. V. Perez-Arribas, *J. Chromatogr. A*, **902**, 3 (2000).

97. A. J. Handley and R. D. McDowall, Solid phase extraction (SPE) in organic analysis, in A. J. Handley, ed., *Extraction Methods in Organic Analysis*, Sheffield Academic Press, Sheffield, Yorkshire, England, 1999, pp. 54–74.

98. Varian Sample Preparation Products, Inc., product literature, 2002. *www.varianinc.com*

99. N. J. K. Simpson, Ion exchange extraction, in N. J. K. Simpson, ed., *Solid-Phase Extraction: Principles, Techniques, and Applications*, Marcel Dekker, New York, 2000, pp. 493–497.

100. D. T. Rossi and N. Zhang, *J. Chromatogr. A*, **885**, 97 (2000).

101. D. Stevenson, *J. Chromatogr. B*, **745**, 39 (2000).

102. D. Stevenson, B. A. Abdul Rashid, and S. J. Shahtaheri, Immuno-affinity extraction, in N. J. K. Simpson, ed., *Solid-Phase Extraction: Principles, Techniques, and Applications*, Marcel Dekker, New York, 2000, pp. 349–360.

103. B. Sellergren, *Anal. Chem.*, **66**, 1578 (1994).

104. J. Olsen, P. Martin, and I. D. Wilson, *Anal. Commun.*, **35**, 13H (1998).

105. L. I. Andersson, *J. Chromatogr. B*, **739**, 163 (2000).

106. L. I. Andersson, *J. Chromatogr. B*, **745**, 3 (2000).

107. O. Ramstrom and K. Mosbach, *Bio/Technology*, **14**, 163 (1996).

108. B. Law, Secondary interactions and mixed-mode extraction, in N. J. K. Simpson, ed., *Solid-Phase Extraction: Principles, Techniques, and Applications*, Marcel Dekker, New York, 2000, pp. 227–242.

109. *Analytichem Int. Curr. Newsl.*, **1**(4) (1982).

110. M. J. M. Wells, D. D. Riemer, and M. C. Wells-Knecht, *J. Chromatogr. A.*, **659**, 337 (1994).

111. T. Suzuki, K. Yaguchi, S. Suzuki, and T. Suga, *Environ. Sci. Technol.*, **35**, 3757 (2001).

112. M. J. M. Wells, General procedures for the development of adsorption trapping methods used in herbicide residue analysis, in *Proceedings of the 2nd Annual International Symposium on Sample Preparation and Isolation Using Bonded Silicas*, Analytichem International, Harbor City, CA, 1985, pp. 63–68.

113. C. F. Poole and S. K. Poole, Theory meets practice, in N. J. K. Simpson, ed., *Solid-Phase Extraction: Principles, Techniques, and Applications*, Marcel Dekker, New York, 2000, pp. 183–226.

114. J. Patsias and E. Papadopoulou-Mourkidou, *J. Chromatogr. A*, **904**, 171 (2000).
115. H. Yuan and M. J. M. Wells, in preparation.
116. Alltech Associates, product literature, 2002.
    www.alltechweb.com/productinfo/Technical/datasheets/205000u.pdf
117. M. J. M. Wells, D. M. Ferguson, and J. C. Green, *Analyst*, **120**, 1715 (1995).
118. Varian Sample Preparation Products, Inc., product literature.
    www.varianinc.com/cgibin/nav?varinc/docs/spp/solphase&cid=
    975JIKPLPNMQOGJQLMMIP#steps
119. M. J. M. Wells, A. J. Rossano, Jr., and E. C. Roberts, *Anal. Chim. Acta*, **236**, 131 (1990).
120. R. C. Denney, *A Dictionary of Chromatography*, Wiley, New York, 1976, pp. 60, 71, 72.
121. M. C. Carson, *J. Chromatogr. A*, **885**, 343 (2000).
122. J. R. Dean, Solid phase extraction, in *Extraction Methods for Environmental Analysis*, Wiley, Chichester, West Sussex, England, 1998, pp. 35–61.
123. C. Yu, M. H. Davey, F. Svec, and J. M. J. Frechet, *Anal. Chem.*, **73**, 5088 (2001).
124. I. Ferrer and E. T. Furlong, *Anal. Chem.*, **74**, 1275 (2002).
125. C. L. Arthur and J. Pawliszyn, *Anal. Chem.*, **62**, 2145 (1990).
126. W. M. Mullett, P. Martin, and J. Pawliszyn, *Anal. Chem.*, **73**, 2383 (2001).
127. J. R. Dean, Solid phase microextraction, in *Extraction Methods for Environmental Analysis*, Wiley, Chichester, West Sussex, England, 1998, pp. 63–95.
128. J. Pawliszyn, *Solid Phase Microextraction: Theory and Practice*, Wiley-VCH, New York, 1997, 247 pp.
129. S. A. Scheppers Wercinski and J. Pawliszyn, Solid phase microextraction theory, in S. A. Scheppers Wercinski, ed., *Solid Phase Microextraction: A Practical Guide*, Marcel Dekker, New York, 1999, pp. 1–26.
130. S. Ulrich, *J. Chromatogr. A*, **902**, 167 (2000).
131. N. H. Snow, *J. Chromatogr. A*, **885**, 445 (2000).
132. J. Beltran, F. J. Lopez, and F. Hernandez, *J. Chromatogr. A*, **885**, 389 (2000).
133. G. Theodoridis, E. H. M. Koster, and G. J. de Jong, *J. Chromatogr. B*, **745**, 49 (2000).
134. Z. Penton, Method development with solid phase microextraction, in S. A. Scheppers Wercinski, ed., *Solid Phase Microextraction: A Practical Guide*, Marcel Dekker, New York, 1999, pp. 27–57.
135. R. E. Shirey, SPME fibers and selection for specific applications, in S. A. Scheppers Wercinski, ed., *Solid Phase Microextraction: A Practical Guide*, Marcel Dekker, New York, 1999, pp. 59–110.
136. M. de Fatima Alpendurada, *J. Chromatogr. A*, **889**, 3, 2000.
137. Supelco, Inc., product literature, 2002.
    www.supelco.com

138. Supelco, Inc., *How to Choose the Proper SPME Fiber*, product literature, T499102, 1999/2000. www.supelco.com

139. R. Eisert and J. Pawliszyn, *J. Chromatogr. A*, **776**, 293 (1997).

140. E. H. M. Koster, C. H. P. Bruins, and G. J. de Jong, *Analyst*, **127**(5), 598 (2002).

141. E. Baltussen, P. Sandra, F. David, and C. Cramers, *J. Microcolumn Sep.*, **11**, 737 (1999).

142. B. Tienpont, F. David, C. Bicchi, and P. Sandra, *J. Microcolumn Sep.*, **12**, 577 (2000).

143. C. Bicchi, C. Cordero, C. Iori, P. Rubiolo, and P. Sandra, *J. High-Resolut. Chromatogr.*, **23**, 539 (2000).

144. C. Bicchi, C. Iori, P. Rubiolo, and P. Sandra, *J. Agric. Food Chem.*, **50**, 449 (2002).

145. J. Vercauteren, C. Peres, C. Devos, P. Sandra, F. Vanhaecke, and L. Moens, *Anal. Chem.*, **73**, 1509 (2001).

146. A. G. J. Tredoux, H. H. Lauer, T. Heideman, and P. Sandra, *J. High-Resolut. Chromatogr.*, **23**, 644 (2000).

147. A. Hoffmann, R. Bremer, P. Sandra, and F. David, *LaborPraxis*, **24**(2), 60 (2000).

148. P. Popp, C. Bauer, and L. Wennrich, *Anal. Chim. Acta*, **436**(1), 1 (2001).

149. P. Sandra, B. Tienpont, J. Vercammen, A. Tredoux, T. Sandra, and F. David, *J. Chromatogr. A*, **928**(1), 117 (2001).

150. N. Ochiai, K. Sasamoto, M. Takino, S. Yamashita, S. Daishima, A. Heiden, and A. Hoffmann, *Anal. Bioanal. Chem.*, **373**(1/2), 56 (2002).

151. T. Benijts, J. Vercammen, R. Dams, H. P. Tuan, W. Lambert, and P. Sandra, *J. Chromatography B: Biomedical Sciences and Applications*, **755**(1/2), 137 (2001).

152. N. Ochiai, K. Sasamoto, M. Takino, S. Yamashita, S. Daishima, A. Heiden, and A. Hoffmann, *Analyst*, **126**(10), 1652 (2001).

153. N. J. K. Simpson, A comparison between solid-phase extraction and other sample processing techniques, in N. J. K. Simpson, ed., *Solid-Phase Extraction: Principles, Techniques, and Applications*, Marcel Dekker, New York, 2000, pp. 489–492.

154. J. R. Dean, Comparison of extraction methods, in *Extraction Methods for Environmental Analysis*, Wiley, Chichester, West Sussex, England, 1998, pp. 211–216.

155. J. Pawliszyn, *Can J. Chem.*, **79**, 1403 (2001).

156. Y. Luo and J. Pawliszyn, Solid phase microextraction (SPME) and membrane extraction with a sorbent interface (MESI) in organic analysis, in A. J. Handley, ed., *Extraction Methods in Organic Analysis*, Sheffield Academic Press, Sheffield, Yorkshire, England, 1999, pp. 75–99.

CHAPTER

3

EXTRACTION OF SEMIVOLATILE ORGANIC COMPOUNDS FROM SOLID MATRICES

DAWEN KOU AND SOMENATH MITRA

Department of Chemistry and Environmental Science, New Jersey Institute of
Technology, Newark, New Jersey

3.1. INTRODUCTION

This chapter covers techniques for the extraction of semivolatile organics from solid matrices. The focus is on commonly used and commercially available techniques, which include Soxhlet extraction, automated Soxhlet extraction, ultrasonic extraction, supercritical fluid extraction (SFE), accelerated solvent extraction (ASE), and microwave-assisted extraction (MAE). The underlying principles, instrumentation, operational procedures, and selected applications of these techniques are described. In a given application, probably all the methods mentioned above will work, so it often boils down to identifying the most suitable one. Consequently, an effort is made to compare these methodologies.

The U.S. Environmental Protection Agency (EPA) has approved several methods for the extraction of pollutants from environmental samples. These standard methods are listed under EPA publication SW-846, *Test Methods for Evaluating Solid Waste: Physical/Chemical Methods* [1]. Many of them were approved only in the last decade. Automated Soxhlet was promulgated in 1994, SFE and ASE in 1996, and MAE in 2000. The Association of Official Analytical Chemists (AOAC) has published its own standard extraction methods for the food, animal feed, drug, and cosmetics industries [2]. Some extraction methods have also been approved by the American Society for Testing and Materials (ASTM) [3]. Table 3.1 summarizes the standard methods from various sources.

Sample Preparation Techniques in Analytical Chemistry, Edited by Somenath Mitra
ISBN 0-471-32845-6   Copyright © 2003 John Wiley & Sons, Inc.

Table 3.1. Methods Accepted as Standards for the Extraction of
Semivolatile Organics from Solid Matrices

Technique	Analytes	Standard Method
Soxhlet extraction	Semivolatile and nonvolatile organics	EPA 3540C
	Fat in cacao products	AOAC 963.15
Automated Soxhlet extraction	Semivolatile and nonvolatile organics	EPA 3541
Pressurized fluid extraction (PFE)	Semivolatile and nonvolatile organics	EPA 3545A
Microwave-assisted extraction (MAE)	Semivolatile and nonvolatile organics	EPA 3546
	Total petroleum hydrocarbons, organic compounds	ASTM D-5765 ASTM D-6010
	Fat in meat and poultry products	AOAC 991.36
Ultrasonic extraction	Semivolatile and nonvolatile organics	EPA 3550C
Supercritical fluid extraction (SFE)	Semivolatile petroleum hydrocarbons, PAHs, PCBs, and organochlorine pesticides	EPA 3560 EPA 3561 EPA 3562

3.1.1. Extraction Mechanism

Extraction of organics from solids is a process in which solutes desorb from the sample matrix and then dissolve into the solvent. Extraction efficiency is influenced by three interrelated factors: solubility, mass transfer, and matrix effects. Much of the discussion in Chapter 2 on solvents and solubility is also relevant to solid matrices. The solubility of an analyte depends largely on the type of the solvent, and for a selected solvent, its solubility is affected by temperature and pressure. *Mass transfer* refers to analyte transport from the interior of the matrix to the solvent. It involves solvent penetration into the matrix and removal of solutes from the adsorbed sites. Mass transfer is dependent on the diffusion coefficient as well as on the particle size and structure of the matrix. High temperature and pressure, low solvent viscosity, small particle size, and agitation facilitate mass transfer [4]. It is a more important issue than solubility when the analyte concentration in the extraction solvent is below its equilibrium solubility (i.e., when the analyte is readily soluble in the solvent). Matrix effects are the least understood of the three factors. A highly soluble compound can be "unextractable" because it is locked in the matrix pores, or is strongly bound to its surface. For example, analytes in aged soil bind more strongly than in a clean soil when spiked with the same analyte. Desorption is more difficult and may take longer. Some extraction techniques, such as SFE, are found to be matrix dependent

[5]. Different extraction parameters are employed for different groups of analytes in different matrices.

Solvent selection depends largely on the nature of the analytes and the matrix. Although the discussions in Chapter 2 can be used as a guideline to account for the solvent–analyte interactions, the matrix effects are often unpredictable. There is no single solvent that works universally for all analytes and all matrices. Sometimes, a mixture of water-miscible solvents (such as acetone) with nonmiscible ones (such as hexane or methylene chloride) are used. The water-miscible solvents can penetrate the layer of moisture on the surface of the solid particles, facilitating the extraction of hydrophilic organics. The hydrophobic solvents then extract organic compounds of like polarity. For instance, hexane is efficient in the extraction of nonpolar analytes, and methylene chloride extracts the polar ones.

As temperature and pressure play important roles in extraction kinetics, extraction techniques can be classified based on these parameters. Classical methods include Soxhlet extraction, automated Soxhlet extraction, and ultrasonic extraction. They are operated under atmospheric pressure, with heating or ultrasonic irradiation. These methods consume relatively large volumes of organic solvents, and the extraction may take a long time. The other group consists of SFE, ASE, and MAE, which are performed under elevated pressure and/or temperature. The extraction is faster, more efficient, and sample throughput is high. With relatively less consumption of organic solvents, these methods are more environmentally friendly. Moreover, the costs of solvent purchase and waste disposal are reduced. Despite the high initial equipment cost, these methods may be more economical in the long run, especially for the routine analysis of a large number of samples.

3.1.2. Preextraction Procedures

Most extraction methods perform best on dry samples with small particle size. If possible, samples may be air-dried and ground to a fine powder before extraction. However, this procedure is not recommended if the sample contains volatile analytes and/or worker exposure is a concern. Instead, the sample can be dried by mixing with anhydrous sodium sulfate or palletized diatomaceous earth. In certain applications such as in MAE, water can be used as a part of the solvent mixture [6,7]. Instead of drying, water is added into the sample to maintain a certain moisture level.

3.1.3. Postextraction Procedures

Some extraction techniques generate large volumes of solvent extract. The extract needs to be concentrated to meet the detection limit of the analytical method. Moreover, in most cases, extracts of soil, sludge, and waste samples require some degree of cleanup prior to analysis. The purpose of cleanup is to remove interfering compounds and high-boiling materials that may cause error in quantification, equipment contamination, and deterioration of chromatographic resolution. The details of postextraction techniques have been discussed in Chapter 1.

3.2. SOXHLET AND AUTOMATED SOXHLET

Soxhlet extraction and automated Soxhlet extraction are described in this section. Soxhlet extraction was named after Baron Von Soxhlet, who introduced this method in the mid-nineteenth century. It had been the most widely used method until modern extraction techniques were developed in the 1980s. Today, Soxhlet is still a benchmark method for the extraction of semivolatile organics from solid samples. Automated Soxhlet extraction (Soxtec being its commercial name) offers a faster alternative to Soxhlet, with comparable extraction efficiency and lower solvent consumption.

3.2.1. Soxhlet Extraction

A schematic diagram of a typical Soxhlet apparatus is shown in Figure 3.1. The system has three components. The top part is a solvent vapor reflux condenser. In the middle are a thimble holder with a siphon device and a side tube. The thimble holder connects to a round-bottomed flask at the bottom. The sample is loaded into a porous cellulous sample thimble and placed into the thimble holder. Typically, 300 mL of solvent(s) (for a 10-g sample) is added to the flask. A couple of boiling chips are also added, and the flask is gently heated on a heating mantle. Solvent vapor passes through the side tube and goes to the reflux condenser, where it condenses and drips back to the thimble chamber. When the analyte-laden solvent reaches the top of the thimble holder, it is drained back into the bottom flask through the siphon device. This cycle repeats many times for a predetermined time period. Since the extracted analytes have higher boiling points than the extraction solvent, they accumulate in the flask while the solvent recirculates. Consequently, the sample is always extracted with fresh solvents in each cycle.

Because the sample is extracted with cooled, condensed solvents, Soxhlet is slow and can take between 6 to 48 hours. The extract volume is relatively large, so a solvent evaporation step is usually needed to concentrate the analytes prior to extract cleanup and analysis. The sample size is usually 10 g or more. Multiple samples can be extracted on separate Soxhlet units, and the extraction can be run unattended. Soxhlet is a rugged, well-established

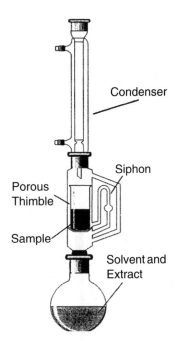

Condenser

Siphon

Porous
Thimble

Sample

Solvent and
Extract

Figure 3.1. Schematic diagram of a Soxhlet apparatus. (Reproduced from Ref. 93, with permission from Nelson Thornes Ltd.)

technique that is often used as the benchmark for comparing other methods. Few parameters can affect the extraction. The main drawbacks are the long extraction time and relatively large solvent consumption. The routine use of Soxhelt is decreasing as faster extraction techniques are finding their way into the analytical arena.

3.2.2. Automated Soxhlet Extraction

In 1994, automated Soxhlet extraction (Soxtec, commercially) was approved by EPA as a standard method. A shematic diagram of Soxtec is shown in Figure 3.2. The extraction is carried out in three stages: boiling, rinsing, and solvent recovery. In the first stage, a thimble containing the sample is immersed in the boiling solvent for about 60 minutes. Extraction here is faster than Soxhlet, because the contact between the solvent and the sample is more vigorous, and the mass transfer in a high-temperature boiling solvent is more rapid. In the second stage, the sample thimble is lifted above the boiling solvent. The condensed solvent drips into the sample, extracts the organics, and falls back into the solvent reservoir. This rinse–extract process is similar to Soxhlet and is usually set for 60 minutes. The third stage is a concentration step for 10 to 20 minutes. The solvent is evaporated to 1 to

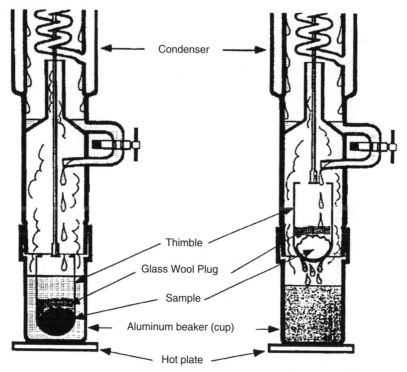

Figure 3.2. Schematic diagram of an automatic Soxhlet extraction device (Soxtec).

2 mL, as would occur in a Kuderna–Danish concentrator. Since the concentration step is integrated in Soxtec, the extract is ready for cleanup and analysis.

Lopez-Avila et al. [8] published a study in 1993 that evaluated the Soxtec extraction of 29 target compounds (seven nitroaromatic compounds, three haloethers, seven chlorinated hydrocarbons, and 12 organochlorine pesticides) from spiked sandy clay loam and clay loam. Among the five factors investigated (matrix type, spike level, anhydrous sodium sulfate addition, total extraction time, and immersion/extraction time ratio), matrix type, spike level, and total extraction time had the most pronounced effects on method performance at the 5% significance level for 16 of the 29 target compounds. The two solvent mixtures, hexane–acetone (1 : 1) and methylene chloride–acetone (1 : 1), performed equally well. Four compounds were not recovered at all, and apparently were lost from the spike matrix. Limited experimental work was performed with 64 base–neutral–acidic compounds spiked onto clay loam, and with three standard reference materials certified for polycyclic aromatic hydrocarbons (PAHs). For the 64 compounds spiked onto clay loam at 6 mg/kg, 20 had recoveries more than 75%, 22 between 50 and 74%, 12 between 25 and 49%, and 10 less than 25%.

3.2.3. Comparison between Soxtec and Soxhlet

Soxhlet can be applied universally to almost any sample. It is not uncommon to use Soxhlet as the benchmark method for validating other extraction techniques. Soxtec reduces the extraction time to 2 to 3 hours as compared to 6 to 48 hours in Soxhlet. It also decreases solvent use from 250 to 500 mL per extraction to 40 to 50 mL per extraction. Two to six samples can be extracted simultaneously with a single Soxtec apparatus.

Recent studies comparing Soxtec with Soxhlet show comparable or even better results for Soxtec. Brown et al. [9] compared the efficiency of the standard Soxhlet method against three different protocols using the Soxtec extractor (Tecator, Inc. Silver Spring, MD). Organic mutagens were extracted from municipal sewage sludge using MeOH and CH_2Cl_2 as solvents. Both the Soxtec (with 5 minutes of boiling time and 55 minutes of rinsing time), and Soxhlet procedures yielded reproducible mutagenic responses within the variability of the bioassay. The data indicate that the Soxtec extraction, which was faster and required less solvent, provided adequate extraction of organic mutagens from sewage sludge.

Foster and Gonzales [10] reported a collaborative study by 11 laboratories of Soxtec and Soxhlet methods for the determination of total fat in meat and meat products. Each lab analyzed six samples: canned ham, ground beef, frankfurters, fresh pork sausage, hard salami, and beef patties with added soy. In general, results for the Soxtec system showed improved performance. The method was first adopted by AOAC International for the extraction of fat from meat. Membrado et al. [11] tested Soxtec against Soxhlet extraction for the extraction of coal and coal-derived products. Optimization of Soxtec operating conditions reduced the total extraction time to 10% of what was needed by Soxhlet extraction. The recovery and precision by the two methods were comparable.

3.3. ULTRASONIC EXTRACTION

Ultrasonic extraction, also known as *sonication*, uses ultrasonic vibration to ensure intimate contact between the sample and the solvent. Sonication is relatively fast, but the extraction efficiency is not as high as some of the other techniques. Also, it has been reported that ultrasonic irradiation may lead to the decomposition of some organophosphorus compounds [12].

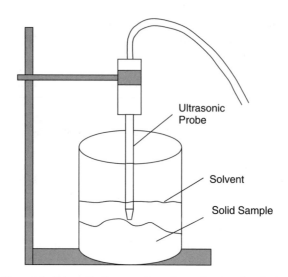

Figure 3.3. Schematic diagram of an ultrasonic extraction device.

Thus, the selected solvent system and the operating conditions must demonstrate adequate performance for the target analytes in reference samples before it is implemented for real samples. This is particularly important for low-concentration [parts per billion (ppb) level] samples.

Figure 3.3 shows a schematic diagram of a sonication device. It is a horn-type ultrasonic disruptor equipped with a titanium tip. There are two types of disruptors. A $\frac{3}{4}$-in. horn is typically used for low-concentration samples and a $\frac{1}{8}$-in. tapered microtip attached to a $\frac{1}{2}$-in. horn for medium/high-concentration samples. The sample is usually dried with anhydrous sodium sulfate so that it is free flowing. For trace analysis, the sample size is typically 30 g. Then a certain volume (typically, 100 mL) of selected solvents are mixed with the sample. The most common solvent system is acetone–hexane (1 : 1 v/v) or acetone–methylene chloride (1 : 1 v/v). For nonpolar analytes such as polychlorinated biphenyls (PCBs), hexane can also be used. The extraction is performed in the pulsed mode, with ultrasonic energy being on and off rather than continuous. The disruptor horn tip is positioned just below the surface of the solvent, yet above the sample. Very active mixing between the sample and the solvent should be observed. Extraction can be carried out in duration as short as 3 minutes. Since it is a fast procedure, it is important that one strictly follow the specific operating conditions. For low-concentration samples, the sample needs to be extracted two or more times, each time with the same amount of fresh solvents. Then the extracts from the different extractions are combined. For high-concentration (over 20 ppm) samples, approximately 2 g of sample is needed, and a single extraction with 10 mL of solvents may be adequate. After extraction, the extract is filtrated or centrifuged, and some form of cleanup is generally needed prior to analysis.

3.3.1. Selected Applications and Comparison with Soxhlet

Like Soxhlet, sonication is also recognized as an established conventional method, although it is not as widely used. Limited research has focused on sonication per se or its comparison with Soxhlet. Qu et al. [13] developed a method using sonication with methanol for the extraction of linear alkylbenzene sulfonate (LAS) in plant tissues (rice stems and leaves). Both efficiency and accuracy were found to be high. The mean recovery was 89% (84 to 93% for LAS concentration of 1 to 100 mg/kg), and the relative standard deviation (RSD) was 3% for six replicate analyses. Its advantages over Soxhlet extraction were speed (1 hour), less solvent consumption, and smaller sample requirement (2 to 3 g).

Marvin et al. [14] compared sonication with Soxhlet for the extraction of PAHs from sediments, and from an urban dust standard reference material (SRM 1649). The sonication method required less than 5 g of sample. The amount of organic materials extracted by sonication with two solvents was $2.53 \pm 0.10\%$ of the sediment samples (w/w), while $2.41 \pm 0.14\%$ was extracted by Soxhlet. Sequential sonicaion with two solvents was much faster (45 minutes) than Soxhlet (2 days), with practically the same extraction efficiency. The variation of PAH extracted by sonication from the urban dust SRM was within 15%.

Haider and Karlsson [15] developed a simple procedure for the determination of aromatic antioxidants and ultraviolet stabilizers in polyethylene using ultrasonic extraction. Chloroform was used for the isolation of Chimassorb 944 from 150-μm-thick commerical low-density polyethylene and Irganox 1010 and Irgafos 168 from 25-μm medium-density polyethylene film. The recovery of the additives increased remarkably at higher temperatures and longer extraction times. At 60°C, quantitative recovery was achieved in 15, 45, and 60 minutes for Irgafos 168, Irganox 1010, and Chimassorb 944, respectively.

Eiceman et al. [16] reported the ultrasonic extraction of polychlorinated dibenzo-p-dioxins (PCDDs) and other organic compounds from fly ash from municipal waste incinerators. Ten to 20 grams of sample was extracted with 200 mL of benzene for 1 hour. Results from five replicate analyses yielded averages and RSDs (ng/g) for the tetra- to octachlorinated dibenzo-p-dioxins of 8.6 ± 2.2, 15.0 ± 4.0, 13.0 ± 3.4, 3.2 ± 1.0, and 0.4 ± 0.1, respectively.

Golden and Sawicki [17] studied ultrasonic extraction of almost all of the polar compounds from airborne particulate material collected on Hi-Vol filters. Full recovery of PAH and good reproducibility were achieved. Total analysis time was approximately 1.5 hours. The same research group also reported a sonication procedure for the extraction of total particulate aromatic hydrocarbon (TpAH) from airborne particles collected on glass fiber filters [18]. Significantly higher recovery of TpAH and PAH were achieved by 40 minutes of sonication than by 6 to 8 hours of Soxhlet extraction.

3.4. SUPERCRITICAL FLUID EXTRACTION

Supercritical fluid extraction (SFE) utilizes the unique properties of supercritical fluids to facilitate the extraction of organics from solid samples. Analytical scale SFE can be configured to operate on- or off-line. In the online configuration, SFE is coupled directly to an analytical instrument, such as a gas chromatograph, SFC, or high-performance liquid chromatograph. This offers the potential for automation, but the extract is limited to analysis by the dedicated instrument. Off-line SFE, as its name implies, is a stand-alone extraction method independent of the analytical technique to be used. Off-line SFE is more flexible and easier to perform than the online methods. It allows the analyst to focus on the extraction per se, and the extract is available for analysis by different methods. This chapter focuses on off-line SFE.

The discovery of supercritical fluids by Baron Cagniard de la Tour dates back to 1822 [19]. In 1879, Hannay and Hogarth demonstrated the solvating power of supercritical ethanol [20]. Between 1964 and 1976, Zosel filed several patents on decaffeination of coffee, which signified a major development in SFE. In 1978, a decaffeination plant was opened by the Maxwell House Coffee Division. Since then, SFE has found many industrial applications. The use of supercritical fluids for analytical purposes started with capillary supercritical fluid chromatography (SFC), which was introduced by Novotny et al. in 1981 [21]. Analytical scale SFE became commercially available in the mid-1980s. In 1996, EPA approved two SFE methods, one for the extraction of total petroleum hydrocarbons (TPHs) and the other for PAHs. Another SFE method was promulgated by EPA in 1998 for the extraction of PCBs and organochlorine pesticides (OCPs).

3.4.1. Theoretical Considerations

A supercritical fluid is a substance above its critical temperature and pressure. Figure 3.4 shows a phase diagram of a pure substance, where curve

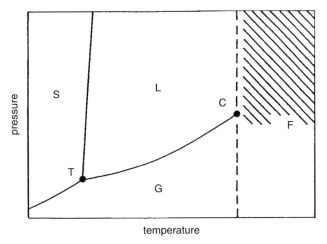

Figure 3.4. Phase diagram of a pure substance. (Reproduced from Ref. 24, with permission from Kluwer Academic Publishers.)

$T-C$ is the interface between gas and liquid. Each point on the line corresponds to a certain temperature and the pressure needed to liquefy the gas at this temperature. Point C is the critical point. Beyond the critical temperature, a gas does not liquefy under increasing pressure. Instead, it is compressed into a supercritical fluid. The critical point is substance-specific. Table 3.2 shows the supercritical conditions of some selected solvents.

Table 3.2. Critical Parameters of Select Substances

Substance	Critical Temperature (°C)	Critical Pressure (atm)	Critical Density (10^3 kg/m^3)
CO_2	31.3	72.9	0.47
N_2O	36.5	72.5	0.45
SF_6	45.5	37.1	0.74
NH_3	132.5	112.5	0.24
H_2O	374	227	0.34
$n\text{-}C_4H_{10}$	152	37.5	0.23
$n\text{-}C_5H_{12}$	197	33.3	0.23
Xe	16.6	58.4	1.10
CCl_2F_2	112	40.7	0.56
CHF_3	25.9	46.9	0.52

Reproduced from Ref. 24, with permission from Kluwer Academic Publishers.

Table 3.3. Physical Properties of Gases, Supercritical Fluids, and Liquids

State	Conditions[a]	Density (10^3 kg/m^3)	Viscosity (mPa·s)	Self-Diffusion Coefficient $(10^4 \text{ m}^2/\text{s})$
Gas	30°C, 1 atm	$0.6–2 \times 10^3$	$1–3 \times 10^2$	0.1–0.4
Supercritical fluid	Near T_c, p_c	0.2–0.5	$1–3 \times 10^{-2}$	0.7×10^{-3}
	Near T_c, $4p_c$	0.4–0.9	$3–9 \times 10^{-2}$	0.2×10^{-3}
Liquid	30°C, 1 atm	0.6–1.6	0.2–3	$0.2–2 \times 10^{-5}$

Reproduced from Ref. 24, with permission from Kluwer Academic Publishers.
[a] T_c, critical temperature; p_c, critical pressure.

Table 3.3 presents the approximate physical properties of gases, super-critical fluids, and liquids. It shows that the densities of supercritical fluids are close to that of a liquid, whereas their viscosities are gaslike. The diffusion coefficients are in between. Due to these unique properties, supercritical fluids have good solvating power (like liquid), high diffusivity (better than liquid), low viscosity, and minimal surface tension (like gas). With rapid mass transfer in the supercritical phase and with better ability to penetrate the pores in a matrix, extraction is fast in SFE, along with high extraction efficiency.

The solubility of a supercritical fluid is influenced by its temperature, pressure, and density. Solubility correlates better to density than to pressure. An empirical equation can be used to predict solubility [22]:

$$\ln(s) = aD + bT + c \tag{3.1}$$

where s is the solubility in mole or weight percent, D the density in g/mL, T the temperature in kelvin, and a, b, and c are constants. Figure 3.5 depicts the change in analyte solubility in supercritical fluids as a function of temperature and pressure. The predicted solubility using equation (3.1) shows good agreement with the experimental data.

Carbon dioxide (CO_2) has a low supercritical temperature (31°C) and pressure (73 atm). It is nontoxic and nonflammable and is available at high purity. Therefore, CO_2 has become the solvent of choice for most SFE applications. Being nonpolar and without permanent dipole moment, supercritical CO_2 is a good solvent for the extraction of nonpolar and moderately polar compounds. However, its solvating power for polar solutes is rather poor. Moreover, when the solutes bind strongly to the matrix, the solvent strength of CO_2 is often inadequate to break the solute–matrix bond.

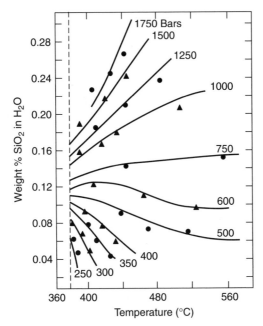

Figure 3.5. Solubility of SiO_2 in supercritical H_2O. (Reproduced from Ref. 22, with permission from Preston Publications.)

This is true even if it is capable of dissolving the solutes. Supercritical solvents such as N_2O and $CHClF_2$ are more efficient in extracting polar compounds, but their routine use is uncommon due to environmental concerns. The extraction efficiency of polar compounds by CO_2 can be improved by the addition of small quantities (1 to 10%) of polar organic solvents, referred to as *modifiers*. This is a common practice in SFE. Table 3.4 lists some common modifiers for supercritical CO_2.

Table 3.4. Commonly Used Modifiers for Supercritical CO_2

Oxygen containing	Methanol, ethanol, isopropyl alcohol, acetone, tetrahydrofuran
Nitrogen containing	Acetonitrile
Sulfur containing	Carbon disulfide, sulfur dioxide, sulfur hexafluoride
Hydrocarbons and halogenated organics	Hexane, toluene, methylene chloride, chloroform, carbon tetrachloride, trichlorofluoromethane
Acids	Formic acid

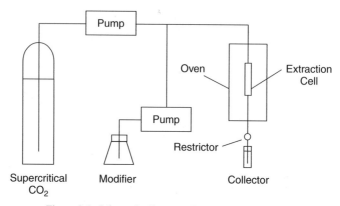

Figure 3.6. Schematic diagram of an off-line SFE system.

3.4.2. Instrumentation

The schematic diagram of an SFE system is shown in Figure 3.6. The basic components include a tank of CO_2, a high-pressure pump, an extraction cell, a heating oven, a flow restrictor, and an extract collector. A source of organic modifier and a pump for its delivery may also be needed. High-purity CO_2 is generally supplied in a cylinder with a dip tube (or eductor tube). The function of the dip tube is to allow only liquefied CO_2 to be drawn into the pump, as the liquid stays at the bottom of the vertically placed cylinder while the gaseous CO_2 is at the top. Aluminum cylinders are generally preferred over steel cylinders. Impurities in CO_2 may cause interference during analysis. The extraction cells, frits, restrictors, and multiport valves may also carry-over analytes from high-concentration samples. It has been found that contamination is more likely to be caused by SFE instrumentation and associated plumbing than by the CO_2 itself [23]. All connections in the SFE system should be metal to metal, and the use of lubricants should be avoided. The extraction system should also be cleaned after each extraction.

The basic requirement for a SFE pump is the ability to deliver constant flow (at least 2 mL/min) in the pressure range 3500 to 1000 psi. Reciprocating and syringe pumps are most common. To maintain CO_2 in a liquid state, the pump head is cooled by using a recirculating bath. There are several ways to add a modifier to the CO_2. One is to add it directly to the extraction cell, but the modifier is exhausted with the flow of extraction fluid. Another approach is to add the modifier to the CO_2 tank (i.e., it is premixed with CO_2). However, it has been reported that the ratio of modifier to CO_2 in the mixture changes with time [24]. Moreover, the modifier may contaminate the CO_2 pump. A better alternative is to use a second pump for modifier delivery. The modifier and the CO_2 are mixed at a point after the pump but before the extraction cell. This way, the type of the modifier and its concentration can easily be controlled, and the CO_2 pump is free of modifier contamination.

The extraction cell is usually made of stainless steel, PEEK (polyether ether ketone), or any other suitable material that can withstand high pressure (up to 10,000 psi). It is fitted with fingertight frits, which eliminate use of a wrench and reduces the wear and tear that can result from overtightening. Research indicates that the shape of the cell has little impact on the extraction efficiency [24]. Short squat cells are preferred because they are easier to fill than the long thin ones. The extraction cell is placed in an oven that can heat up to 200°C.

The pressure of the supercritical fluid is controlled by the restrictor. Restrictors can be broadly classified into two types: fixed and variable. *Fixed* (diameter) *restrictors* are typically made of fused silica or metal tubing. They are inexpensive and easy to replace, but are subject to plugging problems. A common cause of plugging is water freezing at the restrictor tip because of the rapid expansion of the released supercritical fluid. Plugging can also happen when the matrix has high concentrations of extractable materials such as elemental sulfur, bulk hydrocarbons, or fats. *Variable restrictors* have an orifice or nozzle that can be adjusted electronically. They are free from plugging, and a constant flow rate can be maintained. Although variable restrictors are more expensive, they are necessary for real-world applications.

The extract is collected by depressurizing the fluid into a sorbent trap or a collection solvent. A trap may retain the analytes selectively, which may then be selectively washed off by a solvent. This can offer high selectivity, but requires an additional step. The trap can be cryogenically cooled to avoid the loss of analytes. Using a collection solvent is more straightforward. The choice of solvents often depends on the analytical instrumentation. For example, tetrachloroethene is suitable for infrared determination, while methylene chloride and isooctane are appropriate for gas chromatographic separations.

3.4.3. Operational Procedures

The sample is loaded into an extraction cell and placed into the heating oven. The temperature, pressure, flow rate, and the extraction time are set, and the extraction is started. The extract is collected either by a sorbent trap, or by a collection vial containng a solvent. Typical EPA-recommended operating conditions for the extraction of PAHs, pesticides, and PCBs are

Table 3.5. EPA-Recommended SFE Methods for Environmental Samples

	Total Recoverable Petroleum Hydrocarbons	Volatile PAHs	Less Volatile PAHs	Organochlorine Pesticides	PCBs
Extraction fluid	CO_2	CO_2	CO_2–CH_3OH–H_2O (95:1:4 v/v/v)[a]	CO_2	CO_2
Pressure (psi)	6100	1750	4900	4330	4417
Density (g/mL)	0.785	0.3	0.63	0.87	0.75
Temperature (°C)	80	80	120	50	80
Static equilibration time (min)	0	10	10	20	10
Dynamic extraction time (min)	30	10	30	30	40
Flow rate (mL/min)	1.1–1.5	2.0	4.0	1.0	2.5

[a] For HPLC determination only. CO_2–methanol–dichloromethane (95:1:4 v/v/v) should be used for GC.

presented in Table 3.5. Supercritical fluid extraction can be operated in two modes: static or dynamic. In *static extraction* the supercritical fluid is held in an extraction cell for a certain amount of time and then released to a collection device. In *dynamic extraction*, the supercritical fluid flows continuously through the extraction cell and out into a collection device.

3.4.4. Advantages/Disadvantages and Applications of SFE

SFE is fast (10 to 60 minutes) and uses minimum amount of solvents (5 to 10 mL) per sample. CO_2 is nontoxic, nonflammable, and environmentally friendly. Selective extraction of different groups of analytes can be achieved by tuning the strength of the supercritical fluids with different modifiers and by altering operating conditions. In addition, the extract from SFE does not need additional filtration, as the extraction cell has frits.

On the down side, analytical-scale SFE has limited sample size (<10 g), and the instrument is rather expensive. Furthermore, SFE has been found to be matrix dependent. Different methods have to be developed and validated for different sample matrices and for different groups of analytes. For example, Kim et al. [25] conducted an investigation on the effect of plant matrix on the SFE recovery of five schisandrin derivatives. At 60°C and 34.0 MPa, the compounds extracted from the leaves of *Schisandra chinensis* by supercritical CO_2 were 36.9% of what were obtained by organic solvent extraction. However, under the same SFE conditions, extraction from the stem and fruits yielded more than 80% of that by organic solvents. Although the addition of 10% ethanol to CO_2 increased the yield from leaves four times, it had little effect on the extraction of stems and fruits.

SFE has a wide range of applications, which include the extraction of PAHs, PCBs, phenols, pesticides, herbicides, and hydrocarbons from environmental samples, contaminants from foods and feeds, and active gradients from cosmetics and pharmaceutical products. Table 3.6 lists some examples from the literature.

3.5. ACCELERATED SOLVENT EXTRACTION

Accelerated solvent extraction (ASE) is also known as pressurized fluid extraction (PFE) or pressurized liquid extraction (PLE). It uses conventional solvents at elevated temperatures (100 to 180°C) and pressures (1500 to 2000 psi) to enhance the extraction of organic analytes from solids. ASE was introduced by Dionex Corp. (Sunnyvale, CA) in 1995. It evolved as a consequence of many years of research on SFE [45]. SFE is matrix dependent and often requires the addition of organic modifiers. ASE was developed to overcome these limitations. It was expected that conventional solvents would be less efficient than supercritical fluids, which have higher diffusion coefficients and lower viscosity. However, the results turned out to be quite the opposite. In many cases, extraction was faster and more complete with organic solvents at elevated temperature and pressure than with SFE. Extensive research has been done on the extraction of a variety of samples with ASE. ASE was approved by EPA as a standard method in 1996.

3.5.1. Theoretical Considerations

The elevated pressure and temperature used in ASE affects the solvent, the sample, and their interactions. The solvent boiling point is increased under high pressure, so the extraction can be conducted at higher temperatures. The high pressure also allows the solvent to penetrate deeper into the sample matrix, thus facilitating the extraction of analytes trapped in matrix pores. At elevated temperatures, analyte solubility increases and the mass transfer is faster. The high temperature also weakens the solute–matrix bond due to

Table 3.6. Selected SFE Applications

Analytes	Matrix	Reference
Polycyclic aromatic hydrocarbons (PAHs)	Standard reference materials (SRMs)	5
	Wastewater sludge	26
	Soils	27
	Liver samples	28
	Toasted bread	29
Polychlorinated biphenyls (PCBs)	Wastewater sludge	26
	Chicken liver	30
Organochlorine pesticides (OCPs)	Wastewater sludge	26
	Chinese herbal medicines	31
Carbamate pesticides (carbaryl, aldicarb, and carbofuran)	Filter paper and silica gel matrixes	32
Insecticides carbosulfan and imidacloprid	Process dust waste	33
Ten triazine herbicide residues	Eggs	34
Cyanazine and its seven metabolites	Spiked silty clay loam soil	35
Aromatic acids, phenols, pesticides	Soils	27
4-Nonylphenol	Municipal sewage sludge	36
Petroleum hydrocarbons	Spiked clay–sand soil	37
Nine aliphatic hydrocarbons	Chicken liver	30
Nicarbazin (a drug used principally in poultry)	Poultry feeds, eggs, and chicken tissue	38
Fenpyroximate	Apple samples	39
Vitamins A and E	Milk powder	40
Vitamins D_2 and D_3	Pharmaceutical products	41
p-Aminobenzoate (PABA) and cinnamate, ultraviolet absorbers	Cosmetic products	42
Five of the most common sunscreen agents	Cosmetic products	43
Lanolin	Raw wool fibers	44

van der Waals forces, hydrogen bonding, and dipole attractions. In addition, the high temperature reduces the solvent viscosity and surface tension, which enhances solvent penetration into the matrix. All these factors lead to faster extraction and better analyte recovery.

3.5.2. Instrumentation

A schematic diagram of an ASE system is shown in Figure 3.7. It consists of solvent tank(s), a solvent pump, an extraction cell, a heating oven, a collec-

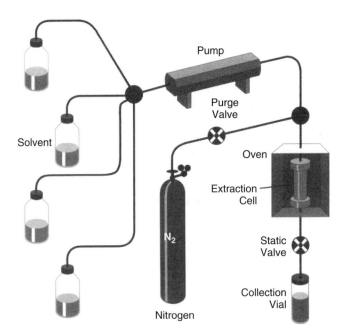

Figure 3.7. Schematic diagram of an ASE system. (Reproduced with permission from Dionex Corp.)

tion vial, and a nitrogen tank. The sample size can be anywhere between 1 and 100 mL. The extraction cells are made of stainless steel that can withstand high temperature and pressure. Each cell has two removable fingertight caps on the ends that allow easy sample loading and cleaning. The caps are fitted with compression seals for high-pressure closure. To load the cell, one end cap is screwed on to fingertightness. Then a filter is introduced into the cell, followed by the sample. The other cap is screwed on to fingertightness for complete closure. The cell is then placed in a carousel that can hold and load multiple cells.

The ASE system is fully automated. An autoseal actuator moves the cell from the carousel into the heating oven. The solvent is delivered from one or more solvent bottles into the extraction cell by a pump. The oven is heated, and the temperature and pressure in the cell rise. When the pressure reaches 200 psi above the preset value, the static valve opens to release the excessive pressure and then closes again. Then the pump delivers fresh solvent to the cell to bring the pressure back to the preset value. The addition of fresh solvent increases the concentration gradient and enhances both mass transfer and extraction efficiency. The extracts are collected in 40 or

60-mL collection vials on a removable vial tray. The vial lids have TFE-coated solvent-resistant septa. The tubing from the extraction cell to the collection vial provides enough heat loss so that additional cooling is not necessary.

An automated solvent controller is available in the latest ASE system. It allows up to four solvents to be mixed and delivered to the extraction cells. This can reduce the time for measuring and mixing solvents and decrease users' exposure to toxic solvents. The solvent controller can be programmed to change solvents between sequential extractions of multiple samples. The same sample can also be reextracted using different solvents. The ASE system has many built-in safety features, which include vapor sensors, liquid-leak detectors, vial overfill monitors, electronic and mechanical over-pressurization prevention systems, solvent flow monitors, and pneumatic source pressure monitors.

3.5.3. Operational Procedures

The steps in the ASE process are shown in Figure 3.8. The sample is loaded into the extraction cell, and then the solvent is pumped in. Then the cell is heated to the desired temperature and pressure. The heat-up time can be 5 to 9 minutes (for up to 200°C). This is referred to as the *prefill method*. Alternatively, the sample can be heated before adding the solvent, which is known as the *preheat method*. However, the preheat method is prone to the loss of volatile analytes. Therefore, the prefill approach is generally preferred [46].

After heating, the extraction can be conducted dynamically, statically, or as a combination of both. In the dynamic mode, the extraction solvent flows through the system, whereas there is no solvent flow in the static mode. Although it may have higher extraction efficiency, dynamic extraction uses more solvents and is not commonly used. Static extraction time is on the order of 5 minutes, although it can be as long as 99 minutes. After extraction, the extract is flushed into the collection vial with fresh solvents. The flush volume can be 5 to 150% of the cell volume, with 60% being the typical choice. As many as five static cycles may be chosen, although a single cycle is the most common option. The total flush volume is divided by the number of cycles, and an equal portion is used in each cycle. After the final solvent flush, the solvent is purged into the collection vial with nitrogen (typically, 1-minute purge at 150 psi). The ASE system can sequentially extract up to 24 samples in one unattended operation. The sequence of introducing and removing the cells to and from the oven can be automated. Extract filtration is not required, but concentration and/or cleanup is often necessary prior to analysis.

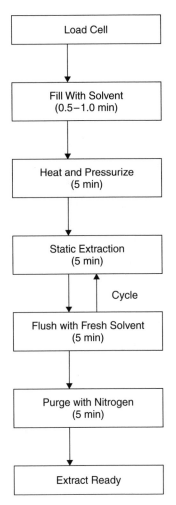

Figure 3.8. Schematic diagram of ASE procedures.

3.5.4. Process Parameters

Typical operating parameters suggested in the EPA standard method are listed in Table 3.7.

Temperature and Pressure

As mentioned before, solubility and mass transfer increase at elevated temperatures. Table 3.8 shows that both recovery and precision improved when the temperature was increased during the extraction of total petroleum

Table 3.7. Suggested System Parameters in EPA Standard Methods for the ASE of Environmental Samples

	Semivoaltiles, Organophosphorus Pesticides, Organochlorine Pesticides, Herbicides, and PCBs	Polychlorinated Dibenzodioxins and Polychlorinated Dibenzofurans	Diesel Range Organics
Oven temperature (°C)	100	150–175	175
Pressure (psi)	1500–2000	1500–2000	1500–2000
Static time (min)	5 (after 5 min preheat time)	5–10 (after 7–8 min preheat time)	5–10 (after 7–8 min preheat time)
Flush volume	60% of the cell volume	60–75% of the cell volume	60–75% of the cell volume
Nitrogen purge[a]	60 s at 150 psi	60 s at 150 psi	60 s at 150 psi
Static cycles	1	2 or 3	1

[a] Purge time may be extended for larger cells.

hydrocarbons from soil [46]. Similar observations were made in other applications as well [47,48]. A certain pressure level is required to keep the solvent in its liquid state when the temperature is above its boiling point at atmospheric pressure. Pressure greater than 1500 psi has no significant influence on the recovery [45]. Typical pressures used in the extraction of environmental samples are in the range 1500 to 2000 psi.

Solvents

The general criteria for the solvent selection are high solubility of the analytes and low solubility of the sample matrix. Solvents used in conventional

Table 3.8. Effects of Temperature on the Recovery of TPHs from Soil Using ASE (1200 mg/kg Certified Value)

Temperature (°C)	Extraction Efficiency (%)	RSD (%)
27	81.2	6.0
50	93.2	5.0
75	99.2	2.0
100	102.7	1.0

Reproduced from Ref. 46, with permission from the American Chemical Society.

Table 3.9. Solvents Recommended by EPA for the ASE of Environmental Samples

Analytes	Solvents
Organochlorine pesticides, semivolatile organics	Acetone–hexane (1:1 v/v) or acetone–methylene chloride (1:1 v/v)
PCBs	Acetone–hexane (1:1 v/v) or acetone–methylene chloride (1:1 v/v) or hexane
Organophosphorus pesticides	Methylene chloride or acetone–methylene chloride (1:1 v/v)
Chlorinated herbicides	Acetone–methylene chloride–phosphoric acid solution (250:125:15 v/v/v) or acetone–methylene chloride–trifluoroacetic acid solution (250:125:15 v/v/v)
Polychlorinated dibenzodioxins and polychlorinated dibenzofurans	Toluene or toluene–acetic acid solution (5% v/v glacial acetic acid in toluene) for fly ash samples
Diesel range organics	Acetone–methylene chloride (1:1 v/v) or acetone–hexane (1:1 v/v) or acetone–heptane (1:1 v/v)

(such as Soxhlet) extraction methods can readily be applied in ASE. However, conventional solvents cannot be used in certain applications, such as the extraction of polymers. This is because the matrix itself can dissolve in the solvent at high temperature and plug the connecting tubing in the system. On the other hand, solvents that are not efficient in Soxhlet extraction may yield high recovery under ASE conditions. For example, hexane was found to be a poor solvent in the Soxhlet extraction of monomers and oligomers from nylon-6 and poly(1,4-butylene terephthalate) (PBT), but it gave satisfactory results in ASE [47]. Table 3.9 lists the solvents recommended in EPA method 3545A for the ASE of different groups of analytes from soils, clays, sediments, sludge, and waste solids.

Small sample size can reduce solvent volume, provided it meets the requirements of sensitivity and homogeneity. Ten to 30 grams of material is usually necessary. The volume of the solvent is a function of the size of the extraction cell rather than the mass of the sample. The solvent volume may vary from 0.5 to 1.4 times that of the cell [1]. Specific solvent/cell volume ratios are usually available in the instrument manufacturer's instructions.

3.5.5. Advantages and Applications of ASE

ASE has many advantages. It uses minimal amount of solvent and is fast (about 15 minutes), fully automated, and easy to use. Filtration is a built-in step, so additional filtration is not needed. While operating at higher temperatures and pressures, ASE can employ the same solvent specified by other existing methods. Therefore, method development is simple. There are more solvents to choose from, because solvents that work poorly in conventional methods may perform well under ASE conditions. In addition, ASE provides the flexibility of changing solvents without affecting the extraction temperature and pressure. Despite high initial equipment cost, cost per sample can be relatively low.

This section is not intended to be a thorough literature survey, but it offers a general description of typical ASE applications. Table 3.10 provides

Table 3.10. Selected ASE Applications

Analytes	Matrix	Reference
PAHs	Soils	49
	Clay loam and soils	50
	Mosses and pine needles	51
	Soils, heap material, and fly ash	52
	Soil	53
PCBs	Mosses and pine needles	52
	Soil	53
Organochlorine pesticides (OCPs)	Soils, heap material, and fly ash	52
	Clay loam and soils	50
Organophosphorus pesticides	Foods	54
Polychlorinated dibenzo-p-dioxins and polychlorinated dibenzofurans	Soils, heap material, and fly ash	52
	Chimney brick, urban dust, and fly ash	55
Hydrocarbons	Wet and dry soils	56
Chlorobenzenes, HCH isomers, and DDX	Soil	53
	Mosses and pine needles	51
Atrazine and alachlor	Soils	57
Diflufenican	Soil	58
Phenols	Spiked soil	59
Chlorophenols	Soil	60
Additive Irganox 1010	Polypropylene	61
	Polypropylene, PVC, and nylon	62
Antioxidant Irganox 1076	Linear low-density polyethylene (LLDPE)	63
Monomers and oligomers	Nylon-6 and poly(1,4-butylene terephthalate)	47
Felodipine	Medicine tablets	64
Active gradients	Medicinal plants	65
Fatty acid and lipids	Cereal, egg yolk, and chicken meat	66

a quick reference to these examples, and more detailed information can be found in some recent reviews [67,68]. In principle, ASE is a universal method that can be used in any solvent extraction. However, majority applications so far have been in the environmental area, such as the extraction of pesticides, herbicides, PAHs, PCBs, base/neutral/acid compounds, dioxins, furans, and total petroleum hydrocarbons. ASE has also been used to extract additives and plasticizers from polymers, additives, and active ingredients from pharmaceuticals, and contaminants/fat from food.

3.6. MICROWAVE-ASSISTED EXTRACTION

It should be noted that microwave-assisted extraction (MAE) discussed in this chapter is different from microwave-assisted acid digestion. The former uses organic solvents to extract organic compounds from solids, while the latter uses acids to dissolve the sample for elemental analysis with the organic contents being destroyed. Microwave-assisted digestion of metals is covered in Chapter 5.

The name *magnetron* (microwave generator) was first used in 1921 by A. W. Hall. In 1946, Percy Spencer discovered the function of microwave as a heating source. Domestic microwave ovens became available in 1967 [69]. In 1975, microwave was first applied to acid digestion for metal analysis by Abu-Samra et al. [70]. Since then much work has been done on microwave-assisted acid digestion, and it has gained widespread acceptance and approval by regulatory agencies as a standard method. Microwave-assisted organic extraction was first carried out in 1986 by Ganzler et al. [71] for the extraction of fats and antinutrients from food and pesticides from soil. In 1992, Pare [72] patented a process called MAP (microwave-assisted process) for the extraction of essential oils from biological materials. This technique was later extended to analytical as well as large-scale applications. In the year 2000, MAE was approved by the EPA as a standard method for the extraction of semivoaltile and nonvolatile compounds from solid samples.

3.6.1. Theoretical Considerations

Microwaves are electromagnetic radiation in the frequency range 0.3 to 300 GHz (corresponding to 0.1 to 100 cm wavelength). They are between the radio frequency and the infrared regions of the electromagnetic spectrum. Microwave is used extensively in radar transmission (1 to 25 cm wavelength) and telecommunications. To avoid interference with communication networks, all microwave heaters (domestic or scientific) are designed to work at either 2.45 or 0.9 GHz. Domestic ovens operate at 2.45 GHz only. When mircowave radiation is applied to molecules in the gas phase, the molecules absorb energy to change their rotational states. The microwave spectrum of molecules shows many sharp bands in the range 3 to 60 GHz. This has been used in microwave spectroscopy to obtain fundamental physical–chemical data such as bond lengths and angels, and to identify gaseous molecules (e.g., molecular species in outer space).

In the liquid and solid states, molecules do not rotate freely in the microwave field; therefore, no microwave spectra can be observed. Molecules respond to the radiation differently, and this is where microwave heating comes in. The mechanism of microwave heating is different from that of conventional heating. In conventional heating, thermal energy is transferred from the source to the object through conduction and convection. In microwave heating, electromagnetic energy is transformed into heat through ionic conduction and dipole rotation. *Ionic conduction* refers to the movement of ions in a solution under an electromagnetic field. The friction between the solution and the ions generates heat. *Dipole rotation* is the reorientation of dipoles under microwave radiation. A polarized molecule rotates to align itself with the electromagnetic field at a rate of 4.9×10^9 times per second. The larger the dipole moment of a molecule, the more vigorous is the oscillation in the microwave field.

The ability of a material to transform electromagnetic energy into thermal energy can be defined as

$$\tan \delta = \frac{\varepsilon''}{\varepsilon'}$$

where $\tan \delta$ is the loss tangent or tangent delta; ε'' is the dielectric loss coefficient, a measure of the efficiency of a material to transform electromagnetic energy to thermal energy; and ε' is the dielectric constant, a measure of the polarizibility of a molecule in an electric field. Table 3.11 lists the physical constants of some selected organic solvents. Polar solvents such as acetone, methanol, and methylene chloride have high $\tan \delta$ values and can be heated rapidly. Nonpolar solvents such as hexane, benzene, and toluene cannot be heated because they lack dipoles and do not absorb microwave.

3.6.2. Instrumentation

In general, organic extraction and acid digestion use different types of microwave apparatus, as these two processes require different reagents and different experimental conditions. A new commercial system, Mars X (CEM Corp., Matthews, NC) offers a duel unit that can perform both proce-

Table 3.11. Physical Constants of Organic Solvents Used in MAE[a]

Solvent	Boiling Point (°C)	Vapor Pressure torr	Vapor Pressure kPa	ε'	Dipole Moment (debye)	$\tan \delta \times 10^4$
Methylene chloride	40	436	58.2	8.93	1.14	—
Acetone	56	184	24.6	20.7	2.69	—
Methanol	65	125	16.7	32.7	2.87	6400
Tetrahydrofuran	66	142	19.0	7.58	1.75	—
Hexane	69	120	16.0	1.88	<0.1	—
Ethyl acetate	77	73	9.74	6.02	1.88	—
Ethanol	78	—	—	24.3	1.69	2500
Methyl ethyl ketone	80	91	12.1	18.51	2.76	—
Acetonitrile	82	89	11.9	37.5	3.44	—
2-Propanol	82	32	4.27	19.92	1.66	6700
1-Propanol	97	14	1.87	20.33	3.09	~2400[b]
Isooctane	99	49	6.54	1.94	0	—
Water	100	760	101.4	78.3	1.87	1570
Methyl isobutyl ketone	116	20	2.67	13.11	—	—
Dimethyl formamide	153	2.7	0.36	36.71	3.86	—
Dimethyl acetamide	166	1.3	0.17	37.78	3.72	—
Dimethyl sulfoxide	189	0.6	0.08	46.68	3.1	—
Ethylene glycol	198	—	—	41.0	2.3	10,000
N-Methyl pyrrolidinone	202	4.0	0.53	32.0	4.09	—

Reproduced from Ref. 85, with permission from the American Chemical Society.
[a] Boiling points were determined at 101.4 kPa; vapor pressures were determined at 25°C, dielectric constants were determined at 20°C; dipole moments were determined at 25°C.
[b] Value was determined at 10°C.

dures. In this chapter only the instrumentation for organic extraction is discussed.

The basic components of a microwave system include a microwave generator (magnetron), a waveguide for transmission, a resonant cavity, and a power supply. For safety and other reasons, domestic microwave ovens are not suitable for laboratory use. There are two types of laboratory microwave units. One uses closed extraction vessels under elevated pressure; the other uses open vessels under atmospheric pressure. Table 3.12 lists the features of some commercial MAE systems.

Closed-Vessel Microwave Extraction Systems

Closed-vessel units were the first commercially available microwave ovens for laboratory use. A schematic diagram of such a system is shown in

Table 3.12. Features of Some Commercial MAE Systems

Model/Manufacturer	Power (W)	Sensors	Max. Pressure (bar)	Vessel Volume (mL)	Vessel Material[a]	Number of Vessels	Max Temp. (°C)
Multiwave/Anton Parr GmbH, Austria	1000	Pressure control and infrared temperature measurement in all vessels	70	100	TFM/ceramics	12	230
			70	100	TFM/ceramics	6	280
			130	50	TFM/ceramics	6	280
			130	50	Quartz	8	300
			130	20	Quartz	8	300
Mars-8/CEM, United States	1500	Infrared temperature measurement in all vessels	35	100	TFM	14	300
			100	100	TFM	12	300
Ethos 900/1600, Milestone, United States	1600	Pressure control and temperature measurement in all vessels	30	120	TFM/PFA	10	240
			100	120	TFM	6	280
			30	120	TFM/PFA	12	240
			100	120	TFM	10	280
Soxwave 100/3.6, Prolabo, France	250	Temperature control	Open vessel	250	Quartz	1	—
			Open vessel	100 or 260	Quartz	1	—

[a]TFM, tetrafluoromethoxyl polymer; PFA, perfluoroalkoxy.

166

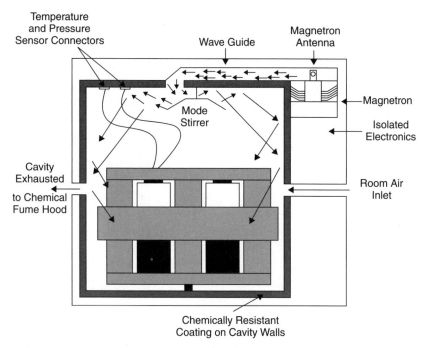

Figure 3.9. Schematic diagram of a closed-vessel cavity MAE system. (Reproduced from Ref. 85, with permission from the American Chemical Society.)

Figure 3.9. In the oven cavity is a carousel (turntable or rotor) that can hold multiple extraction vessels. The carousel rotates 360° during extraction so that multiple samples can be processed simultaneously. The vessels and the caps are constructed of chemically inert and microwave transparent materials such as TFM (tetrafluoromethoxyl polymer) or polyetherimide. The inner liners and cover are made of Teflon PFA (perfluoroalkoxy). The vessels can hold at least 200 psi of pressure. Under elevated pressures, the temperature in the vessel is higher than the solvent's boiling point (see Table 3.11), and this enhances extraction efficiency. However, the high pressure and temperature may pose safety hazards. Moreover, the vessels need to be cooled down and depressurized after extraction.

One of the extraction vessels is equipped with a temperature and pressure sensor/control unit. Figure 3.10 shows the schematic diagram of a control vessel as well as a standard vessel. A fiber-optic temperature probe is built into the cap and the cover of the control vessel. The standard EPA method requires the microwave extraction system to be capable of sensing the temperature to within ±2.5°C and adjusting the microwave field output power

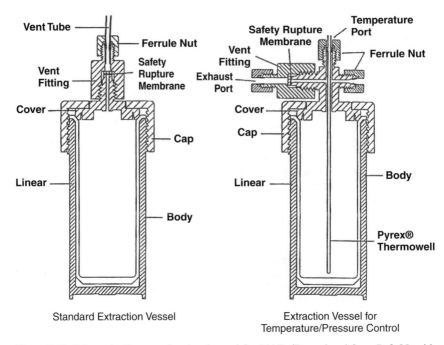

Figure 3.10. Schematic diagram of a closed vessel for MAE. (Reproduced from Ref. 85, with permission from the American Chemical Society.)

automatically within 2 seconds of sensing. The temperature sensor should be accurate to $\pm 2°C$.

Safety features are essential to a microwave apparatus. An exhaust fan draws the air from the oven to a solvent vapor detector. Should solvent vapors be detected, the magnetron is shut off automatically while the fan keeps running. Each vessel has a rupture membrane that breaks if the pressure in the vessel exceeds the preset limit. In the case of a membrane rupture, solvent vapor escapes into an expansion chamber, which is connected to the vessels through vent tubing. To prevent excessive pressure buildup, some manufacturer use resealable vessels. A spring device allows the vessel to open and close quickly, releasing the excess pressure.

Additional features can be found in newer systems. Some have a built-in magnetic stir bar with variable speed control for simultaneous stirring in all the vessels. Stirring enhances contact between the sample and the solvents. This reportedly results in significant reduction in extraction time and improvement in analytes recoveries [68]. The stir bar is made of Weflon, a proprietary polytetrafluoroethylene (PTFE) compound that can absorb microwave. This allows the use of nonpolar solvents for extraction since

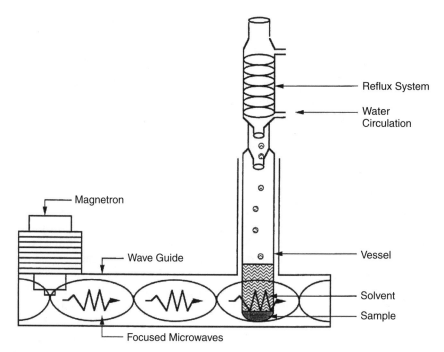

Figure 3.11. Schematic diagram of an open-vessel, waveguide-type MAE system. (Reproduced from Ref. 6, with permission from Elsevier Science.)

heating is done through the stir bar. The same solvents used in conventional methods (both polar and nonpolar) may be adopted here, thus reducing the time for method development.

Open-Vessel Microwave Extraction Systems

Open-vessel systems are also known as *atmospheric pressure microwave* or *focused microwave systems*. An example is Soxwave 100 (Prolabo Ltd., France). A schematic diagram of such a system is shown in Figure 3.11. It uses a "focused" waveguide, that directs the microwave energy into a single-vessel cavity. This provides greater homogeneity of the radiation than in closed-vessel units, where microwave is dispersed into the multivessel cavity. However, only one vessel can be heated at a time, and multiple vessels are to be processed sequentially. The vessel, typically made of glass or quartz, is connected with an air (or a water) condenser to reflux the volatile analytes and solvents. Operating somewhat like Soxhlet extraction, this type of system has been referred to as *microwave-assisted Soxhlet extraction.*

3.6.3. Procedures and Advantages/Disadvantages

In a typical application, 2 to 20 g of sample is dried, weighed, and loaded into an extraction vessel. A certain amount (less than 30 mL) of select solvents is also added. Then parameters such as temperature, pressure, and extraction time are set according to the instructions from the microwave manufacturer. A preextraction heating step (typically, 1 to 2 minutes) is needed to bring the system to the preset values. Subsequently, the samples are extracted for about 10 to 20 minutes. After the extraction, the vessels are cooled, and this normally takes less than 20 minutes. Finally, the extract is filtered, concentrated, and analyzed.

High efficiency is the major advantage of microwave extraction over conventional methods such as Soxhlet. It can achieve the same recovery in a shorter time (20 to 30 minutes) and with less solvent (30 mL). The throughput is high (up to 12 samples per hour for closed-vessel system). On the other hand, MAE has several limitations. Solvents used in Soxhlet extraction cannot readily be applied to microwave extraction because some of them do not absorb microwave. Method development is generally necessary for MAE applications. Moreover, cooling and filtration after extraction prolongs the overall process. Since MAE is quite exhaustive, normally the extract contains interfering species that require cleanup prior to analysis.

3.6.4. Process Parameters

The efficiency of MAE can be influenced by factors such as the choice of solvent, temperature, extraction time, matrix effects, and water contents. In general, some optimization of these conditions is necessary. Typical microwave conditions suggested in a standard EPA method are listed in Table 3.13.

Table 3.13. EPA Standard Procedure for MAE of Environmental Samples

Solvents	25 mL of acetone–hexane (1:1 v/v)
Temperature	100–150°C
Pressure	50–150 psi
Time at temperature	10–20 min
Cooling	To room temperature
Filtering/rinsing	With the same solvent system

Choice of Solvent

The proper choice of solvent is the key to successful extraction. In general, three types of solvent system can be used in MAE: solvent(s) of high ε'' (dielectric loss coefficient), a mixture of solvents of high and low ε'', and a microwave transparent solvent used with a sample of high ε''. Pure water was used for the extraction of triazines from soils [73], and the recovery was comparable to those using organic solvents. In the extraction of organochlorine pesticides (OCPs) from marine sediments, terahydrofuran (THF) yielded better recovery than either acetone or acetone–hexane (1:1) [74]. It was reported that dichloromethane (DCM)–methanol (9:1) was the most efficient solvent for the extraction of phenylurea herbicides (linuron and related compounds) from soils. Other solvent systems, including DCM, DCM–water (5:1), methanol–water (7:3), and methanol–water (9:1) gave poor performance [75]. For the extraction of felodipine and its degradation product H152/37 from medicine tablets [76], acetonitrile–methanol (95:5) was found to be the optimum solvent composition. Methanol was capable of dissolving the tablet's outer covering layer, while acetonitrile broke the inner matrix into small pieces. Hexane–acetone (typically 1:1) has proven to be an efficient solvent system for the extraction of PAHs, phenols, PCBs, and OCPs from environmental samples [77,78].

Temperature and Pressure

Generally, recovery increases with the increase in temperature and then levels off after a certain point. For thermally labile compounds, analyte degradation occurs at high temperatures and results in low recovery. Excessively high temperatures lead to matrix decomposition in polymer extractions and should be avoided. In general, pressure is not a critical parameter in MAE. It changes with the solvent system and the temperature used and is acceptable below a preset limit.

It was reported that the recoveries of 17 PAHs from six certified reference marine sediments and soils [77] increased from 70 to 75% when the temperature was increased from 50°C to 115°C, and remained at 75% from 115 to 145°C. In the extraction of OCPs from sediments, recovery was unchanged from 100 to 120°C [74]. In the extraction of phenylurea herbicides from soils, the recovery peaked in the range 60 to 80°C and decreased at lower or higher temperatures [75]. In the extraction of sulfonylurea herbicides from soils, recovery dropped from 70 to 80% to 1 to 30%, due to decomposition when temperature increased from 70°C to 115°C [79]. The recovery of oligomers from poly(ethyleneterephthalate) increased as temperature rose from 70°C to 140°C [80]. However, polymer fusion occurred at temperatures above 125°C; therefore, 120°C was chosen as the optimum.

Extraction Time

Many microwave extractions can reach maximum recovery in 10 to 20 minutes. Longer extraction time is not necessary and may lead to the decomposition of thermolabile analytes. It was reported that the recovery of sulfonylurea from soil was not affected by extraction time in the range 5 to 30 minutes [79]. Similar observation was made in the extraction of PAHs from soils and sediments [6]. In the extraction of PAHs and LAHs (linear aliphatic hydrocarbons) from marine sediments, the extraction time was found to be dependent on the irradiation power and the number of samples extracted per run [81]. When the irradiation power was 500 W, the extraction time varied from 6 minutes for one sample to 18 minutes for eight samples [74]. The recovery of OCPs from spiked marine sediments increased from 30% at 5 and 10 minutes to 60% at 20 minutes and to 74 to 99% at 30 minutes [82].

Matrix Effects and Water Content

Matrix effects have been observed in MAE applications. It was reported that recoveries of OCPs from aged soils (24 hours of aging) were lower than those from freshly spiked samples [78]. Similar matrix effects were also reported in the extraction of sulfonylurea herbcides from aged soils [79]. In another study, the average recoveries of 17 PAHs from six different standard reference materials (marine sediments and soils) varied from 50 to 100% [77].

Because water is a polar substance that can be heated by microwave irradiation, it can often improve analyte recovery. In a study of focused MAE of PAHs from soil and sediments [6], sample moisture level showed significant influence on extraction efficiency, and 30% water in the sample provided the highest recovery. Similarly, the maximum recovery of phenyl-urea herbicides was obtained with 10% water in soils [75]. In the extraction of triazines from soil, water content in the range 10 to 15% yielded the highest recovery [7].

Microwave power output and sample weight seem to have minor effects on extraction efficiency. It was reported that the increase in oven power gave higher recovery of PAHs from atmospheric particles [82]. The reason could be that the microwave system used in that study had no temperature control.

For an extraction conducted at a controlled temperature, the oven power output may have less influence on recovery.

3.6.5. Applications of MAE

Majority MAE applications have been in the extraction of PAHs, PCBs, pesticides, phenols, and total petroleum hydrocarbons (TPHs) from environmental samples. MAE has also been used in the extraction of contaminants and nutrients from foodstuffs, active gradients from pharmaceutical products, and organic additives from polymer/plastics. Table 3.14 lists some typical applications. Readers interested in the details of MAE applications can find more information in some recent reviews [85–87].

3.7. COMPARISON OF THE VARIOUS EXTRACTION TECHNIQUES

Table 3.15 summarizes the advantages and disadvantages of various extraction techniques used in the analysis of semivolatile organic analytes in solid samples. They are compared on the basis of matrix effect, equipment cost, solvent use, extraction time, sample size, automation/unattended operation, selectivity, sample throughput, applicability, filtration requirement, and the need for evaporation/concentration. The examples that follow show the differences among these techniques in real-world applications.

Example 1

Lopez-Avila et al. [88] compared MAE, Soxhlet, sonication, and SFE in their ability to extract 95 compounds listed in the EPA method 8250. Freshly spiked soil samples and two SRMs were extracted by MAE and Soxhlet with hexane–acetone (1:1), by sonication with methylene chloride–acetone (1:1), and by SFE with supercritical carbon dioxide modified with 10% methanol. Table 3.16 shows the number of compounds in different recovery ranges obtained by the various techniques. Sonication yielded the highest recoveries, followed by MAE and Soxhlet, whose performances were similar. SFE gave the lowest recoveries. MAE demonstrated the best precision: RSDs were less than 10% for 90 of 94 compounds. Soxhlet extraction showed the worst precision; only 52 of 94 samples gave RSDs less than 10%. No technique produced acceptable recoveries for 15 polar basic compounds. The recoveries of these compounds by MAE with hexane–acetone at 115°C for 10 minutes (1000 W power) were poor. Consequently, their extraction with MAE was investigated using acetonitrile at 50 and 115°C. Ten of the 15 compounds were recovered quantitatively (>70%) at 115°C.

Table 3.14. Selected MAE Applications

Analytes	Matrix	Vessel Type	Solvents	Extraction Conditions	Recovery (%)	RSD (%)	Reference
PAHs	SRMs, spiked, and real soil samples	Open	20 mL of acetone–hexane (1:1)	1-g sample, 10 min	96–100	<7	83
	Soil and sediments	Open	30 mL of dichloromethane	0.1- to 1-g sample, 30% water, 10 min	85–90	<15	6
	Atmospheric particles	Closed	15 mL of acetone–hexane (1:1)	2.6-g sample, 20 min, 400 W	96–103 compared with Soxhlet	<5 for 12 of 16 compounds	82
Semivolatiles, PCBs, OCPs, OPPs	Freshly spiked soils	Closed	30 mL of acetone–hexane (1:1)	5-g sample, 115°C, 10 min	80–120 for 152 of 187 compounds, 7% higher than Soxhlet and sonication	1–39	78
OCPs	Spiked and natural sediments	Closed	30 mL of terahydrofuran	5-g sample, 100°C, 30 min	74–99	1–10	74

174

Atrazine, OPPs	Orange peel	Closed	10 mL of acetone–hexane (1:1)	1.5- to 2.5-g sample, 90°C, 9 min	93–101	1–3	84
Triazines	Aged spiked soil	Closed	30 mL of water	1-g sample, 0.5 MPa, 4 min	88–91	6–7	73
Sulfonylurea herbicides	Freshly spiked and aged sandy soils	Closed	20 mL of dichloromethane–methanol (9:1)	10-g sample, 60°C, 100 psi, 10 min	70–100	1–10	79
Phenylurea herbicides	Freshly Spiked soils (FSS) and aged soils (AS)	Closed	20 mL of dichloromethane–methanol (9:1)	5-g sample, 10% water, 690 kPa, 70°C, 10 min	80–120 for FSS, 41–113 for AS	<12 for SFS, 1–35 for AS	75
Felodipine, H152/37	Tablets	Closed	10 mL of methanol–acetonitrile (5:95)	Whole tablet, 80°C, 10 min	99–100	2–5	76
Oligomers	Poly(ethylene-terephthalate) (PET)	Closed	40 mL of dichloromethane	8-g sample, 120°C, 150 psi, 120 min	94 compared with Soxhlet	5	80

Table 3.15. Advantages and Disadvantages of Various Extraction Techniques

Technique	Advantages	Disadvantages
Soxhlet extraction	Not matrix dependent Very inexpensive equipment Unattended operation Rugged, benchmark method Filtration not required	Slow extraction (up to 24–48 hrs) Large amount of solvent (300–500 mL) Mandatory evaporation of extract
Automated Soxhlet extraction	Not matrix dependent Inexpensive equipment Less solvent (50 mL) Evaporation integrated Filtration not required	Relatively slow extraction (2 hours)
Ultrasonic extraction	Not matrix dependent Relatively inexpensive equipment Fast extraction (10–45 min) Large amount of sample (2–30 g)	Large amount of solvent (100–300 mL) Mandatory evaporation of extract Extraction efficiency not as high Labor intensive Filtration required
Supercritical fluid extraction (SFE)	Fast extraction (30–75 min) Minimal solvent use (5–10 mL) CO_2 is nontoxic, nonflammable, environmentally friendly Controlled selectivity Filtration not required Evaporation not needed	Matrix dependent Small sample size (2–10 g) Expensive equipment Limited applicability
Accelerated solvent extraction (ASE)	Fast extraction (12–18 min) Small amount of solvent (15–40 mL) Large amount of sample (up to 100 g) Automated Easy to use Filtration not required	Expensive equipment Cleanup necessary
Microwave-assisted extraction (MAE)	Fast extraction (20–30 min) High sample throughput Small amount of solvent (30 mL) Large amount of sample (2–20 g)	Polar solvents needed Cleanup mandatory Filtration required Moderately expensive equipment Degradation and chemical reaction possible

Table 3.16. Number of Compounds in Different Recovery Ranges Obtained by Various Extraction Techniques

Technique	Recovery			
	>80%	50–79%	29–49%	<19%
Sonication	63	25	4	2
MAE	51	33	8	2
Soxhlet	50	32	8	4
SFE	37	37	12	8

Example 2

A study compared ASE and SFE to Soxhlet and sonication in the determination of long-chain trialkylamines (TAMs) in marine sediments and primary sewage sludge [89]. The recoveries of these compounds by SFE at 50°C and 30 MPa with CO_2 (modified dynamically with methanol or statically with triethylamine) were 10 to 77% higher than those by Soxhlet or sonication with dichloromethane–methanol (2:1). ASE at 150°C and 17 MPa with the same solvent mixture as Soxhlet showed the highest extraction efficiency among the extraction methods evaluated. SFE exhibited the best precision because no cleanup was needed, whereas Soxhlet, sonication, and ASE extracts required an alumina column cleanup prior to analysis. SFE and ASE used less solvent and reduced the extraction time by a factor of 3 and a factor of 20 compared to sonication and Soxhlet, respectively.

Example 3

Heemken et al. [90] compared ASE and SFE with Soxhlet, sonication, and methanolic saponificaion extraction (MSE) for the extraction of PAHs, aliphatic and chlorinated hydrocarbons from a certified marine sediment samples, and four suspended particulate matter (SPM) samples. Average PAH recovery in three different samples using SFE was between 96 and 105% of that by Soxhlet, sonication, and MSE; for ASE the recovery was between 97 and 108%. Compared to the certified values of sediment HS-6, the average recoveries of SFE and ASE were 87 and 88%; for most compounds the results were within the limits of confidence. For alkanes, SFE recovery was between 93 and 115%, and ASE recovery was between 94 and 107% of that by Soxhlet, sonication, and MSE. While the natural water content of the SPM sample (56%) led to insufficient recovery by ASE and SFE, quantitative extractions were achieved in SFE after addition of anhydrous sodium sulfate to the sample.

Example 4

Llompart et al. [91] compared SFE and MAE with the EPA sonication protocol, for the extraction of phenolic compounds (phenol, o-cresol, m-cresol, and p-cresol) from soil. The samples were five artificially spiked soil matrices with carbon content ranging from 2 to 10%, and a real phenol-contaminated soil with a high carbon content (18%). The extracts from SFE and MAE were analyzed directly by a gas chromatography/mass spectrometry method without cleanup or preconcentration. These two methods showed no significant difference in precision, with RSDs in the range 3 to 15%. They were more efficient than sonication, with at least twice the recovery in both spiked and real soil samples. MAE showed the best recoveries (>80%) for the five spiked matrixes, except for o-cresol in soils with carbon content higher than 5%. Although SFE provided satisfactory recovery from low-carbon (<5%) soils, recoveries were low in more adsorptive (high-carbon-content) soils. Extraction efficiency improved significantly when a derivatization step was combined to SFE. However, in the real soil samples, the recoveries achieved by both SFE and MAE derivatization were lower than those by SFE and MAE without derivatization.

Example 5

Vandenburg et al. [92] compared extraction of additive Irganox 1010 from freeze-ground polypropylene polymer by pressurized fluid extraction (PFE) and MAE with reflux, ultrasonic, shake-flask, and Soxhlet extraction. PFE and MAE were faster than any conventional method with comparable extraction efficiency. The times to reach 90% recovery by PFE using propan-2-ol at 150°C and acetone at 140°C were 5 and 6 minutes, respectively. Reflux with chloroform was found to be the fastest method performed under atmospheric pressure with 90% recovery in 24 minutes. Reflux with cyclohexane–propan-2-ol (1:1) required 38 minutes. Ultrasonic, shake-flask, and Soxhlet extraction required about 80 minutes (90% extraction). The total sample preparation time for PFE was 15 minutes, MAE 28 minutes, and reflux with chloroform was 45 minutes.

REFERENCES

1. EPA publication SW-846, *Test Methods for Evaluating Solid Waste: Physical/Chemical Methods.*
   www.epa.gov/epaoswer/hazwaste/test/sw846.htm
2. *Standard Methods of AOAC International*, Vols. 1 and 2, AOAC International, Arlington, VA, 1999.

3. *Book of ASTM Standards: Water and Environmental Technology*, Sec. 11, Vol. 11.02, American Society for Testing and Materials, Philadelphia, PA.

4. J. Pawliszyn, *J. Chromatogr. Sci.*, **31**, 31–37 (1993).

5. B. A. Benner, *Anal. Chem.*, **70**, 4594–4601 (1998).

6. H. Budzinski, M. Letellier, P. Garrigues, and K. Le Menach, *J. Chromatogr. A*, **837**(1/2), 187–200 (1999).

7. G. Xiong, B. Tang, X. He, M. Zhao, Z. Zhang, and Z. Zhang, *Talanta*, **48**(2), 333–339 (1999).

8. V. Lopez-Avila, K. Bauer, J. Milanes, and W. F. Beckert, *J. AOAC Int.*, **76**(4), 864–880 (1993).

9. K. W. Brown, C. P. Chisum, J. C. Thomas, and K. C. Donnelly, *Chemosphere*, **20**(1/2), 13–20 (1990).

10. M. L. Foster, Jr. and S. E. Gonzales, *J. AOAC Int.*, **75**(2), 288–292 (1992).

11. G. Membrado, J. Vela Rodrigo, N. Ferrando, C. Ana, and V. L. Cebolla Burillo, *Energy Fuels*, **10**(4), 1005–1011 (1996).

12. A. Kotronarou et al., *Environ. Sci. Technol.*, **26**, 1460–1462 (1992).

13. Z. Q. Qu, L. Q. Jia, H. Y. Jin, A. Yediler, T. H. Sun, and A. Kettrup, *Chromatographia*, **44**(7/8), 417–420 (1997).

14. C. H. Marvin, L. Allan, B. E. McCarry, and D. W. Bryant, *Int. J. Environ. Anal. Chem.*, **49**(4), 221–230 (1992).

15. N. Haider and S. Karlsson, *Analyst*, **124**(5), 797–800 (1999).

16. G. A. Eiceman, A. C. Viau, and F. W. Karasek, *Anal. Chem.*, **52**(9), 1492–1496 (1980).

17. C. Golden and E. Sawicki, *Anal. Lett.*, **A11**(12), 1051–1062 (1978).

18. C. Golden and E. Sawicki, *Int. J. Environ. Anal. Chem.*, **4**(1), 9–23 (1975).

19. C. de la Tour, *Ann. Chim. Phys.*, **21**, 127 (1822).

20. J. B. Hannay and J. Hogarth, *Proc. R. Soc.*, **29**, 324 (1879).

21. M. Novotny, S. R. Springton, P. A. Peaden, J. C. Fjeldsted, and M. L. Lee, *Anal. Chem.*, **53**, 407A (1981).

22. S. Mitra and N. Wilson, *J. Chromatogr. Sci.*, **29**, 305–309 (1991).

23. B. A. Charpentier and M. R. Sevenants, eds., *Supercritical Fluid Extraction and Chromatography*, American Chemical Society, Washington, DC, 1988.

24. S. A. Westwood, ed., *Supercritical Fluid Extraction and Its Use in Chromatographic Sample Preparation*, Chapman & Hall, New York, 1993.

25. Y. Kim, Y. H. Choi, Y. Chin, Y. P. Jang, Y. C. Kim, J. Kim, J. Y. Kim, S. N. Joung, M. J. Noh, and K. Yoo, *J. Chromatogr. Sci.*, **37**(12), 457–461 (1999).

26. J. D. Berset and R. Holzer, *J. Chromatogr. A*, **852**(2), 545–558 (1999).

27. F. Guo, Q. Li, X., and J. P. Alcantara-Licudine, *Anal. Chem.*, **71**(7), 1309–1315 (1999).

28. S. G. Amigo, M. S. G. Falcon, M. A. L. Yusty, and J. S. Lozano, *Fresenius' J. Anal. Chem.*, **367**(6), 572–578 (2000).

29. M. N. Kayali-Sayadi, S. Rubio-Barroso, R. Garcia-Iranzo, and L. M. Polo-Diez, *J. Liq. Chromatogr. Relat. Technol.*, **23**(12), 1913–1925 (2000).

30. T. J. L. Y. Lopez-Leiton, M. A. L. Yusty, M. E. A. Pineiro, and J. S. Lozano, *Chromatographia*, **52**(1/2), 109–111 (2000).

31. Y.-C. Ling, H.-C. Teng, and C. Cartwright, *J. Chromatogr. A*, **835**(1/2), 145–157 (1999).

32. M. Lee Jeong and D. J. Chesney, *Anal. Chim. Acta*, **389**(1/3), 53–57 (1999).

33. C. S. Eskilsson and L. Mathiasson, *J. Agric. Food Chem.*, **48**(11), 5159–5164 (2000).

34. J. W. Pensabene, W. Fiddler, and D. J. Donoghue, *J. Agric. Food Chem.*, **48**(5), 1668–1672 (2000).

35. D. M. Goli, M. A. Locke, and R. M. Zablotowicz, *J. Agric. Food Chem.*, **45**(4), 1244–1250 (1997).

36. J. Lin, R. Arunkumar, and C. Liu, *J. Chromatogr. A*, **840**(1), 71–79 (1999).

37. L. Morselli, L. Setti, A. Iannuccilli, S. Maly, G. Dinelli, and G. Quattroni, *J. Chromatogr. A*, **845**(1/2), 357–363 (1999).

38. D. K. Matabudul, N. T. Crosby, and S. Sumar, *Analyst*, **124**(4), 499–502 (1999).

39. B. L. Halvorsen, C. Thomsen, T. Greibrokk, and E. Lundanes, *J. Chromatogr. A*, **880**(1/2), 121–128 (2000).

40. C. Turner and L. Mathiasson, *J. Chromatogr. A*, **874**(2), 275–283 (2000).

41. L. Gamiz-Gracia, M. M. Jimenez-Carmona, and C. de Luque, *Chromatographia*, **51**(7/8), 428–432, (2000).

42. S.-P. Wang and W.-J. Chen, *Anal. Chim. Acta*, **416**(2), 157–167 (2000).

43. S. Scalia, *J. Chromatogr. A*, **870**(1/2), 199–205 (2000).

44. R. Alzaga, E. Pascual, P. Erra, and J. M. Bayona, *Anal. Chim. Acta*, **381**(1), 39–48 (1999).

45. B. E. Richter, *LC-GC*, **17**, S22–S28 (1999).

46. B. E. Richter, B. A. Jones, J. L. Ezzell, and N. L. Porter, *Anal. Chem.*, **68**, 1033–1039 (1996).

47. X. Lou, J. Hans-Gerd, and C. A. Cramers, *Anal. Chem.*, **69**(8), 1598–1603 (1997).

48. I. Windal, D. J. Miller, E. De Pauw, and S. B. Hawthorne, *Anal. Chem.*, **72**, 3916–3921 (2000).

49. N. Saim, J. R. Dean, M. P. Abdullah, and Z. Zakaria, *Anal. Chem.*, **70**(2), 420–424 (1998).

50. J. A. Fisher, M. J. Scarlett, and A. D. Stott, *Environ. Sci. Technol.*, **31**(4), 1120–1127 (1997).

51. K. Wenzel, A. Hubert, M. Manz, L. Weissflog, W. Engewald, and G. Schueuermann, *Anal. Chem.*, **70**(22), 4827–4835 (1998).

52. P. Popp, P. Keil, M. Moeder, A. Paschke, and U. Thuss, *J. Chromatogr. A*, **774**(1/2), 203–211 (1997).

53. A. Hubert, K. Wenzel, M. Manz, L. Weissflog, W. Engewald, and G. Schueuermann, *Anal. Chem.*, **72**(6), 1294–1300 (2000).

54. H. Obana, K. Kikuchi, M. Okihashi, and S. Hori, *Analyst*, **122**(3), 217–220 (1997).

55. B. E. Richter, J. L. Ezzell, D. E. Knowles, F. Hofler, and J. Huau, *Spectra Anal.*, **27**(204), 21–24 (1998).

56. B. E. Richter, *J. Chromatogr. A*, **874**(2), 217–224 (2000).

57. J. Gan, S. K. Papiernik, W. C. Koskinen, and S. R. Yates, *Environ. Sci. Technol.*, **33**(18), 3249–3253 (1999).

58. M. Giulia and A. Franco, *J. Chromatogr. A*, **765**(1), 121–125 (1997).

59. J. R. Dean, A. Santamaria-Rekondo, and E. Ludkin, *Anal. Commun.*, **33**(12), 413–416 (1996).

60. A. Kreisselmeier and H. W. Duerbeck, *J. Chromatogr. A*, **775**(1/2), 187–196 (1997).

61. H. J. Vandenburg, A. A. Clifford, K. D. Bartle, S. A. Zhu, J. Carroll, I. Newton, and L. M. Garden, *Anal. Chem.*, **70**(9), 1943–1948 (1998).

62. H. J. Vandenburg, A. A. Clifford, K. D. Bartle, R. E. Carlson, J. Carroll, and I. D. Newton, *Analyst*, **124**(11), 1707–1710 (1999).

63. M. Waldeback, C. Jansson, F. J. Senorans, and K. E. Markides, *Analyst*, **123**(6), 1205–1207 (1998).

64. E. Bjorklund, M. Jaremo, L. Mathiasson, L. Karlsson, J. T. Strode III, J. Eriksson, and A. Torstensson, *J. Liq. Chromatogr. Relat. Technol.*, **21**(4), 535–549 (1998).

65. B. Benthin, H. Danz, and M. Hamburger, *J. Chromatogr. A*, **837**, 211–219 (1999).

66. K. Schafer, *Anal. Chim. Acta*, **358**(1), 69–77 (1998).

67. K. Giergielewicz-Mozajska, L. Dabrowski, and J. Namiesnik, *Crit. Rev. Anal. Chem.*, **31**(3), 149–165 (2001).

68. V. Lopez-Avilla, *Crit. Rev. Anal. Chem.*, **29**(3), 195–230 (1999).

69. D. J. E. Ingram, *Radio and Microwave Spectroscopy*, Wiley, New York, 1976.

70. A. Abu-Samra, J. S. Morris, and S. R. Koirtyohann, *Anal. Chem.*, **47**, 1475 (1975).

71. K. Ganzler, A. Salgo, and K. Valko, *J. Chromatogr.*, **371**, 299–306 (1986).

72. J. Pare, Can. Pat. Appl., 1992, 35 pp. Application: CA 91-2055390 19911113. Priority: JP 90-310139 19901115.

73. G. Xiong, J. Liang, S. Zou, and Z. Zhang, *Anal. Chim. Acta*, **371**(1), 97–103 (1998).

74. I. Silgoner, R. Krska, E. Lombas, O. Gans, E. Rosenberg, and M. Grasserbauer, *Fresenius' J. Anal. Chem.*, **362**(1), 120–124 (1998).

75. C. Molins, E. A. Hogendoorn, E. Dijkman, H. A. G. Heusinkveld, and R. A. Baumann, *J. Chromatogr. A*, **869**(1/2), 487–496 (2000).

76. C. S. Eskilsson, E. Bjorklund, L. Mathiasson, L. Karlsson, and A. Torstensson, *J. Chromatogr. A*, **840**(1), 59–70 (1999).

77. V. Lopez-Avila, R. Young, and W. F. Beckert, *Anal. Chem.*, **66**, 1097–1106 (1994).

78. V. Lopez-Avila, R. Young, J. Benedicto, P. Ho, R. Kim, and W. F. Beckert, *Anal. Chem.*, **67**(13), 2096–2102 (1995).

79. N. Font, F. Hernandez, E. A. Hogendoorn, R. A. Baumann, and P. van Zoonen, *J. Chromatogr. A*, **798**(1/2), 179–186 (1998).

80. C. T. Costley, J. R. Dean, I. Newton, and J. Carroll, *Anal. Commun.*, **34**(3), 89–91 (1997).

81. B. E. Vazquez, M. P. Lopez, L. S. Muniategui, R. D. Prada, and F. E. Fernandez, *Fresenius' J. Anal. Chem.*, **366**(3), 283–288 (2000).

82. M. Pineiro-Iglesias, P. Lopez-Mahia, E. Vazquez-Blanco, S. Muniategui-Lorenzo, D. Prada-Rodriguez, and E. Fernandez-Fernandez, *Fresenius' J. Anal. Chem.*, **367**(1), 29–34 (2000).

83. Y. Y. Shu, C. Chiu, R. Turle, T. C. Yang, and R. C. Lao, *Organohalogen Compounds*, **31**, 9–13 (1997).

84. A. Bouaid, A. Martin-Esteban, P. Fernandez, and C. Camara, *Fresenius' J. Anal. Chem.*, **367**(3), 291–294 (2000).

85. H. M. Kingston and S. J. Haswell, eds., *Microwave-Enhanced Chemistry: Fundamentals, Sample Preparations, and Applications*, American Chemical Society, Washington, DC, 1997.

86. C. S. Eskilsson and E. Bjorklund, *J. Chromatgr. A*, **902**, 227–250 (2000).

87. M. Letellier and H. Budzinski, *Analusis*, **27**, 259–271 (1999).

88. V. Lopez-Avila, R. Young, and N. Teplitsky, *J. AOAC Int.*, **79**(1), 142–156 (1996).

89. R. Alzaga, C. Maldonado, and J. M. Bayona, *Int. J. Environ. Anal. Chem.*, **72**(2), 99–111 (1998).

90. O. P. Heemken, N. Theobald, and B. W. Wenclawiak, *Anal. Chem.*, **69**(11), 2171–2180 (1997).

91. M. P. Llompart, R. A. Lorenzo, R. Cela, K. Li, J. M. R. Belanger, and J. R. J. Pare, *J. Chromatogr. A*, **774**(1/2), 243–251 (1997).

92. H. J. Vandenburg, A. A. Clifford, K. D. Bartle, J. Carroll, and I. D. Newton, *Analyst*, **124**(3), 397–400 (1999).

93. S. Mitra and B. Kebbekus, *Environmental Chemical Analysis*, Blackie Academic Press, London, 1998.

CHAPTER

4

EXTRACTION OF VOLATILE ORGANIC COMPOUNDS FROM SOLIDS AND LIQUIDS

GREGORY C. SLACK

Department of Chemistry, Clarkson University, Potsdam, New York

NICHOLAS H. SNOW

*Department of Chemistry and Biochemistry, Seton Hall University,
South Orange, New Jersey*

DAWEN KOU

*Department of Chemistry and Environmental Science,
New Jersey Institute of Technology, Newark, New Jersey*

4.1. VOLATILE ORGANICS AND THEIR ANALYSIS

From an analytical point of view, volatile organic compounds (VOCs) can be defined as organic compounds whose vapor pressures are greater than or equal to 0.1 mmHg at 20°C. For regulatory purposes, VOCs are defined by the U.S. Environmental Protection Agency (EPA) as "any compound of carbon, excluding carbon monoxide, carbon dioxide, carbonic acid, metallic carbides or carbonates, and ammonium carbonate, which participates in atmospheric photochemical reactions" [1]. Many VOCs are environmental pollutants. They are not only toxic but are also important ozone precursors in the formation of smog.

An important feature of VOC analysis is that in most cases the analytes are first transferred to a gas–vapor phase and then analyzed by an instrument. Gas chromatography (GC) is the instrumental method of choice for the separation and analysis of volatile compounds. GC is mature, extremely reliable, and there is a wealth of literature regarding analysis of volatile compounds by GC [2–6]. In general, the analysis of pure volatile compounds is not difficult and can be accomplished via direct injection of the analyte into a gas chromatograph [7,8]. However, the analytical task

Sample Preparation Techniques in Analytical Chemistry, Edited by Somenath Mitra
ISBN 0-471-32845-6   Copyright © 2003 John Wiley & Sons, Inc.

becomes challenging when the analytes of interest are dissolved or sorbed in a complex matrix such as soil, food, cosmetics, polymers, or pharmaceutical raw materials. The challenge is to extract the analytes from this matrix reproducibly, and to accurately determine their mass or concentration. There are several approaches to this, including static headspace extraction (SHE), dynamic headspace extraction (purge and trap), solid-phase microextraction (SPME), membrane extraction, and liquid extraction, possibly combined with large-volume GC injection for enhanced sensitivity. The choice of technique depends on the type of sample matrix, information required (quantitative or qualitative), sensitivity required, need for automation, and budget.

In this chapter, techniques for the extraction of volatile compounds from various matrices are described. Details are provided on the basic theory and applications of each technique with a focus on providing useful information to the analyst working on the analysis of volatile analytes from difficult matrices. Since the analytes are volatile, most of the techniques are geared toward preparation of samples for gas chromatography, although they are appropriate for many instrumental methods. The chapter is heavily referenced and the reader should refer to the appropriate references for more details on a particular technique or application.

4.2. STATIC HEADSPACE EXTRACTION

Static headspace extraction is also known as *equilibrium headspace extraction* or simply as *headspace*. It is one of the most common techniques for the quantitative and qualitative analysis of volatile organic compounds from a variety of matrices. This technique has been available for over 30 years [9], so the instrumentation is both mature and reliable. With the current availability of computer-controlled instrumentation, automated analysis with accurate control of all instrument parameters has become routine. The method of extraction is straightforward: A sample, either solid or liquid, is placed in a headspace autosampler (HSAS) vial, typically 10 or 20 mL, and the volatile analytes diffuse into the headspace of the vial as shown in Figure 4.1. Once the concentration of the analyte in the headspace of the vial reaches equilibrium with the concentration in the sample matrix, a portion of the headspace is swept into a gas chromatograph for analysis. This can be done by either manual injection as shown in Figure 4.1 or by use of an autosampler.

Figure 4.2 shows a typical schematic diagram for a headspace gas chromatographic (HSGC) instrumental setup. Typically, the analyte is

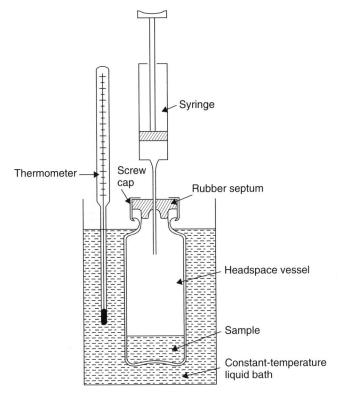

Figure 4.1. Typical static headspace vial, showing the location of the analytical sample and vial headspace. (Reprinted with permission from Ref. 10.)

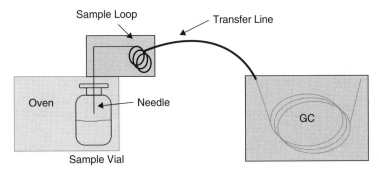

Figure 4.2. Schematic diagram of headspace extraction autosampler and GC instrument.

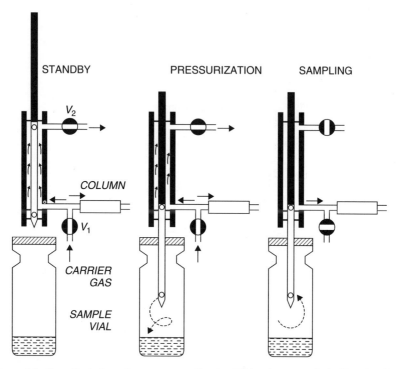

Figure 4.3. Steps for balanced pressure sampling in GC headspace analysis. [Reprinted with permission from Ref. 11 (Fig. 6, p. 208). Copyright John Wiley & Sons.]

introduced as a result of balanced pressure sampling, as demonstrated in Figure 4.3. In this example, the sample vial is brought to a constant temperature and pressure, with both typically being elevated from ambient conditions. Once equilibrium is reached, the vial is connected to the GC column head through a heated transfer line, which is left connected for a given period of time while the sample is transferred to the column by a pressure drop between the vial and the GC inlet pressure. Following this transfer, the vial is again isolated. For automated systems this sampling process can be repeated with the same or the next vial.

4.2.1. Sample Preparation for Static Headspace Extraction

The ease of initial sample preparation is one of the clear advantages of static headspace extraction. Often, for qualitative analysis, the sample can be placed directly into the headspace vial and analyzed with no additional preparation. However, for quantitation, it may be necessary to understand and optimize the matrix effects to attain good sensitivity and accuracy. For quantitative analysis of volatile compounds from solid particles, equilibrium between the analyte concentration in the headspace and in the sample matrix must be reached in a sensible period of time, typically a matter of minutes. For large solid samples it may be necessary to change the physical state of the sample matrix. Two common approaches are crushing or grinding the sample and dissolving or dispersing the solid into a liquid. The first approach increases the surface area available for the volatile analyte to partition into the headspace. However, the analyte is still partitioning between a solid and the headspace. The second approach is preferred since liquid or solution sample matrices are generally easier to work with than solids since the analyte partitioning process into the headspace usually reaches equilibrium faster. Also, analyte diffusion in liquids eliminates unusual diffusion path problems, which often occur with solids and can unpredictably affect equilibration time.

Solid Sample Matrices

One example of suspending or dissolving a solid in solution is seen in USP method 467, which provides an approach for the analysis of methylene chloride in coated tablets. The sample preparation procedure calls for the disintegration of 1 g of tablets in 20 mL of organic-free water via sonication. The solution is centrifuged after sonication, and 2 mL of the supernatant solution is transferred to a HSAS vial and then analyzed by HSGC [12].

Preparation of Liquids for Static Headspace Extraction

In static headspace extraction, sample preparation for liquid samples is usually quite simple—most often, the sample can just be transferred to the headspace sample vial and sealed immediately following collection of sample to minimize storage and handling losses [13].

4.2.2. Optimizing Static Headspace Extraction Efficiency and Quantitation

There are many factors involved in optimizing static headspace extraction for extraction efficiency, sensitivity, quantitation, and reproducibility. These include vial and sample volume, temperature, pressure, and the form of the matrix itself, as described above. The appropriate choice of physical conditions may be both analyte and matrix dependent, and when there are multiple analytes, compromises may be necessary.

Liquid Sample Matrices

The major factors that control headspace sensitivity are the analyte partition coefficient (K) and phase ratio (β). This was demonstrated by Ettre and Kolb [14]:

$$A \approx C^G = \frac{C^0}{K + \beta} \qquad (4.1)$$

where A is the GC peak area for the analyte, C^G the concentration of the analyte in the headspace, C^0 the initial concentration of the analyte in the liquid sample, K the partition coefficient, and β the phase volume ratio. The effect of the parameters K, controlled by the extraction temperature and β, controlled by the relative volume of the two phases, on static headspace extraction analysis sensitivity depends on the solubility of the analyte in the sample matrix. For analytes that have a high partition coefficient, temperature will have a greater influence than the phase ratio. This is because the majority of the analyte stays in the liquid phase, and heating the vial drives the volatile into the headspace. For volatile analytes with a low partition coefficient, the opposite will be true. The volumes of sample and headspace have a greater influence on sensitivity than does the temperature. Essentially, the majority of the volatile analyte is already in the headspace of the vial and there is little analyte left to drive out of the liquid matrix. This is illustrated in Figure 4.4, where a plot of detector response versus temperature for a headspace analysis shows that in an aqueous matrix, increasing the temperature increases the area counts for polar analytes, while the area for nonpolar analytes remains essentially the same [15].

The influence of analyte solubility in an aqueous matrix can also be seen in Figure 4.5, where the influence of sample volume is presented. For a polar analyte in an aqueous matrix, the sample volume will have minimal effect on the area response and a dramatic effect on less polar analytes. The example presented in Figure 4.5 shows the effect of increasing the sample volume from 1 (a) to 5 (b) mL on area response for analytes cyclohexane and 1,4-dioxane [15]. Salt may also be added to both direct immersion and headspace SPME (discussed later) samples to increase extraction recovery by the classical "salting-out" effect. This effect is demonstrated in Figure 4.5(b) and (c). Typically, sodium chloride is added to generate a salt concentration of over 1 M. When examining Figure 4.5, one must remember that the concentration of the analytes has not changed, only the volume in the sample and the amount of salt added. Adding salt results in an increase in peak area of 1,4-dioxane (peak 2) and no change in cyclohexane (peak 1). Meanwhile, the result of changing sample volume is an increase in the

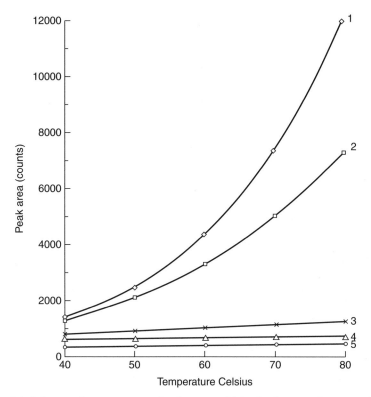

Figure 4.4. Influence of temperature on headspace sensitivity (peak area values, counts) as a function of the partition coefficient K from an aqueous solution with $\beta = 3.46$. The volatiles plotted above are ethanol (1), methyl ethyl ketone (2), toluene (3), n-hexane (4), and tetrachloroethylene (5). [Reprinted with permission from Ref. 15 (p. 26). Copyright John Wiley & Sons.]

area for cyclohexane (peak 1) and no change in 1,4-dioxane (peak 2). For an analyte with a large partition coefficient, the impact of β is insignificant on the area. For example, ethanol has a K value around 1000. For a 10-mL headspace vial filled with 1 or 5 mL of the analyte solution, $C_G = C_0/(1000 + 9)$ or $C_G = C_0/(1000 + 1)$, respectively. The difference in the results of these two calculations will be negligible. One can also see that for analytes where K is small, the effect of β will be significant. This phenomenon is extremely useful for the development chemist when method robustness is more important than sensitivity for a quantitative method. By choosing a matrix solvent that has a high affinity for the volatile analytes, problems with sample and standard transfer from volumetric flasks to the headspace vials are eliminated. Also, in the event that a second analysis of

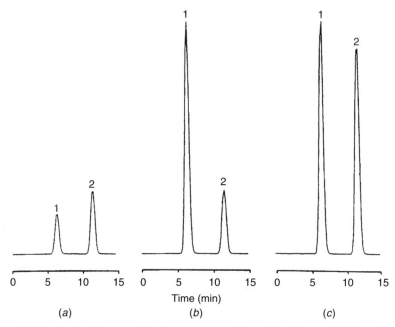

Figure 4.5. Analysis of three samples of an aqueous solution of cyclohexane (0.002 vol %) and 1,4-dioxane (0.1 vol %) in a 22.3-mL vial: (*a*) 1.0 mL of solution ($\beta = 21.3$); (*b*) 5.0 mL of solution ($\beta = 3.46$); (*c*) 5.0 mL of solution ($\beta = 3.46$ to which 2 g of NaCl was added. Headspace conditions: equilibration at 60°C, with shaker. Peaks: 1, cyclohexane; 2, 1,4-dioxane. [Reprinted with permission from Ref. 15 (p. 30). Copyright John Wiley & Sons.]

the analytes in the headspace vial is necessary, the drop in signal from the first to the second injection will be minimal. To determine the impact of β when K values are not readily available, simply prepare the analytes in the desired matrix (aqueous or organic) and determine the area counts versus sample volume.

4.2.3. Quantitative Techniques in Static Headspace Extraction

The four most common approaches to quantitative HSGC calibration are classical external standard, internal standard, standard addition, and multiple headspace extraction (MHE). The choice of technique depends on the type of sample being analyzed.

External Standard Calibration

External standard quantitation involves the preparation of a classical calibration curve, as shown in Figure 4.6*a*. Standard samples are prepared at various concentrations over the desired range and analyzed. A calibration

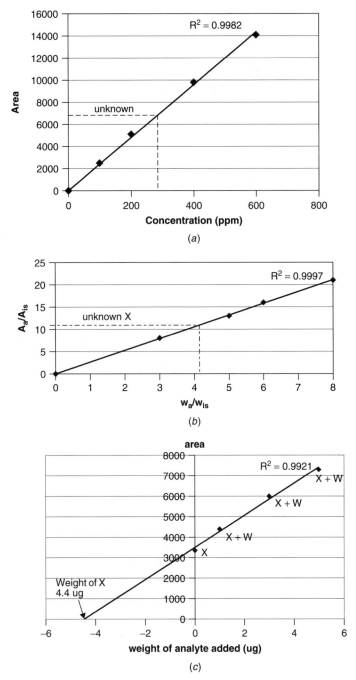

Figure 4.6. Types of calibration curves: (*a*) external standard; (*b*) internal standard; (*c*) standard addition.

curve is then generated, with raw GC peak area plotted versus standard concentration. Peak areas of each analyte are then determined and compared to the curve to generate analyte concentration. This method is best for analytes in liquid samples where the analytes are soluble in the sample matrix and the matrix has no effect on the analyte response. If the analyte has a low solubility in the sample matrix, preparation of standards via serial dilution can be difficult. It is important to match the standard and sample matrix as closely as possible and to demonstrate equivalence in the response between the standards and samples. For solid samples, dissolving or dispersing in a liquid and demonstrating equivalence between standards and samples is preferred to matrix matching, since this simplifies standard preparation. The main difficulty with external standard calibration is that is does not compensate for any variability due to the GC injection or due to variation in the analyte matrix.

Internal Standard Calibration

Internal standard calibration can be used to compensate for variation in analyte recovery and absolute peak areas due to matrix effects and GC injection variability. Prior to the extraction, a known quantity of a known additional analyte is added to each sample and standard. This compound is called an *internal standard*. To prepare a calibration curve, shown in Figure 4.6b, the standards containing the internal standard are chromatographed. The peak areas of the analyte and internal standard are recorded. The ratio of areas of analyte to internal standard is plotted versus the concentrations of the known standards. For the analytes, this ratio is calculated and the actual analyte concentration is determined from the calibration graph.

Although internal standard calibration compensates for some errors in external standard quantitation, there are several difficulties in method development. First, choosing an appropriate internal standard can often be difficult, as this compound must be available in extremely pure form and it must never appear in the samples of interest. Second, it cannot interfere in either the extraction or the chromatography of the analytes. Finally, it must be structurally similar to the analytes, so that it undergoes similar extraction and chromatography, otherwise, the compensation will be lost.

Standard Addition

In standard addition calibration, an additional known quantity of the analyte is added directly to the samples, following an initial analysis. By adding one or more aliquots of standard, a calibration curve can be prepared.

The concentration of analyte in the sample can then be determined by extrapolating the calibration curve, as shown in Figure 4.6*c*. For this method, analyte response must be linear throughout the range of concentrations used in the calibration curve. A practical approach to standard addition is to divide up the sample into several equal portions, then add increasing levels of standard. The samples are analyzed and area response versus the final concentration is plotted. The final concentration of the standard is the concentration of the standard after it is added to the sample. The original concentration is then determined by extrapolation to the *x*-axis. Alternatively, a single additional sample can be prepared and the original concentration the analyte can be determined from the following equation:

$$\frac{\text{original concentration of analyte}}{\text{final concentration of analyte (sample + standard)}}$$

$$= \frac{\text{area from original sample}}{\text{area from (sample + standard)}} \tag{4.2}$$

To calculate the original concentration of the sample using Figure 4.6*c*, the final (diluted) concentration of the sample is expressed in terms of the initial concentration of the sample. Then the initial concentration of the sample is determined [16]. It is important to remember that the sample and the standard are the same chemical compound.

Multiple Headspace Extraction

Multiple headspace extraction (MHE) is used to find the total peak area of an analyte in an exhaustive headspace extraction, which allows the analyst to determine the total amount of analyte present in the sample. This technique, along with the mathematical models behind it, was originally presented by McAuliffe [17] and Suzuki et al. [18]. Kolb and Ettre have an in-depth presentation of the mathematics of MHE in their book [15], and the reader is encouraged to reference that work for further information on the mathematical model.

The advantage to MHE is that sample matrix effects (which are mainly an issue only with solid samples) are eliminated since the entire amount of analyte is examined. This examination is done by performing consecutive analyses on the same sample vial. With the removal of each sample aliquot from the vial, the partition coefficient K will remain constant; however, the total amount of analyte remaining in the sample will decline as each analysis is performed and more of the analyte is driven up into the vial headspace for removal and analysis. Chromatograms of each injection of sample show declining peak areas as the amount of analyte declines in the sample, and when the peak area eventually falls to zero, one knows that the amount of analyte in the sample has been completely exhausted.

The process described above is, however, not in common practice. MHE has been simplified through laboratory use, and in practice, a limited number of consecutive extractions, usually three to four [15], are taken. Then a linear regression analysis is used to determine mathematically the total amount of analyte present in the sample.

4.3. DYNAMIC HEADSPACE EXTRACTION OR PURGE AND TRAP

For the analysis of trace quantities of analytes, or where an exhaustive extraction of the analytes is required, *purge and trap*, or *dynamic headspace extraction*, is preferred over static headspace extraction. Like static headspace sampling, purge and trap relies on the volatility of the analytes to achieve extraction from the matrix. However, the volatile analytes do not equilibrate between the gas phase and matrix. Instead, they are removed from the sample continuously by a flowing gas. This provides a concentration gradient, which aids in the exhaustive extraction of the analytes.

Purge and trap is used for both solid and liquid samples, which include environmental (water and soil) [19–21], biological [21,22] industrial, pharmaceutical, and agricultural samples. This technique is used in many standard methods approved by the EPA [23–25]. Figure 4.7 shows a chromatogram obtained using a purge and trap procedure described in EPA method 524.2 [26]. The detection limits suggested by the EPA are listed in Table 4.1 [23]. Quantitation is easily performed by external standard calibration.

4.3.1. Instrumentation

Figure 4.8 shows a schematic diagram of a typical purge and trap system [27]. It consists of a purge vessel, a sorbent trap, a six-port valve, and transfer lines. The water sample is placed in the purge vessel. A purge gas (typically, helium) passes through the sample continuously, sweeping the volatile organics to the trap, where they are retained by the sorbents. Once the purging is complete, the trap is heated to desorb the analytes into the GC for analysis.

Three types of purge vessels are most prevalent: frit spargers, fritless spargers, and needle spargers. Frit spargers create uniformed fine bubbles with large surface area that facilitate mass transfer (Figure 4.8*a*). However, these spargers can be used only for relatively clean water samples, not for complex samples that may foam or have particles that can clog the frits.

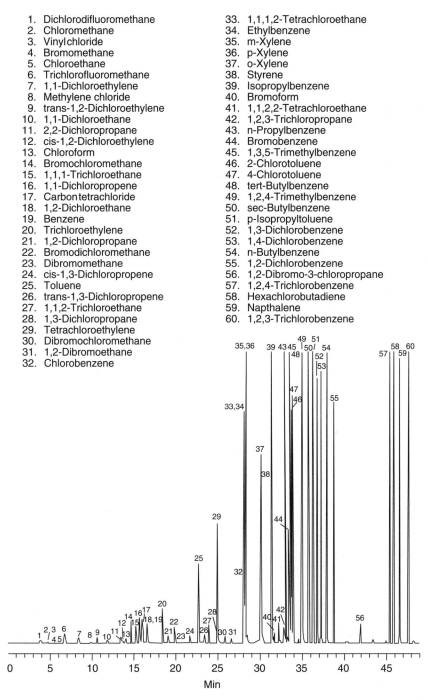

1. Dichlorodifluoromethane
2. Chloromethane
3. Vinylchloride
4. Bromomethane
5. Chloroethane
6. Trichlorofluoromethane
7. 1,1-Dichloroethylene
8. Methylene chloride
9. trans-1,2-Dichloroethylene
10. 1,1-Dichloroethane
11. 2,2-Dichloropropane
12. cis-1,2-Dichloroethylene
13. Chloroform
14. Bromochloromethane
15. 1,1,1-Trichloroethane
16. 1,1-Dichloropropene
17. Carbon tetrachloride
18. 1,2-Dichloroethane
19. Benzene
20. Trichloroethylene
21. 1,2-Dichloropropane
22. Bromodichloromethane
23. Dibromomethane
24. cis-1,3-Dichloropropene
25. Toluene
26. trans-1,3-Dichloropropene
27. 1,1,2-Trichloroethane
28. 1,3-Dichloropropane
29. Tetrachloroethylene
30. Dibromochloromethane
31. 1,2-Dibromoethane
32. Chlorobenzene

33. 1,1,1,2-Tetrachloroethane
34. Ethylbenzene
35. m-Xylene
36. p-Xylene
37. o-Xylene
38. Styrene
39. Isopropylbenzene
40. Bromoform
41. 1,1,2,2-Tetrachloroethane
42. 1,2,3-Trichloropropane
43. n-Propylbenzene
44. Bromobenzene
45. 1,3,5-Trimethylbenzene
46. 2-Chlorotoluene
47. 4-Chlorotoluene
48. tert-Butylbenzene
49. 1,2,4-Trimethylbenzene
50. sec-Butylbenzene
51. p-Isopropyltoluene
52. 1,3-Dichlorobenzene
53. 1,4-Dichlorobenzene
54. n-Butylbenzene
55. 1,2-Dichlorobenzene
56. 1,2-Dibromo-3-chloropropane
57. 1,2,4-Trichlorobenzene
58. Hexachlorobutadiene
59. Napthalene
60. 1,2,3-Trichlorobenzene

Figure 4.7. Chromatogram obtained using a purge-and-trap procedure as described in EPA method 524.2. (Reproduced from Ref. 26, with permission from Supelco Inc.)

195

Table 4.1. Detection Limits of the Volatile Organics in EPA Method 524.2[a]

Analyte	MDL (µg/L)	Analyte	MDL (µg/L)
Benzene	0.04	1,3-Dichloropropane	0.04
Bromobenzene	0.03	2,2-Dichloropropane	0.35
Bromochlorobenzene	0.04	1,1-Dichloropropane	0.10
Bromodichlorobenzene	0.08	cis-1,2-Dichloropropene	N/A
Bromoform	0.12	trans-1,2-Dichloropropene	N/A
Bromomethane	0.11	Ethylbenzene	0.06
n-Butylbenzene	0.11	Hexachlorobutadiene	0.11
sec-Butylbenzene	0.13	Isopropylbenzene	0.15
tert-Butylbenzene	0.14	4-Isopropyltoluene	0.12
Carbon tetrachloride	0.21	Methylene chloride	0.03
Chlorobenzene	0.04	Naphthalene	0.04
Chloroethane	0.10	n-Propylbenzene	0.04
Chloroform	0.03	Styrene	0.04
Chloromethane	0.13	1,1,1,2-Tetrachloroethane	0.05
2-Chlorotoluene	0.04	1,1,2,2-Tetrachloroethane	0.04
4-Chlorotoluene	0.06	Tetrachloroethene	0.14
Dibromochloromethane	0.05	Toluene	0.11
1,2-Dibromo-3-chloropropane	0.26	1,2,3-Trichlorobenzene	0.03
1,2-Dibromoethane	0.06	1,2,4-Trichlorobenzene	0.04
Dibromoethane	0.24	1,1,1-Trichloroethane	0.08
1,2-Dichlorobenzene	0.03	1,1,2-Trichloroethane	0.10
1,3-Dichlorobenzene	0.12	Trichloroethene	0.19
1,4-Dichlorobenzene	0.03	Trichlorofluoromethane	0.08
Dichlorodifluoromethane	0.10	1,2,3-Trichloropropane	0.32
1,1-Dichloroethane	0.04	1,2,4-Trimethylbenzene	0.13
1,2-Dichloroethane	0.06	1,3,5-Trimethylbenzene	0.05
1,1-Dichloroethene	0.12	Vinyl chloride	0.17
cis-1,2-Dichloroethene	0.12	o-Xylene	0.11
trans-1,2-Dichloroethene	0.06	m-Xylene	0.05
1,2-Dichloropropane	0.04	p-Xylene	0.13

[a] This method uses purge and trap with GC-MS (with a wide-bore capillary column, a jet separator interface, and a quadrupole mass spectrometer).

Fritless spargers and needle spargers (Figure 4.8b) are recommended for these samples, which include soils, slurries, foaming liquids, polymers, pharmaceuticals, and foods. The purging is less efficient, but clogging and foaming problems are eliminated. The most common sizes of the purge vessel are 25 and 5 mL.

In general, the trap should do the following: retain the analytes of interest, not introduce impurities, and allow rapid injection of analytes into the

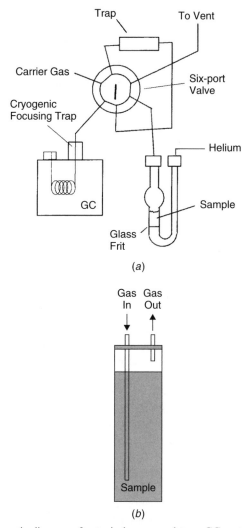

Figure 4.8. (*a*) Schematic diagram of a typical purge and trap–GC system. (Reprinted with permission from Nelson Thornes, Ref. 27.) (*b*) Needle sparger for purge and trap.

column. The trap is usually a stainless steel tube 3 mm in inside diameter (ID) and 25 mm long packed with multiple layers of adsorbents, as shown in Figure 4.9. The sorbents are arranged in layers in increasing trapping capacity. During purging/sorption, the purge gas reaches the weaker sorbent first, which retains only less volatile species. More volatile species break through this layer and are trapped by the stronger adsorbents. During

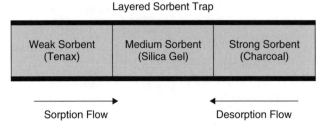

Figure 4.9. Schematic diagram of a multilayer sorbent trap.

desorption, the trap is heated and back-flushed with the GC carrier gas. In this way, the less volatile compounds never come in contact with the stronger adsorbents, so that irreversible adsorption is avoided.

The materials commonly used for trapping volatile organics include Tenax, silica gel, activated charcoal, graphitized carbon black (GCB or Carbopack), carbon molecular sieves (Carbosieve), and Vocarb. Tenax is a porous polymer resin based on 2,6-diphenylene oxide. It is hydrophobic and has a low affinity for water. However, highly volatile compounds and polar compounds are poorly retained on Tenax. To avoid decomposition, Tenax should not be heated to temperatures above 200°C. There are two grades of Tenax: Tenax TA and Tenax GC. The former is of higher purity and is preferred for trace analysis. Silica gel is a stronger sorbent than Tenax. It is hydrophilic and therefore an excellent material for trapping polar compounds. However, water is also retained. Charcoal is another sorbent that is stronger than Tenax. It is hydrophobic and is used mainly to trap very volatile compounds (such as dichlorodifluromethane, a.k.a Freon 12) that can break through Tenax and silica gel. Conventional traps usually contain Tenax, silica gel, and charcoal in series. If the boiling points of the analytes are above 35°C, Tenax itself will suffice, and silica gel and charcoal can be eliminated. Graphitized carbon black (GCB) is hydrophobic and has about the same trapping capacity as Tenax. It is often used along with carbomolecular sieves, which serve as an alternative to silica gel and charcoal for trapping highly volatile species. Vocarb is an activated carbon that is very hydrophobic. It minimizes water trapping and can be dry purged quickly. Vocarb is often used with an ion-trap mass spectrometer, which can be affected by trace levels of water or methanol. GCB, carbon molecular sieves, and Vocarb have high thermal stability and can be operated at higher desorption temperatures than traps containing Tenax.

The transfer line between the trap and the GC is made of nickel, deactivated fused silica, or silica-lined stainless steel tubing. Active sites that can interact with the anlaytes are eliminated on these inert materials. The line is maintained at a temperature higher than 100°C to avoid the condensation of water and volatile organics. The six-port valve that controls the gas flow path is also heated above 100°C to avoid condensation.

4.3.2. Operational Procedures in Purge and Trap

A purge and trap cycle consists of several steps: purge, dry purge, desorb preheat, desorb, and trap bake. Each step is synchronized with the operation of the six-port valve and the GC [or GC-MS (mass spectrometer)]. First, a sample is introduced into the purge vessel. Then the valve is set to the purge position such that the purge gas bubbles through the sample, passes through the trap, and then is vented to the atmosphere. During purge, dry purge, and preheat, the desorb (carrier) gas directly enters the GC. Typically, the purge time is 10 to 15 minutes, and the helium flow rate is 40 mL/min. The trap is at the ambient temperature. After purging, the purge gas is directed into the trap without going through the sample, called *dry purge*. The purpose of dry purging is to remove the water that has accumulated on the trap. Dry purge usually takes 1 to 2 minutes. Then the purge gas is turned off, and the trap is heated to about 5 to 10°C below the desorption temperature. Preheat makes the subsequent desorption faster. Once the preheat temperature is reached, the six-port valve is rotated to the desorb position to initiate the desorption step. The trap is heated to 180 to 250°C and back-flushed with the GC carrier gas. Desorption time is about 1 to 4 minutes. The flow rate of the desorb gas should be selected in accordance with the type of GC column used. After desorption, the valve is returned back to the purge position. The trap is reconditioned/baked at (or 15°C above) the desorption temperature for 7 to 10 minutes. The purpose of trap baking is to remove possible contamination and eliminate sample carryover. After baking, the trap is cooled, and the next sample can be analyzed. The operational parameters (temperature, time, flow rate, etc.) in each step should be the same for all the samples and calibration standards.

4.3.3. Interfacing Purge and Trap with GC

The operational conditions of the purge and trap must be compatible with the configuration of the GC system. A high carrier gas (desorb gas) flow rate can be used with a packed GC column. The trap desorption time is short at the high flow rate, producing a narrowband injection. The optimum flow is about 50 mL/min. Capillary columns are generally preferred over packed columns for better resolution, but these columns require lower flow rate.

Megabore capillary columns (0.53 mm ID or larger) are typically used at a flow rate of 8 to 15 mL/min. Desorption is slower at such flow rates, and the column is often cooled to subambient temperature (typically, 10°C or lower) at the beginning of the GC run to retain the highly volatile species. Sub-ambient cooling may be avoided by using a long (60- to 105-m) column with a thick-film stationary phase (3 to 5 μm). Nevertheless, this flow rate is still too high for a GC-MS. A jet separator or an open split interface can be used at the GC/MS interface to reduce the flow into the MS. However, an open split interface decreases the analytical sensitivity because only a portion of the analytes enters the detector.

Narrow-bore capillary columns (0.32 mm ID or smaller) with MS detector are typically operated at a lower flow rate (less than 5 mL/min). There are two ways to couple purge and trap with this type of column. One is to desorb the trap at a high flow rate and then split the flow into the GC using a split injector. A fast injection is attained at the expense of loss in analytical sensitivity. The other approach is to use a low desorb flow rate, which makes desorption time too long for a narrow bandwidth injection. The desorbed analytes need to be refocused on a second trap, usually by cryogenic trapping (Figure 4.8a). A cryogenic trap is made of a short piece of uncoated, fused silica capillary tubing. It is cooled to −150°C by liquid nitrogen. After refocusing, the cryogenic trap is heated rapidly to 250°C to desorb the analytes into the GC. Cryogenic trapping requires a dedicated cryogenic module and a liquid-nitrogen Dewar tank.

Without a moisture control device, water can go into the GC from purge and trap. The gas from the purge vessel is saturated with water, which can be collected on the trap and later released into the GC during trap heating. Water reduces column efficiency and causes interference with certain detectors (especially PID and MS), resulting in distorted chromatograms. The column can also be plugged by ice if cryofocusing is used. Therefore, water needs to be removed before entering the GC. Two water management methods are commonly used. One is to have a dry purge step prior to the desorption. However, some hydrophilic sorbents (such as silica gel) are not compatible with dry purging. The other approach is to use a condenser between the trap and the GC. The condenser is made of inert materials such as a piece of nickel tubing. It is maintained at ambient temperature, serving as a cold spot in the heated transfer line. During desorption, water is condensed and removed from the carrier gas. After desorption is complete, the condenser is heated and water vapor is vented.

4.4. SOLID-PHASE MICROEXTRACTION

Solid-phase microextraction (SPME) is a relatively new method of sample introduction, developed by Pawliszyn and co-workers in 1989 [28,29] and made commercially available in 1993. This technique has already been described in Chapter 2. The additional discussion here pertains mainly to the analysis of volatile organics. SPME is a solventless extraction method that employs a fused silica fiber coated with a thin film of sorbent, to extract volatile analytes from a sample matrix. The fiber is housed within a syringe needle that protects the fiber and allows for easy penetration of sample and GC vial septa. Most published SPME work has been performed with manual devices, although automated systems are also available.

There are two approaches to SPME sampling of volatile organics: direct and headspace. In *direct sampling* the fiber is placed directly into the sample matrix, and in *headspace sampling* the fiber is placed in the headspace of the sample [30,31]. Figure 4.10 illustrates the two main steps in a typical SPME analysis, analyte extraction (adsorption or absorption, depending on the fiber type) and analysis (thermal desorption into a GC inlet). To extract the analytes from a sample vial, the needle containing the fiber is placed in the sample by piercing the septa, the fiber is exposed to the sample matrix (extraction step), retracted into the housing, and removed from the vial. The injection process is similar: Pierce the GC septum with the needle, expose the fiber (desorption step), and then retract the fiber and remove the needle. A high-performance liquid chromatograph interface for SPME is available [32] and SPME has been interfaced to capillary electrophoresis [33] and FT-IR [34]. These have also been described in Chapter 2.

SPME has several advantages in the analysis of volatile organics. First, no additional instruments or hardware are required. Second, the cost of fibers is low compared to the cost of other methods for volatile analyte extraction. Fibers can be reused from several to thousands of times, depending on extraction and desorption conditions. SPME requires minimal training to get started, although there may be many variables involved in a full-method development and validation. SPME is also easily portable, and field sampling devices are readily available. Finally, with a variety of fiber coating chemistries available, SPME can be applied to a wide variety of volatile organic analytes. Table 4.2 shows a list of available SPME fibers, with their usual applications. A complete bibliography of SPME applications has been published by Supelco [35]. SPME has been used to extract volatile organic compounds from a wide variety of sample matrixes, such as air, foods, beverages, pharmaceuticals, natural products, and biological fluids [35].

4.4.1. SPME Method Development for Volatile Organics

The simplest way to begin developing an SPME method is to consult the applications guide provided by Supelco. This allows the analyst to quickly

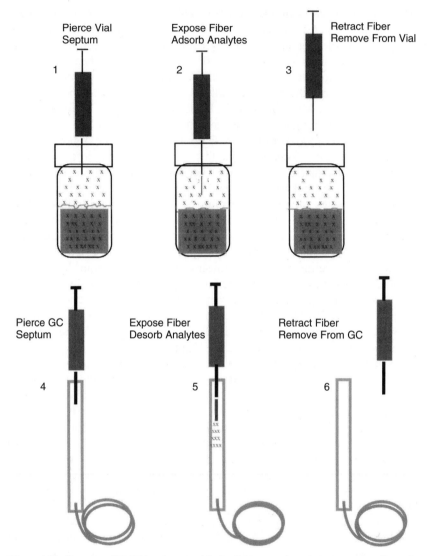

Figure 4.10. Steps in a SPME headspace analysis: 1–3, extraction; 4–6, desorption. (Drawings courtesy of Supelco, Inc.)

Table 4.2. Commercially Available SPME Fibers and Applications

Coating Material	Coating Thickness (μm)	Applications
Polydimethyl siloxane (PDMS)	100	GC/HPLC for volatiles
PDMS	30	GC/HPLC for nonpolar semi-volatiles
PDMS	7	GC/HPLC for nonpolar high-molecular-weight compounds
PDMS/divinylbenzene (PDMS/DVB)	65	GC/HPLC for volatiles, amines, notroaromatics
Polyacrylate (PA)	85	GC/HPLC for polar semivolatiles
Carbowax/divinylbenzene (CW/DVB)	65, 70	GC/HPLC for alcohols and polar compounds
Carboxen/PDMS	75, 85	GC/HPLC for gases and low-molecular-weight compounds
Divinylbenzene/Carboxen	50/30	GC/HPLC for flavor compounds
PDMS/DVB	60	HPLC for amines and polar compounds
Carbowax/templated resin	50	HPLC for surfactants

determine initial extraction and chromatographic conditions for several hundred frequently analyzed compounds from a wide variety of sample matrices [35]. For unique compounds or sample matrices, there are three basic steps to be considered when developing a SPME method analyte extraction, injection into the GC, and chromatographic conditions. A complete list of variables involved in SPME analysis is given in Table 4.3. Not all of these are usually considered by all method developers, but they may become issues in validation, transfer, or troubleshooting. The discussion that follows centers on optimizing the most important variables in SPME extractions of volatile organics and GC analysis.

The optimization of the extraction process, along with SPME extraction theory for both direct and headspace SPME extraction has been described thoroughly by Louch and co-workers [37]. The key issues involved in developing an extraction procedure include: extraction mode (direct or headspace), choice of fiber coating, agitation method, length of extraction, extraction temperature, and matrix modification. Choosing between direct immersion SPME and headspace SPME is relatively straightforward. Direct immersion SPME is warranted for liquid samples or solutions for which other solid-phase or liquid–liquid extraction methods would be considered.

Table 4.3. Variables Involved in Generating Reproducible SPME Results

Extraction	Desorption
Volume of the fiber coating	Geometry of the GC inlet
Physical condition of the fiber coating (cracks, contamination)	GC inlet liner type and volume
Moisture in the needle	Desorption temperature
Extraction temperature	Initial GC column temperature and column dimensions
Sample matrix components (salt, organics, moisture, etc.)	Fiber position in the GC inlet
Agitation type	Contamination of the GC inlet
Sampling time (especially important if equilibrium is not reached)	Stability of GC detector
Sample volume and headspace volume	Carrier gas flow rate
Vial shape	
Time between extraction and analysis	
Adsorption on sampling vessel or components	

Source: Adapted from Ref. 36.

Headspace SPME would be considered for the same analytes as static headspace extraction or purge and trap. Therefore, headspace SPME should be considered for extracting volatile compounds from solid or liquid samples, in which the normal boiling point of the analyte(s) of interest is less than about 200°C. For higher-boiling analytes, direct immersion SPME will probably be necessary. Also, the nature of the sample matrix should be considered. Headspace SPME is preferred for especially complex or dirty samples, as these may foul the fiber coating in a direct immersion analysis. However, SPME fibers have been shown to be usable for about 50 direct immersions into urine [38]. Some laboratories have reported using a fiber for thousands of extractions from drinking water.

4.4.2. Choosing an SPME Fiber Coating

SPME fibers have different coatings for the same reason that GC capillary columns have different coatings: There is no single coating that will extract and separate all volatile organics from a sample, therefore, different types of coatings with different polarities are used on SPME fibers. Currently, three classes of fiber polarity coatings are commercially available: nonpolar, semipolar, and polar coatings [39]. There are several advantages of using different fiber polarities. For one, using a matched-polarity fiber (i.e., polar-coated for a polar analyte) offers enhanced selectivity. Also, there is less of a chance of extracting interfering compounds along with the analyte of interest, and an organic matrix is not a problem—polar compounds can still be extracted [39].

As shown in Table 4.2, there are several SPME fiber coatings commercially available. These range in polarity from polydimethylsiloxane (PDMS), which is nonpolar, to Carbowax–divinylbenzene (CW-DVB), which is highly polar. The overall application of each is shown in the table. Throughout the literature, about 80% of SPME work is done using PDMS fibers, which are versatile and selective enough to obtain some recovery of most organic compounds from water. In most method development schemes, a PDMS fiber is attempted first, followed by a more polar fiber if necessary. Figure 4.11 provides a graphical scheme for choosing a SPME fiber based on analyte polarity and volatility. The nonpolar fibers are more commonly used for headspace SPME as the majority of volatile analytes tend to be non- or slightly polar. Also, as described below, the fiber coating thickness affects extraction recovery in both direct immersion and headspace SPME. The PDMS fiber is the only one available in more than one thickness.

Fiber coating thickness is a second consideration in selecting a fiber for both direct immersion and headspace SPME. The PDMS coating is avail-

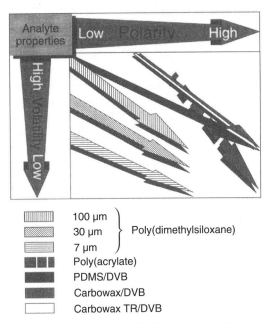

Figure 4.11. Graphical scheme for choosing a SPME fiber coating. [Reprinted with permission from Ref. 36 (Fig. 4.3, p. 99). Copyright John Wiley & Sons.]

able in three thicknesses: 100, 30, and 7 μm. The 100-μm fiber is generally used for highly volatile compounds or when a larger organic extraction volume is needed to improve recovery. Oppositely, the 7-μm-thick fiber is used for less volatile compounds that may present difficulty in thermal desorption in the GC inlet. The 30-μm fiber represents a compromise. For headspace work, the 100-μm fiber is most commonly used, as the larger organic volume enhances partitioning from the headspace.

4.4.3. Optimizing Extraction Conditions

Once the fiber is chosen, extraction conditions must be optimized. As shown in Table 4.3, there are many variables, with extraction time, sample volume, agitation, temperature, and modification of the sample matrix being most important. Extraction time is optimized by extracting a standard using a range of extraction times and plotting the analyte GC peak area versus the extraction time. As extraction time is increased, a plateau in peak area is reached. This represents the time required for the system to reach equilibrium and is the optimized extraction time. This has been presented in detail in Chapter 2. If the extraction time can be controlled carefully, and if sensitivity is adequate, shorter extraction time can be used without fully reaching equilibrium. Due to more rapid kinetics, headspace SPME generally reaches equilibrium faster than does direct immersion SPME. Most SPME headspace extractions are completed in less than 5 minutes, while direct immersion may require more than 30 minutes, although this is highly matrix dependent.

The sample volume also has an effect on both the rate and recovery in SPME extractions, as determined by extraction kinetics and by analyte partition coefficients. The sensitivity of a SPME method is proportional to n, the number of moles of analyte recovered from the sample. As the sample volume (V_s) increases, analyte recovery increases until V_s becomes much larger than the product of K_{fs}, the distribution constant of the analyte, and V_f, the volume of the fiber coating (i.e., analyte recovery stops increasing when $K_{fs}V_f \ll V_s$) [41]. For this reason, in very dilute samples, larger sample volume results in slower kinetics and higher analyte recovery.

As with any extraction, the agitation method will affect both the extraction time and recovery and should be controlled as closely as is practical. In direct-immersion SPME, agitation is usually accomplished using magnetic stirring, so the stirring rate should be constant. Also, the fiber should not be centered in the vial, as there is little to no liquid velocity there; the fiber should always be off-centered so that liquid is moving quickly around it. Agitation can also be achieved by physical movement of the fiber or by movement of the sample vial. Sonication is also used. Typically, headspace SPME sample vials are not agitated.

Extraction temperature can also be an important factor, especially in headspace SPME analyses. However, in SPME, unlike in GC headspace analysis, increasing the temperature in SPME can result in a maximum usable temperature for the method (i.e., going from 25°C to 30°C may result in a reduction in sensitivity [42].

The sample matrix may also be modified to enhance extraction recovery. This is typically done by either dissolving a solid sample in a suitable solvent, usually water or a strongly aqueous mixture, or by modifying the pH or salt content of a solution. Modifying the pH to change the extraction behavior works the same way in SPME as it does for classical liquid–liquid extraction. At low pH, acidic compounds will be in the neutral form and will be extracted preferentially into the fiber coating; at high pH, basic compounds are extracted favorably. Neutral compounds are not affected appreciably by solution pH.

4.4.4. Optimizing SPME–GC Injection

The GC injection following SPME is typically performed under splitless conditions. Since no solvent is present, the GC inlet liner does not need to have a large volume to accommodate the sample solvent, so special small-internal-diameter glass liners are often used. Optimizing SPME–GC injections has been discussed in detail by Langenfeld et al. [43] and Okeyo and Snow [44]. The main considerations involve transferring the analytes in the shortest possible time out of the fiber coating, through the inlet and onto the capillary GC column and in focusing the analytes into the sharpest bands possible. Thus, both inlet and chromatographic conditions play roles.

For semivolatile compounds, inlet optimization is very simple. Classical splitless inlet conditions, followed by an initial column temperature cool enough to refocus the analyte peaks following the desorption, work well. Thus, a typical condition would be a temperature of about 250°C, a head pressure sufficient to maintain optimum GC column flow and an initial column temperature at least 100°C below the normal boiling point of the analyte. For semivolatile analytes, a classical splitless inlet liner can be used, as the cool column will refocus these peaks. The desorption time in the inlet must be determined by experimentation, but typically, runs between 1 and 5 minutes.

For volatile analytes, optimizing the inlet is more difficult, as making the initial column temperature low enough to refocus these analytes is often not possible without cryogenics. The inlet must therefore be optimized to provide the fastest-possible desorption and transfer to the GC column, while the GC column is maintained as cool as possible to achieve any focusing that is possible. First, a low-volume SPME inlet liner should be used in place of the classical splitless liner. Second, a pulsed injection, with the inlet pressure higher than usual during the desorption, should be used to facilitate rapid analyte transfer. With an electronically controlled inlet, the pressure can be returned to the optimum for the GC column following the desorption. Finally, it may be necessary to use a thicker-film GC column to aid in retaining the volatile analytes.

As an example, Figure 4.12 shows the effect of inlet liner diameter on the separation of a hydrocarbon sample. In the first chromatogram, a 0.75-mm-ID liner was used and all of the peaks are sharp. In the second and third chromatograms, 2- and 4-mm liners are used. Significant peak broadening of the early peaks is seen in the 4-mm case especially. Also in the 4-mm case, however, the later eluting peaks are not significantly broadened, indicating that the liner diameter is not important for these compounds.

4.5. LIQUID–LIQUID EXTRACTION WITH LARGE-VOLUME INJECTION

Classical liquid–liquid and liquid–solid extractions are recently receiving additional examination, as new injection techniques for GC have made very simple, low-volume extractions feasible. Recently, several commercial systems for large-volume liquid injections (up to 150 µL all at once, or up to 1 to 2 mL over a short period of time) have become available. When combined with robotic sampling systems, these have become powerful tools in the trace analysis of a variety of sample types. Due to its simplicity, classical liquid–liquid extraction is often the method of choice for sample preparation. Some of the robotic samplers available for this type of analysis, such as the LEAP Technologies Combi-PAL robotic sampler, which has been licensed by several instrument vendors, are also capable of performing automated SPME and SHE.

4.5.1. Large-Volume GC Injection Techniques

The techniques for injecting large volumes into a capillary GC column were developed in the 1970s but not widely commercialized until the 1990s, when electronic control of the GC pneumatics became available. Two methods are used for large-volume injection: solvent vapor exit (SVE), which is based on a classical on-column inlet and programmed temperature vaporization (PTV), which was originally built into a split/splitless inlet. For relatively clean samples, both are capable of satisfactory large-volume injections,

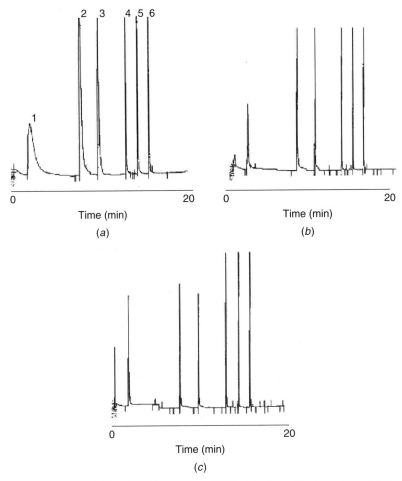

Figure 4.12. Effect of inlet liner diameter on SPME injection of hydrocarbons. (*a*) 4-mm-diameter liner; (*b*) 2-mm-diameter liner; (*c*) 0.75-mm-diameter liner. Analytes: 1, octane; 2, decane; 3, undecane; 4, tridecane; 5, tetradecane; 6, pentadecane. [Reprinted with permission from Ref. 44 (Fig. 3). Copyright Advanstar Communications.]

while for dirty samples, the SVE inlet is prone to fouling. These two inlets are pictured schematically in Figure 4.13.

The SVE configuration begins with a classical cool on-column inlet. A retention gap consisting of a length (usually about 5 m) of uncoated fused silica tubing is connected to the inlet. Following the retention gap is a short length (2 m) of coated analytical column that serves as a retaining pre-

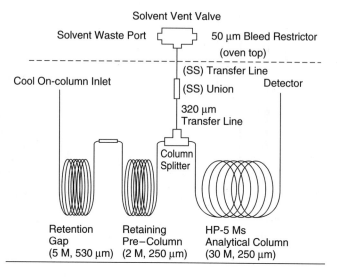

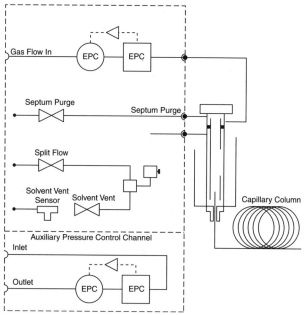

Figure 4.13. Schematic diagrams of large-volume injection systems. Top: on-column configuration with solvent vapor exit. (Drawing courtesy of Agilent Technologies.) Bottom: programmed temperature vaporization configuration. (Drawing courtesy of ATAS, International.)

column. Following the retaining precolumn, the flow is split to the analytical column and to the solvent vapor exit. The solvent vapor exit consists of a transfer line (uncoated tubing) and an electronically controlled solenoid valve that opens and closes. A restrictor is used to maintain a small permanent flow through the vapor exit so that back-flushing of solvent does not occur. Prior to injection, the vapor exit valve is opened and it remains opened during the injection process. Following injection, liquid solvent enters the retention gap, where it is evaporated and ejected through the vapor exit. After evaporation of about 95% of the solvent vapor, with the analytes being retained in the retaining precolumn, the vapor exit is closed and the analytical run is started. This allows the injection of sample amounts of up to 100 μL all at once, or up to several milliliters of sample using a syringe pump. SVE large-volume injection is generally used for relatively "clean" samples, such as drinking water or natural water extracts, since as in on-column injection, the entire sample reaches the retention gap, making fouling a common occurrence. Commercial systems generally include software that assists in optimizing the many injection variables.

The PTV large-volume inlet is, essentially, a temperature-programmable version of the classical split/splitless GC inlet. The main design change is that the glass liner within the inlet and the inlet itself is of low thermal mass, so that the temperature can be programmed rapidly. The PTV inlet can operate in several modes, including the classical split and splitless, cold split solvent vent, and hot split solvent vent. In the cold injection modes, the inlet begins at a relatively low temperature, below the normal boiling point of the sample solvent. The sample is injected, usually into a packed glass sleeve within the inlet. The solvent vapor is then vented through the open split vent, while the inlet is cool and the analytes remain behind in the liner. When about 95% of the solvent vapor has exited through the vent, the vent is closed, the inlet is heated rapidly, and the analytes are thermally desorbed into the GC column. This method also allows rapid injection of up to 150 μL of liquid sample, with the benefit that nonvolatile or reactive material will remain in the inlet sleeve rather than in the GC column or retention gap. The analysis of a lake water extract using liquid–liquid extraction followed by PTV injection is shown in Figure 4.14. A thorough and readable manual for PTV large-volume injection that is freely available on the Internet has been written by Janssen and provided by Gerstel [45].

4.5.2. Liquid–Liquid Extraction for Large-Volume Injection

The ability to inject 100 or more microliters of a liquid sample rapidly and automatically into a capillary gas chromatograph necessitates another look at liquid–liquid extraction. Sensitivity of the analysis is a common problem

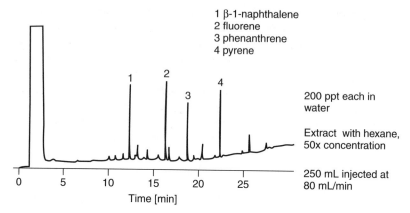

Figure 4.14. Chromatogram of lake water extract analyzed using liquid–liquid extraction with large-volume injection. (Drawing courtesy of ATAS, International.)

with all extraction methods, as sample concentration is often difficult. In SPME, sample concentration occurs automatically. In liquid–liquid extraction, however, an evaporation step is often required, which greatly increases the possibility of contamination and sample losses. For example, in a trace analysis, 1.0 L of water is often extracted with several hundred milliliters of organic solvent, which is then evaporated down to 1 mL prior to classical splitless injection of 1 µL of the remaining extract. If a 100-µL large-volume injection is available, the same concentration amount can be achieved by extracting 10 mL of water with 1 mL of solvent and injecting 100 µL of the extract, without an evaporation step. The same 1000-fold effective sample concentration is achieved without the potentially counterproductive concentration and with over a 99% reduction in solvent use and with less sample requirement.

4.6. MEMBRANE EXTRACTION

Membrane extraction has emerged as a promising alternative to conventional sample preparation techniques. It has undergone significant developments in the last two decades and is still evolving. It has been used for the extraction of a wide variety of analytes from different matrices. Only the extraction of volatile organics is discussed in this chapter. Figure 4.15 shows the concept of membrane separation. The sample is in contact with one side of the membrane, which is referred to as the *feed* (or *donor*) *side*. The membrane serves as a selective barrier. The analytes pass through to the other

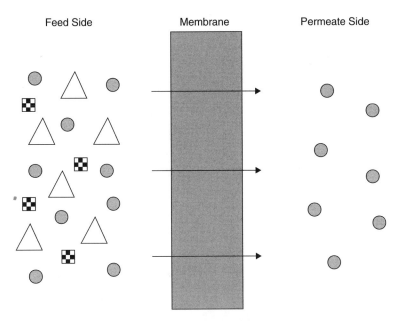

Figure 4.15. Concept of membrane separation; the circles are the analytes.

side, referred to as the *permeate side*. Sometimes, the permeated species are swept by another phase, which can be either a gas or a liquid.

A major advantage of membrane extraction is that it can be coupled to an instrument for continuous online analysis. Typically, a mass spectrometer [46–56] or gas chromatograph [57–66] is used as the detection device. Figure 4.16 shows the schematic diagrams of these systems. In membrane introduction mass spectrometry (MIMS), the membrane can be placed in the vacuum compartment of the MS. The permeates enter the ionization source of the instrument directly. In membrane extraction coupled with gas chromatography (Figure 4.16*b*), a sorbent trap is used to interface the membrane to the GC. The analytes that have permeated across the membrane are carried by a gas stream to the trap for preconcentratin. The trap is heated rapidly to desorb the analytes into the GC as a narrow injection band. For complex samples, GC has been the method of choice, due to its excellent separation ability. Tandem MS is emerging as a faster alternative to GC separation, but such instruments are more expensive. Detection limits of the membrane-based techniques are typically in the ppt to ppb range.

Membrane pervaporation (permselective "evaporation" of liquid molecules) is the term used to describe the extraction of volatile organics from an aqueous matrix to a gas phase through a semipermeable membrane.

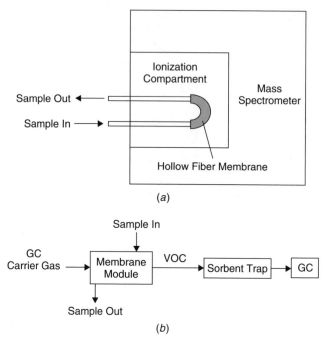

Figure 4.16. (*a*) Mass introduction mass spectrometry. (*b*) Hyphenation of membrane extraction with online GC.

The extraction of volatiles from a gas sample to a gaseous acceptor across the membrane is called *permeation*, which is the mechanism of extraction from the headspace of an aqueous or solid sample. For both pervaporation and permeation, the transport mechanism can be described by the solution–diffusion theory [67]. In pervaporation, the organic analytes first move through the bulk aqueous sample to the membrane surface and then dissolve/partition into it. After diffusing through the membrane to the permeate side, the analytes evaporate into the gas phase. In headspace sampling, an additional step of transporting the analytes from the bulk aqueous phase into the headspace is involved. In both cases, the extraction is driven by the concentration gradient across the membrane.

Steady-state permeation is governed by *Fick's first law:*

$$J = -AD\frac{dC}{dx} = AD\frac{\Delta C}{l} \tag{4.3}$$

where J is the analyte flux, A the membrane surface, D the diffusion coefficient, C the solute concentration, x the distance along the membrane wall, and l the membrane thickness. It can be seen from the equation that mass transfer is faster across a thin, large-surface-area membrane. In pervaporation, the overall mass transfer resistance is the sum of the mass transfer resistance of the bulk aqueous phase on the feed side, the membrane, and the gas on the permeate side. In headspace sampling, the overall mass transfer resistance is the sum of the mass transfer resistance of the bulk aqueous sample, the liquid–gas interface, the gas phase on the feed side, the membrane, and the gas on the permeate side. Non-steady-state permeation can be described by *Fick's second law:*

$$\frac{dC(x,t)}{dt} = -D\frac{d^2C(x,t)}{dt^2} \tag{4.4}$$

where $C(x,t)$ is the solute concentration at position x and time t.

4.6.1. Membranes and Membrane Modules

Membranes can be classified as porous and nonporous based on the structure or as flat sheet and hollow fiber based on the geometry. Membranes used in pervaporation and gas permeation are typically hydrophobic, nonporous silicone (polydimethylsiloxane or PDMS) membranes. Organic compounds in water dissolve into the membrane and get extracted, while the aqueous matrix passes unextracted. The use of mircoporous membrane (made of polypropylene, cellulose, or Teflon) in pervaporation has also been reported, but this membrane allows the passage of large quantities of water. Usually, water has to be removed before it enters the analytical instrument, except when it is used as a chemical ionization reagent gas in MS [50]. It has been reported that permeation is faster across a composite membrane, which has a thin (e.g., 1 μm) siloxane film deposited on a layer of microporous polypropylene [61].

As the name suggests, flat-sheet membranes are flat, like a sheet of paper, and can be made as thin as less than 1 μm. However, they need special holders to hold them in place. Hollow-fiber membranes are shaped like tubes (200 to 500 μm ID), allowing fluids to flow inside as well as on the outside. Hollow fibers are self-supported and offer the advantage of larger surface area per unit volume and high packing density. A large number of parallel fibers can be packed into a small volume.

There are two ways to design a membrane module [66]. The membrane can be introduced into the sample, referred to as *membrane in sample* (MIS), or the sample can be introduced into the membrane, referred to as *sample in membrane* (SIM). Figure 4.17a is a schematic diagram of the MIS configu-

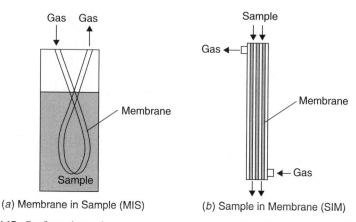

(a) Membrane in Sample (MIS)   (b) Sample in Membrane (SIM)

Figure 4.17. Configurations of membrane modules using hollow-fiber membranes. (a) Membrane in sample (MIS). (b) Sample in membrane (SIM).

ration. A hollow-fiber membrane is shown here, although a flat membrane fitted on the tip of a probe can also be used. The membrane is submerged in the sample, and the permeated analytes are stripped by a flowing gas (or vacuum) on the other side of the membrane. At any time, only a small fraction of the sample is in direct contact with the membrane. The ratio of membrane surface area to sample volume is quite low. The sample is usually stirred to enhance analyte diffusion through the aqueous phase. The membrane can also be placed in the headspace of a sample. The analytes first vaporize and then permeate through the membrane. In the MIS configuration, the time to achieve exhaustive extraction can be rather long. On the other hand, this configuration is simple and does not require the pumping of samples. It can also be used for headspace extraction where the membrane is not in direct contact with the sample. In this way, possible contamination of the membrane can be avoided, and the extraction can be applied to solid samples as well.

Figure 4.17b shows a schematic diagram of the SIM configuration. The membrane module has the classical shell-and-tube design. The aqueous sample is either made to "flow through" or "flow over" the hollow fiber, while the stripping gas flows countercurrent on the other side. In both cases, the sample contact is dynamic, and the contact surface/volume ratio is much higher than in the MIS extraction. Consequently, extraction is more efficient. The flow-through mode provides higher extraction efficiency than the flow-over mode. This is because tube-side volume is smaller than the shell-side volume, which results in higher surface/volume ratio for the aqueous sample. Comparison studies show that under similar experimental conditions, flow-through extraction provides the highest sensitivity among all available membrane module configurations [59].

4.6.2. Membrane Introduction Mass Spectrometry

The use of membrane introduction mass spectrometry (MIMS) was first reported in 1963 by Hoch and Kok for measuring oxygen and carbon dioxide in the kinetic studies of photosynthesis [46]. The membrane module used in this work was a flat membrane fitted on the tip of a probe and was operated in the MIS mode. The permeated anaytes were drawn by the vacuum in the MS through a long transfer line. Similar devices were later used for the analysis of organic compounds in blood [47]. Memory effects and poor reproducibility plagued these earlier systems. In 1974, the use of hollow-fiber membranes in MIMS was reported, which was also operated in the MIS mode [48]. Lower detection limits were achieved thanks to the larger surface area provided by hollow fibers. However, memory effects caused by analyte condensation on the wall of the vacuum transfer line remained a problem.

In the late 1980s, Bier and Cooks [49] introduced a new membrane probe design, which was operated in the SIM mode. The schematic diagram of such a system is shown in Figure 4.16a. The sample flowed though the hollow-fiber membrane, which was inserted directly in the ionization chamber of the mass spectrometer. This eliminated memory effects and increased sensitivity and precision. Sample introduction was accomplished using flow injection, which increased the speed of analysis. Instruments based on this design were commercialized in 1994 by MIMS Technology, Inc. (Palm Bay, FL). MIMS in its modern forms has several advantages. Sample is directly introduced into the MS through the membrane, without additional preparation. The sensitivity is high, with detection limits in the sub-ppb (parts per billion) range. The analysis is fast, typically from 1 to 6 minutes. This technique is especially attractive for online, real-time analysis. It has been used in environmental monitoring [51–53], bioreactor monitoring [54,55], and chemical reaction monitoring [56].

The absence of chromatographic separation makes MIMS a fast technique. It is advantageous in some applications where only select compounds are to be detected or the total concentration of a mixture is to be determined. For instance, the total concentration of trihalomethanes (THMs, including chloroform, bromoform, bromodichloromethane, and dibromochloromethane) in drinking water can be determined by MIMS in less than

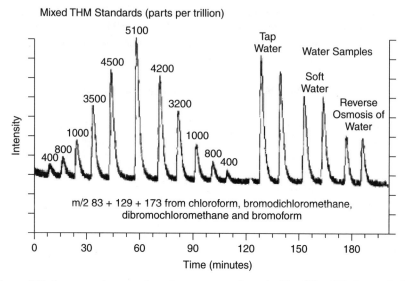

Figure 4.18. Ion current summation chromatogram for m/z $83 + 129 + 173$ from trihalomethane analysis. (Reproduced from Ref. 51, with permission from the American Chemical Society.)

10 minutes, without identifying the individual species [51]. Figure 4.18 shows the ion current chromatogram obtained using this method, where the peak area is proportional to the total THM concentration. MIMS works best for nonpolar, volatile organics with small molecular weight (<300 amu). In recent years efforts have been made to extend the application of MIMS to semivolatiles. This is beyond the scope of this chapter and is not discussed here. More details on MIMS can be found in several review articles [68,69].

4.6.3. Membrane Extraction with Gas Chromatography

The hyphenation of membrane extraction with gas chromatography is more complex. The analytes pervaporate into the GC carrier gas, which is at a positive pressure, thus reducing the partial pressure gradient. A sorbent trap is used to concentrate the analytes prior to GC analysis. Continuous monitoring can be carried out by pumping the water through the membrane module continuously, and heating the sorbent trap intermittently to desorb the analytes into the GC for analysis [57,58]. Although this works for the monitoring of a water stream, discrete, small-volume samples cannot be

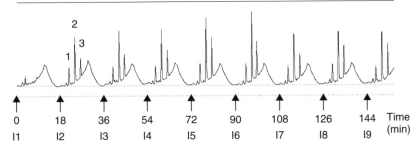

1. 1,1 Dichloroethylene
2. cis 1,2 Dichloroethylene
3. Trichloroethylene

Figure 4.19. Chromatograms obtained during continuous monitoring of a contaminated groundwater well. Sample Injections were made every 18 minutes (I1, injection 1; and so on). (Reproduced from Ref. 65, with permission from Wiley-VCH.)

analyzed in this fashion. Moreover, it may take a relatively long time for the permeation to reach steady state. In other words, the membrane response to the concentration change in the stream can be slow. Any measurement during the transition period provides erroneous results.

A non-steady-state membrane extraction method referred to as pulse introduction membrane extraction (PIME) has been developed to avoid these problems [62]. PIME resembles a flow-injection operation. Deionized water (or an aqueous solution) serves as a carrier fluid, which introduces the sample into the membrane as a pulse. Analyte permeation does not have to reach steady state during extraction. Once the extraction is complete, the analytes are thermally desorbed from sorbent trap into the GC. A chromatogram is obtained for each sample that reflects its true concentration. PIME can be used for the analysis of multiple discrete samples, as well as for the continuous monitoring of a stream by making a series of injections. Figure 4.19 shows chromatograms obtained during continuous monitoring of contaminated groundwater using PIME [65]. The sample injections were made every 18 minutes.

The greatest challenge in membrane extraction with a GC interface has been the slow permeation through the polymeric membrane and the aqueous boundary layer. The problem is much less in MIMS, where the vacuum in the mass spectrometer provides a high partial pressure gradient for mass transfer. The time required to complete permeation is referred to as *lag time*. In membrane extraction, the lag time can be significantly longer than the sample residence time in the membrane. An important reason is the boundary layer effects. When an aqueous stream is used as the carrier fluid, a static boundary layer is formed between the membrane and the aqueous phase. The analytes are depleted in the boundary layer, and this reduces the concentration gradient for mass transfer and increases the lag time. In a typical analytical application, mass transfer through the boundary layer is the rate-limiting step in the overall extraction process [63,64].

Sample dispersion is another cause of the long lag time in flow injection techniques where an aqueous carrier fluid is used [63,64]. Dispersion is caused by axial mixing of the sample with the carrier stream. This increases the sample volume, resulting in longer residence time in the membrane. Dilution reduces the concentration gradient across the membrane, which is the driving force for diffusion. The overall effects are broadened sample band and slow permeation.

Gas Injection Membrane Extraction

Gas injection membrane extraction (GIME) of aqueous samples has been developed to address the issues of boundary layer effects and sample dispersion [66]. This is shown in Figure 4.20. An aqueous sample from the loop

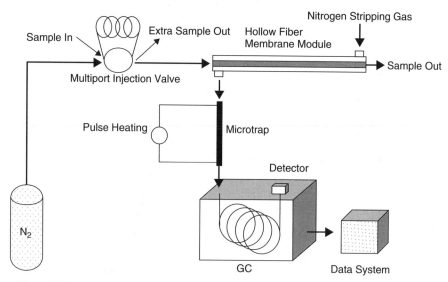

Figure 4.20. Schematic diagram of gas injection membrane extraction. (Reproduced from Ref. 66, with permission from the American Chemical Society.)

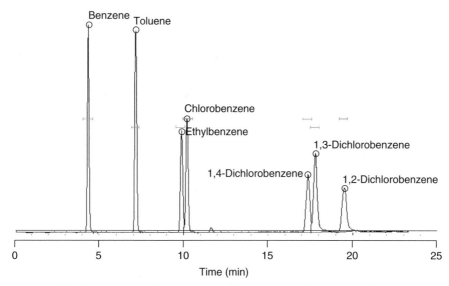

Figure 4.21. Chromatogram of an aqueous sample containing ppb-level purgable aromatics as listed in EPA standard method 602 by GIME. (Reproduced from Ref. 66, with permission from the American Chemical Society.)

of a multiport injection valve is injected into the hollow fiber membrane module by an N_2 stream. The gas pushes the sample through the membrane fibers, while the organic analytes permeate to the shell side, where they are swept by a countercurrent nitrogen stream to a microsorbent trap. After a predetermined period of time, the trap is electrically heated to desorb the analytes into the GC. Figure 4.21 shows a chromatogram of ppb-level volatile organic compounds, as listed in EPA method 602, obtained by GIME [66].

The permeation profiles obtained by aqueous elution and GIME are shown in Figure 4.22. It can be seen that the lag time was reduced significantly by gas injection of aqueous samples. There is no mixing between the eluent gas and the sample; thus dispersion is eliminated. The boundary layer is also greatly reduced, as the gas cleans the membrane by removing any water sticking on the surface. GIME is a pulsed introduction technique that can be used for the analysis of individual samples by discrete injections or for continuous on-line monitoring by sequentially injecting a series of samples. This technique is effective in speeding up membrane extraction. It can significantly increase sample throughput in laboratory analysis and is desirable for online water monitoring.

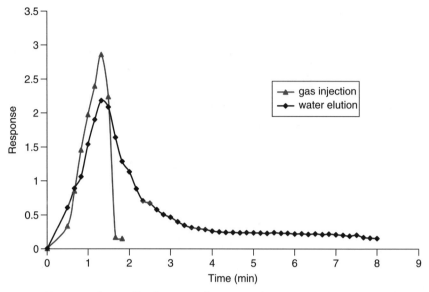

Figure 4.22. Permeation profiles for 1 mL of a 500-ppb benzene sample at an eluent (gas or liquid) flow rate of 1 mL/min. (Reproduced from Ref. 66, with permission from the American Chemical Society.)

4.6.4. Optimization of Membrane Extraction

Several factors affect the efficiency of membrane extraction and hence the sensitivity of the analysis: temperature, membrane surface area, membrane thickness, geometry, sample volume, and sample flow rate. These parameters need to be optimized for specific applications. Higher temperature facilitates mass transfer by increasing diffusion coefficient, but at the same time decreases analyte partition coefficient in the membrane. The temperature of the membrane module needs to be controlled to avoid fluctuation in extraction efficiency and sensitivity. Extraction efficiency can also be improved by using thinner membranes, which provide faster mass transfer. In the case of hollow fiber membranes, extraction efficiency can be increased by using longer membranes and multiple fibers, which provide lager contact area between the membranes and the sample. It has been reported that spiraled membranes provide more efficient extraction than straight membranes, because the former facilitates turbulent flow in the membrane module and reduces the boundary layer effects. The larger the sample volume, the more analytes it has and the higher is the sensitivity. However, larger volumes take longer to extract. Lower sample flow rates increase the extraction efficiency but prolong the extraction time.

4.7. CONCLUSIONS

There are many techniques available for the preparation of volatile analytes prior to instrumental analysis. In this chapter the major techniques, leading primarily to gas chromatographic analysis, have been explored. It is seen that the classical techniques: purge and trap, static headspace extraction, and liquid–liquid extraction still have important roles in chemical analysis of all sample types. New techniques, such as SPME and membrane extraction, offer promise as the needs for automation, field sampling, and solvent reduction increase. For whatever problems may confront the analyst, there is an appropriate technique available; the main analytical difficulty may lie in choosing the most appropriate one.

ACKNOWLEDGMENTS

N.H.S. gratefully acknowledges the Robert Wood Johnson Pharmaceutical Research Institute for support during the sabbatical year in which this chapter was written. Special thanks go to Rebecca Polewczak (Clarkson University), who provided valuable assistance in organizing materials for this chapter.

REFERENCES

1. 40 CFR Part 51 Sec. 51.100.
2. H. M. McNair and E. J. Bonelli, *Basic Gas Chromatography*, Varian Instrument, Palo Alto, CA, 1968.
3. H. M. McNair and J. M. Miller, *Basic Gas Chromatography*, Wiley, New York, 1997.
4. J. M. Miller, *Chromatography: Concepts and Contrasts*, Wiley, New York, 1988.
5. R. L. Grob, *Modern Practice of Gas Chromatography*, 2nd ed., Wiley, New York, 1985.
6. K. J. Hyver and P. Sandra, *High Resolution Gas Chromatography*, 3rd ed., Hewlett-Packard, Palo Alto, CA, 1989.
7. K. Grob, *Split and Splitless Injection in Capillary GC*, 3rd ed., Hüthig, Heidelberg, 1993.
8. K. Grob, *On-Column Injection in Capillary Gas Chromatography*, Hüthig, Heidelberg, 1991.
9. H. Hachenberg and A. P. Schmidt, *Gas Chromatographic Headspace Analysis*, Heyden, London, 1977.
10. Ref. 9, p. 21.

11. B. Kolb and P. Popisil, in P. Sandra, ed., *Sample Introduction in Capillary Gas Chromatography*, Vol. 1, Hüthig, Heidelberg, 1985.

12. USP 24-NF 19, Method 467, *United States Pharmacopoeia Convention*, Rockville, MD, 2000.

13. A. Cole and E. Woolfenden, *LC-GC*, **10**(2), 76–82 (1992).

14. L. S. Ettre and B. Kolb, *Chromatographia*, **32**, 5–12 (1991).

15. B. Kolb and L. S. Ettre, *Static Headspace–Gas Chromatography: Theory and Practice*, Wiley-VCH, New York, 1997.

16. D. C. Harris, *Quantitative Chemical Analysis*, 5th ed., W.H. Freeman, New York, 1999, p. 102.

17. C. McAuliffe, *Chem Technol.*, 46–51 (1971).

18. M. Suzuki, S. Tsuge, and T. Takeuchi, *Anal. Chem.*, **42**, 1705–1708 (1970).

19. I. Silgoner, E. Rosenberg, and M. Grasserbauer, *J. Chromatogr. A*, **768**, 259–270 (1997).

20. Z. Bogdan, *J. High Resolut. Chromatogr.*, **20**, 482–486 (1997).

21. L. Dunemann and H. Hajimiragha, *Anal. Chim. Acta*, **283**, 199–206 (1993).

22. P. Roose and U. A. Brinkman, *J. Chromatogr. A*, **799**, 233–248 (1998).

23. EPA methods 502.2 and 524.2, in *Methods for the Determination of Organic Compounds in Drinking Water*, Supplement III, National Exposure Research Laboratory, Office of Research and Development, U.S. Environmental Protection Agency, Cincinnati, OH, 1995.

24. EPA methods 601 and 602, in *Methods for Organic Chemical Analysis of Municipal and Industrial Wastewater*, 40 CFR Part 136, App. A.

25. EPA methods 8021 and 8260, in EPA Publication SW-846, *Test Methods for Evaluating Solid Waste, Physical/Chemical Methods*. *www.epa.gov/epaoswer/hazwaste/test/sw846.htm*

26. Chromatogram of EPA 524.2 standard. *www.supelco.com*

27. S. Mitra and B. Kebbekus, *Environmental Chemical Analysis*, Blackie Academic Press, London, 1998, p. 270.

28. R. Berlardi and J. Pawliszyn, *Water Pollut. Res. J. Can.*, **24**, 179 (1989).

29. C. Arthur and J. Pawliszyn, *Anal. Chem.*, **62**, 2145 (1990).

30. Z. Zhang and J. Pawliszyn, *Anal. Chem.*, **65**, 1843 (1993).

31. B. Page and G. Lacroix, *J. Chromatogr.*, **648**, 199 (1993).

32. J. Chen and J. Pawliszyn, *Anal. Chem.*, **67**, 2350 (1995).

33. C.-W. Whang, in J. Pawliszyn, ed., *Applications of Solid Phase Micro-extraction*, Royal Society of Chemistry, Cambridge, 1999, pp. 22–40.

34. J. Burck, in Ref. 33, pp. 638–653.

35. *SPME Applications Guide*, Supelco, Bellefonte, PA, 2001.

36. J. Pawliszyn, *Solid Phase Microextraction: Theory and Practice*, Wiley-VCH, New York, 1997.

37. D. Louch, S. Matlagh, and J. Pawliszyn, *Anal. Chem.*, **64**, 1187 (1992).

38. P. D. Okeyo and N. H. Snow, *J. Microcol. Sep.*, **10**(7), 551–556 (1998).

39. V. Mani, in Ref. 33, pp. 60–61.

40. Ref. 36, p. 99.

41. Ref. 36, pp. 117–122.

42. C. Grote and K. Levsen, in Ref. 33, pp. 169–187.

43. J. Langenfeld, S. Hawthorne, and D. Miller, *J. Chromatogr. A*, **740**, 139–145 (1996).

44. P. Okeyo and N. H. Snow, *LC-GC*, **15**(12), 1130–1136 (1997).

45. Guide to injection techniques may be found at: *www.gerstelus.com*

46. G. Hoch and B. Kok, *Arch. Biochem, Biophys.*, **101**, 160–170 (1963).

47. S. Woldring, G. Owens, and D. C. Woolword, *Science*, **153**, 885 (1966).

48. L. B. Westover, J. C. Tou, and J. H. Mark, *Anal. Chem.*, **46**, 568–571 (1974).

49. M. E. Bier and R. G. Cooks, *Anal. Chem.*, **59**, 597–601 (1987).

50. T. Choudhury, T. Kotiaho, and R. G. Cooks, *Talanta*, **39**, 573–580 (1992).

51. S. J. Bauer and D. Solyom, *Anal. Chem.*, **66**, 4422 (1994).

52. M. Soni, S. Bauer, J. W. Amy, P. Wong, and R. G. Cooks, *Anal. Chem.*, **67**, 1409–1412 (1995).

53. V. T. Virkki, R. A. Ketola, M. Ojala, T. Kotiaho, V. Komppa, A. Grove, and S. Facchetti, *Anal. Chem.*, **67**, 1421 (1995).

54. M. J. Hayward, T. Kotiaho, A. K. Lister, R. G. Cooks, G. D. Austin, R. Narayan, and G. T. Tsao, *Anal. Chem.*, **62**, 1798–1804 (1990).

55. N. Srinivasan, N. Kasthurikrishnan, R. G. Cooks, M. S. Krishnan, and G. T. Tsao, *Anal. Chim. Acta*, **316**, 269 (1995).

56. P. Wong, N. Srinivasan, N. Kasthurikrishnan, R. G. Cooks, J. A. Pincock, and J. S. Grossert, *J. Org. Chem.*, **61**(19), 6627 (1996).

57. K. F. Pratt and J. Pawliszyn, *Anal. Chem.*, **64**, 2107–2110 (1992).

58. Y. Xu and S. Mitra, *J. Chromatogr. A*, **688**, 171–180 (1994).

59. M. J. Yang, S. Harms, Y. Z. Luo, and J. Pawliszyn, *Anal. Chem.*, **66**, 1339–1346 (1994).

60. S. Mitra, L. Zhang, N. Zhu, and X. Guo, *J. Microcol. Sep.*, **8**, 21–27 (1996).

61. L. Zhang, X. Guo, and S. Mitra, **44**, 529–540 (1997).

62. X. Guo and S. Mitra, *J. Chromatogr. A*, **826**, 39–47 (1998).

63. X. Guo and S. Mitra, *Anal. Chem.*, **71**, 4587–4593 (1999).

64. X. Guo and S. Mitra, *Anal. Chem.*, **71**, 4407–4413 (1999).

65. A. San Juan, X. Guo, and S. Mitra, *J. Sep. Sci.*, **24**(7), 599–605 (2001).

66. D. Kou, A. San Juan, and S. Mitra, *Anal. Chem.*, **73**, 5462–5467 (2001).

67. H. Lonsdale, U. Merten, and R. Riley, *J. Appl. Polym. Sci.*, **9**, 1341 (1965).

68. S. Bauer, *Trends Anal. Chem.*, **14**(5), 202–213 (1995).

69. N. Srinivasan, R. C. Johnson, N. Kasthurikrishnan, P. Wong, and R. G. Cooks, *Anal. Chim. Acta*, **350**, 257–271 (1997).

CHAPTER

5

PREPARATION OF SAMPLES FOR METALS ANALYSIS

BARBARA B. KEBBEKUS

Department of Chemistry and Environmental Science, New Jersey Institute of Technology, Newark, New Jersey

5.1. INTRODUCTION

Metals contained in samples are determined by a wide variety of analytical methods. Bulk metals, such as copper in brass or iron in steel, can be analyzed readily by chemical methods such as gravimetry or electrochemistry. However, many metal determinations are for smaller, or trace, quantities. These are determined by various spectroscopic or chromatographic methods, such as atomic absorbance spectrometry using flame (FAAS) or graphite furnace (GFAAS) atomization, atomic emission spectrometry (AES), inductively coupled plasma atomic emission spectrometry (ICP-AES), inductively coupled plasma mass spectrometry (ICP-MS), x-ray fluorescence (XRF), and ion chromatography (IC).

Preparation of materials for determination of their metal content serves several purposes, which vary with the type of sample and the demands of the particular analysis. Some of the major functions of sample preparation are:

- To degrade and solubilize the matrix, to release all metals for analysis.
- To extract metals from the sample matrix into a solvent more suited to the analytical method to be used.
- To concentrate metals present at very low levels to bring them into a concentration range suitable for analysis.
- To separate a single analyte or group of analytes from other species that might interfere in the analysis.
- To dilute the matrix sufficiently so that the effect of the matrix on the analysis will be constant and measurable.

Sample Preparation Techniques in Analytical Chemistry, Edited by Somenath Mitra
ISBN 0-471-32845-6 Copyright © 2003 John Wiley & Sons, Inc.

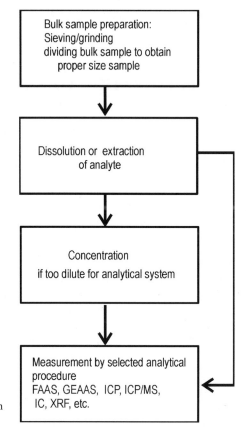

Figure 5.1. Plan for sample preparation for metals determination.

• To separate different chemical forms of the analytes for individual determination of the species present.

Although not all of these functions are needed in every case, most analyses require one or more of them. Figure 5.1 is a schematic of the analysis procedure. The major concerns in selection of sample preparation methods for metal analysis are the analytical method to be used, the concentration range of the analyte, and the type of matrix in which the analyte exists.

A common result of the sample preparation is the dissolution of the entire sample, producing a clear solution. The digestion method must be selected to suit the type of sample, the metals being determined, and finally, the analytical method. Of the methods listed above, most require a liquid sample, except for x-ray fluorescence, which often is used on solid samples. Wet digestion in acid solution, dry ashing, and extraction of the analytes from the sample without total matrix destruction are common sample preparation methods. Dry ashing is useful for moist samples, such as food or botanical samples, because it destroys large amounts of wet organic matter easily and quickly. However, if the analyte metal is present in a volatile form, methyl mercury, for example, dry ashing can cause loss of analyte. Many sample matrices, both organic and inorganic, can be dissolved by heating in a strong oxidizing acid solution. Other samples can be treated by extracting the metals from the matrix. This method is frequently used for water samples, where a chelating agent may be used to complex the metals of interest, enabling their easy separation from the aqueous matrix.

Since many of these analyses are done to determine traces of metals, and because the reagents used for matrix destruction are often quite aggressive, prevention of sample contamination from the containers or from the reagents themselves requires constant care. In addition to contamination of samples, there also exists the possibility of loss of analyte during the preparation of the samples. Metal ions may adsorb on surfaces, especially on glass surfaces. Silica vessels are less likely to sorb metals. Some metals, notably iron, mercury, gold, and palladium, may be removed from samples when they are heated in platinum crucibles. This adsorbed material is not only lost to the analysis, but may remain on the container walls to contaminate future samples. Volatilization is another major cause of analyte loss and can be serious for the more volatile metals when samples are dry ashed. Cadmium, lead, mercury, selenium, and zinc are especially volatile, being lost at temperatures below 500°C. Volatility depends rather strongly on matrix and on the form of the metal present, so metals that have been found to be well recovered at a certain temperature in one type of sample may be subject to losses by volatilization in a different matrix or sample.

Digestion Methods

Many metal analyses are carried out using atomic spectroscopic methods such as flame or graphite furnace atomic absorption or inductively coupled plasma atomic emission spectroscopy (ICP-AES). These methods commonly require the sample to be presented as a dilute aqueous solution, usually in acid. ICP-mass spectrometry requires similar preparation. Other samples may be analyzed in solid form. For x-ray fluorescence, the solid sample may require dilution with a solid buffer material to produce less variation between samples and standards, reducing matrix effects. A solid sample is also preferred for neutron activation analyses and may be obtained from dilute aqueous samples by precipitation methods.

Total matrix dissolution is common and ensures complete availability of the analytes for analysis. However, it is a lengthy process in many cases, and other methods may achieve useful analytical samples with less time and labor. Slurry sampling is one such method. If the sample can be finely powdered and the powder taken up in a fluid slurry, it may give acceptable analytical results. Also, the analyte may be leached or extracted from the matrix without dissolving the entire matrix. Finally, if the metals in a sample are to be speciated in the analysis, that is, if the actual form in which the metal exists in the original sample is to be determined, an entirely different sample preparation scheme is required. Aggressive acid digestion usually renders all the metals into the same form and destroys any information about the species originally present.

5.2. WET DIGESTION METHODS

The common methods used for dissolving samples for metals analysis are digestion in an open flask, digestion in a pressurized, sealed container, and microwave assisted decomposition. Some common solvents used are listed in Table 5.1.

Samples to be analyzed for elemental metal content are usually prepared by digesting the matrix in a strong acid. In the case of organic matrices, an oxidizing mixture is used to destroy the entire organic matrix and solubilize the sample. This yields a clear solution containing the metals for analysis by such techniques as AA, ICP, or ICP-MS. Nitric acid is commonly used, because there is no chance of forming insoluble salts as might happen with HCl or H_2SO_4. Hydrogen peroxide may be added to increase the oxidizing power of the digestion solution.

Inorganic samples, soils, sediments, ores, rocks, and minerals may be digested in dilute or concentrated acids or mixtures of acids, which may be sufficient to leach out the analytes. However, if total dissolution is required,

Table 5.1. Reagents Commonly Used in Sample Dissolution or Digestion

Reagent	Sample Type
Water	Soluble salts
Dilute acids	Dry-ashed sample residues, easily oxidized metals and alloys, salts
Concentrated acid (e.g., HNO_3)	Less readily oxidized metals and alloys, steels, metal oxides
Concentrated acid with added oxidizing agent	Metals, alloys, soils, particulates from air, refractory minerals, vegetable matter
Hydrofluoric acid	Silicates and other rock samples

hydrofluoric acid can be used as a final digestion step to dissolve silicates. Refractory materials such as cements, ceramics, and slags may require fusion or flux digestion, which involves melting the ground sample with a salt such as sodium carbonate or sodium peroxide. The resulting solid is then more easily dissolved for analysis. However, the method is not generally suitable for metals, which tend to be lost by volatilization because of the high temperatures required. In addition, the material of the container becomes more critical as the aggressiveness of the digestion process increases.

5.2.1. Acid Digestion—Wet Ashing

The simplest method for wet digestion is carried out in an open container. Samples are dried, weighed, and placed in a beaker. The digestion reagent is added. The beaker is covered with a watch glass and placed on a hot plate, as shown in Figure 5.2. The sample is allowed to boil very gently to avoid spattering. More solution may be added from time to time to prevent the sample from drying out. Hydrogen peroxide may be added at a point during the digestion to help oxidize organic materials. When the sample has been digested completely, it is evaporated to near dryness and then taken up in a dilute acid solution and diluted to volume for analysis. Samples are generally not allowed to dry completely, as species even less soluble may form. Filtration at this point is often necessary, as many matrices will leave some insoluble matter, such as silica. The filter must be rinsed carefully to avoid the loss of analyte.

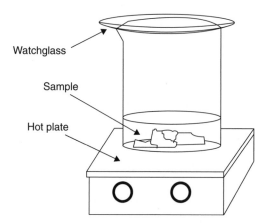

Figure 5.2. Open digestion can be done on a hotplate in a loosely covered beaker.

The choice of acid to be used depends on the sample. Relatively soluble inorganic samples, salts, active metals, or alloys may be dissolved in water or dilute acid. Electropositive metals will also dissolve in dilute acid, although aluminum may need a trace of mercuric chloride added to prevent the formation of an impervious oxide layer.

The mildest solution that will digest the sample is preferred, as stronger acids are more likely to add to the blank, attack digestion vessels, and generally require more care in the laboratory. Concentrated acids may be used individually, in mixtures, or in sequence. Hot concentrated acids will dissolve many metals and alloys. Nitric acid oxidizes the sample and should be used before a stronger oxidizer such as perchloric acid, to remove the more readily oxidized material.

Several progressively more aggressive digestion schemes can be considered. If the sample is not water soluble, a nitric acid digestion may be suitable. In preparation, the glassware should be acid washed and rinsed with distilled water. A general procedure is to place the weighed sample in a conical flask or beaker and add 5 to 10% nitric acid. This is covered with a watch glass and brought to a slow boil on a hot plate. If the sample is prone to bumping, a few boiling chips can be added. The solution is evaporated down to a few milliliters without allowing it to dry. Heating is continued and small quantities of concentrated nitric acid are added until the digested solution is clear and light colored. The beaker walls are washed down. If filtration is necessary, it is carried out at this point. The solution is transferred to a volumetric flask and diluted to volume before analysis. Sometimes, 1:1 hydrochloric acid is added when the nitric acid digestion is complete, and a further digestion is carried out, before filtration and dilution.

If the sample is not digested satisfactorily by nitric acid alone, or by nitric acid followed by HCl, further treatment with sulfuric acid can be done. A 2:1 mixture of sulfuric and nitric acids is added. The sample is evaporated to dense white fumes of SO_3. More nitric acid may be added if the solution does not clear. The solution is again heated to SO_3 fumes. The solution is then cooled, diluted with water, and heated to dissolve any salts. Then it is filtered, if necessary, diluted to volume, and is ready for analysis.

An even more aggressive digestion begins with the nitric acid digestion as described above. After the sample has been digested and boiled down to a few milliliters, the sample is cooled and equal volumes of nitric and perchloric acids are added, cooling the beaker between additions. It is evaporated gently until dense white fumes of perchloric acid are seen. If further digestion is needed, nitric acid can again be added. The cooled sample is diluted with water, filtered if needed, and then diluted to volume.

For samples that show significant losses of analyte due to the retention of metals in the silica residues, the sample is first digested thoroughly with nitric acid in a PTFE (polytetrafluoroethylene) beaker. Then concentrated perchloric acid and a small amount of hydrofluoric acid are added. The sample is boiled until clear and white fumes have appeared. The sample is cooled and diluted to volume.

For even more dissolution power, mixtures of concentrated acids with oxidizing agents or with hydrofluoric acid are used. Aqua regia, a 3:1 mixture of concentrated hydrochloric and nitric acids will dissolve noble metals. Sulfuric acid with hydrogen peroxide is a powerful oxidizer. A mixture of an oxidizing acid with hydrofluoric acid provides acidity, oxidizing power, and complexation to dissolve the sample. These mixtures will dissolve all metals and alloys and most refractory minerals, soils, rocks, and sediments. Figure 5.3 shows the progressively more aggressive acid digestion solutions.

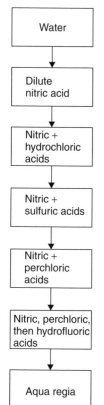

Figure 5.3. Acid mixtures used for digestion. The least aggressive mixture that digests the sample thoroughly should be used.

5.2.2. Microwave Digestion

Digesting a sample in a closed container in a microwave oven has several advantages over open container dissolution methods. The containers are fabricated of high-temperature polymers, which are less likely to contain metal contaminants than are glass or ceramic beakers or crucibles. The sealed container eliminates the chance of airborne dust contamination. The sealed, pressurized containers reduce evaporation, so that less acid digestion solution is required, reducing blanks. The sealed container also eliminates losses of more volatile metal species, which can be a problem in open container sample decomposition, especially in dry ashing. The electronic controls on modern microwave digesters allow very reproducible digestion conditions. Automated systems reduce the need for operator attention. Finally, the controlled exhaust contains the acid fumes, which can then be scrubbed in a neutralizing solution. Otherwise, these fumes tend to corrode exhaust hoods and laboratory fixtures.

A microwave sample digestion system has been described in Chapter 3. It consists of a microwave oven, a rotating carousel holding several sample digestion bombs, and a system for venting these in a controlled fashion. It may also provide monitoring and recording of both temperature and pressure in the containers. The sample containers are relatively high pressure containers, usually made of strong, high-temperature-resistant polymers, often polycarbonate for strength or PTFE for chemical resistance. Systems designed for the strong acid digestion required for metals analysis often include a separate liner, which is more resistant to chemical attack and can be changed as it begins to break down under the very corrosive conditions of high acidity and high temperature in the digestion bomb.

Each bomb has a pressure relief valve, which vents into a manifold. This exhausts the acid fumes into a tube, which should be connected to an acid-scrubbing trap. The relief valves are set so that the sample is heated under pressure, allowing higher temperatures and more rapid digestion than is possible in an open container.

Modern microwave digestion systems monitor both pressure and temperature in the containers. As the temperature or pressure reaches the set point, power to the oven is cut. The oven power as well as the maximum pressure and temperatures can be set. Both digestion time and oven power can be programmed so that each sample is treated in a reproducible manner. The initial digestion is done slowly at low temperature, and the temperature is increased after the majority of the readily digested matrix is dissolved.

Microwave containers for sample digestion are commercially available which can be used for ashing samples at temperatures up to 300°C or pressures to 800 psi, under controlled pressure and temperature. Under these conditions, even refractory samples can be digested successfully in a reasonable time. A method for dissolution of alumina samples uses a high-boiling mixture of H_2SO_4 and H_3PO_4 and digests the sample at 280°C, with the pressure reaching only 40 psi. Similar samples can also be digested at 240°C with HCl, but the pressure reaches as high as 660 psi [1].

Ovens that do not have the facility to monitor and control the sample temperature may need to be calibrated, so that different ovens can be used with similar results. It is not sufficient simply to set the power fraction and make a correction for the difference in wattage of the ovens. The easiest way to calibrate a microwave oven is to measure the temperature rise of 1 L of water at various power settings and times and compare these between ovens.

5.2.3. Comparison of Digestion Methods

In a comparison study of acid digestion in the open and using the microwave oven, plant matter was prepared for determination of chromium [2]. Samples of rye grass, beech leaves, and pine needles were digested in PTFE beakers using two different acid digestion schemes. The digestions were done in nitric acid followed by perchloric and hydrofluoric acids, as well as nitric followed by sulfuric and hydrofluoric acids. Similar sequences of acids were used with closed PTFE vessels in microwave digester. The open digestions took 40 to 90 hours. Microwave sequences were complete in 50 to 75 minutes, with the longer times needed in both cases when sulfuric acid rather than perchloric acid was used.

The results of this study showed that the microwave digestion was equally effective in digesting the samples. In addition, the savings in time were very substantial. Open digestion with perchloric acid resulted in negative bias because of the formation and evaporation of volatile chromyl chlorides. In both methods, there were small differences in the analyte recovery for each of the matrices, indicating that method validation is always a good idea when working with different sample types, even those that appear to be as similar to each other as the two plant materials used in this study.

In laboratories where there must be large sample throughput as well as a large number of analytes, as is the case in food analysis for labeling purposes, the use of a single sample preparation method is highly desirable. Microwave digestion has been tested for such situations. The replacement of a series of separate official methods for different metals at different levels with a single method was examined. A microwave system that measured and controlled temperature and pressure in each vessel simultaneously was used. This allowed foods of different types to be digested together without danger of rupturing either the seals or vessels. Foods were ground in a blender and weighed into the digestion vessels. Five milliliters of ultrapure concentrated nitric acid was added to each and the vessels were sealed. They all were processed under a program that ramped the power from 100 to 600 W over

5 minutes, held it at 600 W for 5 minutes, and at 1000 W for an additional 10 minutes. After 15 minutes of cooling, the samples were opened and diluted to 50 or 100 mL with deionized water, before analysis with ICP-AES. The method was tested on a wide variety of foodstuffs, including cream, nuts, oysters, tuna salad, liver, spinach, corn, and eggs. Acceptable results were obtained in all cases with all spike recoveries within the legislated $\pm 20\%$ limits [3]. This study demonstrates that a streamlined method for metals can be developed if required accuracy limits are not overly stringent. Table 5.2 shows the extraction efficiencies for several metals obtained using microwave digestions of a variety of reference materials in three different laboratories.

In the interests of efficiency and reduction of laboratory waste solvents,

Table 5.2. Some Extraction Efficiencies Using Microwave Methods

Matrix	Analyte	Mean (μg/g)	S.D.	Number of Replicates	Certified Value	Ref.
Corn	Ca	62.2	1.1	3	42 ± 5	3
	Mg	1060	32	3	990 ± 82	3
	Mn	5.24	0.45	3	4.0 ± 0.45	3
Spinach	Fe	549	0.05	3	550 ± 20	3
	K	39100	0.01	3	35600 ± 300	3
	Zn	54.3	0.07	3	50 ± 2	3
Oyster tissue	Ca	1690	85	3	1960 ± 190	3
	Cu	57.2	1.6	3	66.3 ± 4.3	3
	Mn	11.3	0.42	3	12.3 ± 1.5	3
Bovine liver	Ca	112	9.2	3	120 ± 7	3
	Mg	568	49	3	600 ± 15	3
	Na	2050	340	3	2430 ± 130	3
	Zn	126	15	3	123 ± 8	3
Soil (CRM S-1)	Na	4620	110	4	4440 ± 140	93
	K	12080	240	4	12050 ± 580	93
	Ca	2360	60	4	2600 ± 600	93
	Mn	266	21		266 ± 18	93
	Zn	33.3	1.8	4	35 ± 3.3	93
	Ni	11	1.5	4	13	93
	Pb	16.4	2.2	4	15 ± 3.6	93
Soil (CRM 142R)	Pb	35.2	2.6	~20	40.2 ± 1.9	133
	Ni	60.6	1.9	~20	64.5 ± 2.5	133
	Cu	69.7	1.8	~20	69.7 ± 1.3	133
	Cr	111.8	2.8	~20	113 ± 4	133
	Zn	96.1	1.6	~20	101 ± 6	133
	Cd	0.37	0.04	~20	0.34 ± 0.04	133

the EPA developed a method for total recoverable metals using microwave digestion, which reduced the amount of sample from 100 mL to 25 mL and the amount of acid required for digestion from 10 mL to 5 mL. Samples are microwave digested and analyzed by ICP-MS, which requires less sample and gives excellent specificity and accuracy. This method eliminates the use of hotplates, hoods, beakers, and watchglasses, requiring less time, less laboratory space, and much less cleanup of glassware between samples (EPA laboratory method SW 846 3015).

In general, the use of microwave digestion is preferable for practical reasons. Microwave energy is delivered into the sample efficiently without heating containers, hotplates, and so on. The energy can readily be controlled and programmed automatically, ensuring better reproducibility. Sample digestion times are reduced significantly, and the amount of reagent required is usually less. Additionally, there is less chance of volatilization of some analytes, and sample contamination is less likely than when an open container is used. Finally, microwaves provide an excellent opportunity for automation. A review of microwave digestion procedures for an array of environmental samples has been published [4]. Some reagents used in digestions of biological samples are summarized in Table 5.3.

5.2.4. Pressure Ashing

Pressure ashing is also applicable to acid digestion of samples. In this method the weighed samples are placed into small quartz vessels with the appropriate acid digestion solution. These are sealed with PTFE and quartz caps, placed in a heating block, and the apparatus closed and pressurized with nitrogen. The nitrogen serves to support the digestion vessels by equalizing pressure inside and outside the vessels, as they are heated. As in microwave sample dissolution, wet digestion in a sealed container eliminates losses of analytes through volatilization.

Although the sample is protected from losses by volatilization, unwanted materials, especially carbon, are also not removed, and these can cause problems in some cases. For samples containing much organic material, the carbon remaining in the samples after this wet ashing can interfere with the determination of several metals especially arsenic and selenium by ICP-MS [92].

5.2.5. Wet Ashing for Soil Samples

Mineral samples such as rock, soil, and sediments require more aggressive digestion. Total sample dissolution may be done by several methods, and

Table 5.3. Microwave Digestion Reagents

Reagents	Elements Determined	References
Marine Biological Tissues		
HNO$_3$	Al, As, Ca, Cd, Co, Cr, Cu, Fe, Hg, Mg, Mn, Ni, Pb, Se, Sr, Zn	5–21
HNO$_3$, V$_2$O$_5$ catalyst	As	22
HNO$_3$, H$_2$O$_2$	Ag, Al, As, B, Cd, Cr, Cu, Hg, Mg, Mn, Ni, Pb, Se, Sr, Zn	24–35
HNO$_3$, HF	Ag, Al, As, Cd, Co, Cr, Cu, Fe, Hg, Mn, Ni, Pb, Se, Sn, Th, Zn	29, 36
HNO$_3$, H$_2$SO$_4$, H$_2$O$_2$, NH$_4$-EDTA	Ca, Cd, Cu, Fe, K, Mg, Mn, P, Sr, Zn	37
Methanolic KOH	Hg, Methylmercury	38
Other Biological Tissues		
HNO$_3$	Ag, As, Cd, Co, Cr, Cu, Fe, Hg, Mg, Mn, Mo, Po, Pb, Rb, Se, V, Zn	6, 12, 39–44
HNO$_3$, HClO$_4$	Cd, Cu, Pb, Se	45, 46, 47
HNO$_3$, H$_2$O$_2$	B, Bi, Ca, Cd, Co, Ce, Cu, Fe, Hg, K, Mg, Mn, Mo, Na, P, Pb, Rb, Ru, Sb, Se, Sn, Sr, Tl, Zn	25, 30, 48–57
Botanical Samples		
HNO$_3$	Al, As, B, Ba, Be, Bi, Ca, Ce, Cd, Co, Cr, Cu, Eu, Fe, Hg, K, La, Mg, Mn, Mo, Na, Ni, Pb, Po, Rb, Se, Sm, Sr, Tb, Te, Th, U, V, Zn	13, 21, 40, 42, 44, 58–69
HNO$_3$, H$_2$O$_2$	Al, As, B, Ca, Ce, Cd, Co, Cu, Eu, Fe, Hg, K, Mg, Mn, Na, Ni, Pb, Se, Sm, Sr, Tb, Th, U, Zn	25, 34, 49, 53, 56, 57, 70–79
HNO$_3$, HCl	Ca, Co, Cu, Fe, K, Mg, Mn, Na, Ni, Pb, Zn	69, 80–84
HNO$_3$, HClO$_4$	Al, Ba, Ca, Cd, Cu, Fe, K, Mg, Mn, Pb, Zn	47, 85–89
NO$_3$, V$_2$O$_5$ catalyst	As, Cd, Cu, Fe, Pb, Se	90, 91

Source: Ref. 4.

leaching to remove the analytes without dissolving the matrix completely is also possible. In one study [93] several methods were compared on a set of samples of contaminated soils. In each case, the solid samples were air dried and sieved to recover particles below 1 mm in diameter. The procedures were as follows:

1. A weighed 1-g sample was heated in a platinum crucible with 10 mL of concentrated HF and 7.5 mL of concentrated $HClO_4$. After evaporation, a second treatment with the same acids was carried out. Then 20 mL of 4% H_3BO_3 was added and evaporated. The residue was dissolved in 2 mL of concentrated HCl and diluted to volume with distilled water.

2. A weighed 1-g sample was heated in a platinum crucible with 10 mL of concentrated HF and 7.5 mL of concentrated $HClO_4$. After evaporation, a second treatment with the same acids was carried out. Residue was mixed with $LiBO_2$ and melted at 900°C in a muffle furnace. The glassy melt was dissolved in dilute HNO_3 and diluted to volume.

3. A 0.25-g sample was digested in a microwave apparatus with 4 mL of HF, 3 mL of HCl and 3 mL of HNO_3. The microwave was operated at 250 W for 10 minutes, 400 W for 5 minutes, and 500 W for 10 minutes. After venting and cooling, the digest was diluted to 35 mL.

4. The sample was mineralized at 450°C for 8 hours and a 5-g sample was weighed into a platinum dish. Then 15 mL of HNO_3 and 10 mL of $HClO_4$ were added and the sample was leached for 24 hours at room temperature, followed by heating to dryness. The sample was taken up in 25 mL of dilute HCl and digested on a water bath for an hour. The silica residue was filtered off and washed with 1% hot HCl and diluted to 100 mL.

Comparison of the results of analysis by FAAS of the solutions produced by each of these methods indicated that procedures 1 and 3 were preferable to the others with respect to precision and accuracy. Procedure 1 was less accurate for chromium than procedure 3. In addition, procedure 3, which employed microwave digestion, took considerably less time to complete. Method 4, the leaching process, produced acceptable results only for Fe, Mg, Zn, and Cu.

Experiments of this type indicate the importance of properly designed sample preparation schemes and the necessity of running a standard reference material that closely resembles the sample by a new method to ensure accurate results.

5.3. DRY ASHING

For samples that contain much organic matter, which are being analyzed for nonvolatile metals, dry ashing is a relatively simple method of removing the organic matter that can be used for relatively large samples and requires little of the analyst's time. In the open vessel method, the sample is placed in a suitable crucible and is ignited in a muffle furnace. Crucibles used for ashing are usually made of silica, porcelain, platinum, or Pyrex glass.

The major drawbacks of the method are the possible loss of some elements by volatilization, contamination of the sample by airborne dust, as it must be left open to the atmosphere, and irreversible sorption of analyte into the walls of the vessel. It is important to do blanks with each batch of samples. Particles generated within the muffle furnace may be the cause of high or variable blanks. In this case the applicability of the method will depend on the level of analyte expected in the samples. A variable blank can be tolerated when the analyte level is substantially higher than the blank but not when the concentration analyte found in the blank and the sample are similar.

Losses from volatilization of the analyte can be minimized by restricting the temperature at which ashing takes place. For determination of lead, copper, zinc, cadmium, and iron in foodstuffs, for example, good recoveries of the analytes were obtained by heating the samples slowly to 450°C and holding this temperature for 1 hour. A collaborative study showed no significant losses of the analytes under these ashing conditions [94].

Dry ashing is suitable for nutritional elements in foods, such as Fe, K, Ca, Mg, and Mn, which are present in substantial quantity and are stable at the high temperatures required. Fats and oils, however, can pose a problem, as they may ignite and cause losses in smoke particles. These require pretreatment before ignition. Additives such as sulfuric acid or salts may aid in dry ashing. Sulfuric acid has a chemical charring effect, and salts such as magnesium nitrate, sodium carbonate, and magnesium oxide aid in the retention of some elements. These salts leave a soluble alkaline inorganic residue. Silica remaining after destruction of much of the sample matrix can occlude metals and render them insoluble in the acid used to dissolve the residue. If this is a major difficulty with certain samples, further treatment with hydrofluoric acid may be needed to dissolve the silica entirely.

A general procedure is to place the weighed sample into a platinum or silica glass crucible and heat it in a muffle furnace to a white ash. The temperature should be kept at 400 to 450°C if any of the more volatile metals are being determined. Salts or sulfuric acid may be added, if needed, and a final ashing step can be done with hydrofluoric acid if required. The residue is then dissolved in concentrated nitric acid and warm water, and diluted to volume. The final concentration of acid should be between 1 and 5%.

Extraction, Separation, and Concentration

It is not always necessary or required to digest the entire sample in order to free the metals for analysis. In some cases it is not even desirable. In studies of contaminated soils, for instance, the analyte of interest may be present as a soluble salt from a pollution source, as well as also being present in the structure of the mineral crystals. The soluble form is of concern, as it is available to biota and may eventually contaminate groundwater. That in the insoluble particles is not of interest. In such cases, where the analyte is much more soluble than the matrix or where the metals included in the matrix are not of interest, an extraction process rather than complete solubilization is preferred. This is treated further in Section 5.10.

5.3.1. Organic Extraction of Metals

Organic extraction is carried out for recovery of dissolved metals from water samples. Ionic species, including metallic ions, are quite insoluble in organic solvents. If the charge on the metal ion is neutralized or the ion is bound to a larger organic moiety, the metal become soluble in an organic solvent and, consequently, can be extracted from the aqueous phase. This can be achieved either by formation of metal chelates, metal–organic complexes, or by ion pairing.

The formation of metal chelates is the most common extraction technique for metals. A complex formed between a metal and a chelating agent is hydrophobic in nature and soluble in organic solvents. The partition coefficient of the metal complex in an organic solvent such as chloroform or methyl isobutyl ketone (MIBK) is quite high, enabling recovery of the metal by liquid–liquid extraction. In a typical extraction, the chelating agent and the organic solvent are added to the aqueous sample and shaken together. The chelate formed partitions into the organic phase. The extraction involves four different equilibria, which are shown in Figure 5.4. The chelating agent, HA, a weak acid, dissociates in the aqueous phase:

$$HA \rightleftharpoons H^+ + A^-$$

$$K_1 = \frac{[H^+]_{aq}[A^-]_{aq}}{[HA]_{aq}}$$

It also forms a complex with metal ion M^{n+}:

$$nA^- + M^{n+} \rightleftharpoons MA_n$$

$$K_2 = \frac{[MA_n]}{[M^{n+}][A^-]^n}$$

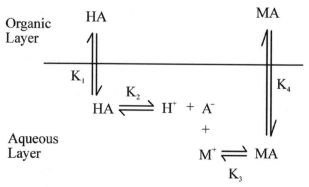

Figure 5.4. Equilibria involved in extraction of a chelated metal from an aqueous phase into an organic solvent. (From Ref. 95.)

The complex is then distributed between the two phases:

$$(MA_n)_{aq} \rightleftharpoons (MA_n)_{org}$$

$$K_3 = \frac{[MA_n]_{org}}{[MA_n]_{aq}}$$

The undissociated chelating agent also is distributed between the organic and aqueous phases:

$$(HA)_{aq} \rightleftharpoons (HA)_{org}$$

$$K_4 = \frac{[HA]_{org}}{[HA]_{aq}}$$

The distribution ratio, D, is defined as

$$D = \frac{\text{concentration of metal in organic phase}}{\text{concentration of metal in aqueous phase}}$$

$$= \frac{[MA_n]_{org}}{[M^{+n}]_{aq} + [MA_n]_{aq}}$$

This ratio should be as large as possible to maximize the efficiency of the extraction.

Assuming that $[M^{+n}] \gg [MA_n]_{aq}$ and substituting from the equations above for K_1, K_2, and K_3, the following is obtained:

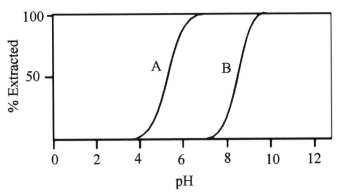

Figure 5.5. pH affects the stability of chelates and may be used to discriminate in an extraction between the desired analyte and interfering metals.

$$D = \frac{K_2 K_3 (K_1)^n [\text{HA}]_{\text{org}}^n}{(K_4)^n [\text{H}^+]_{\text{aq}}^n}$$

If the amount of chelating agent in the organic phase is fixed, then

$$D = \frac{(\text{constant}) K_1}{[\text{H}^+]_{\text{aq}}}$$

Thus, D is a function of the equilibrium constant for the chelate formation, K_1, the number of chelating agent molecules bonded to the metal ion and the pH. The first two of these are fixed by the system chosen. By controlling the pH, both the extraction efficiency and the selectivity can be controlled. For example, Figure 5.5 shows the formation of the chelates of two metals as the pH is varied. When the chelate is extracted into an organic solvent at a pH of 6, metal A is quantitatively extracted, while metal B remains in the aqueous layer.

The most common chelating agents used to extract metals from water samples are ammonia pyrolidine dithiocarbamate (APDC) and 8-hydroxyquinone. Methyl isobutyl ketone (MIBK) is generally used as a solvent. In a typical extraction, 1 mL of APDC is added to 50 to 100 mL of aqueous sample in a volumetric flask. The pH of the aqueous sample is adjusted for maximum extraction of the analyte of interest. Then 10 mL of MIBK is added (the volumetric ratio of sample to MIBK is usually less than 40) and the mixture is vigorously shaken for 30 seconds. The metal chelate partitions into the organic phase, which floats on the water. More water can be added to raise the organic level into the neck of the flask so that it can be

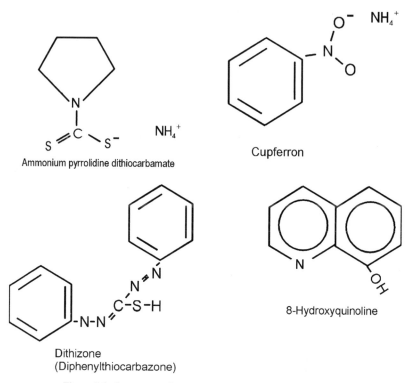

Figure 5.6. Structures of some commonly used chelating agents.

aspirated directly into the analytical instrument. Figure 5.6 shows the structures of some common chelating agents.

5.3.2. Extraction with Supercritical Fluids

Since commercial supercritical fluid extraction apparatus has become available, use of these materials as extractants has become attractive. Solvent evaporation and disposal are eliminated, and the extractions may be very efficient because of the low viscosity of supercritical fluids, which allows them to penetrate readily into the solid sample particles. Carbon dioxide, with or without modifiers such as methanol, is the most commonly used solvent.

For extraction with supercritical carbon dioxide, metals are first chelated with a ligand such as a derivative of dithiocarbamate. It has been found [96] that while the solubility of chelates of metals with sodium diethyl dithiocarbamate in carbon dioxide is quite low, the solubility can be increased significantly by substitution of a longer chain alkyl group for the ethyl groups on the dithiocarbamate. Even better extractions were obtained when the ethyl groups of the diethyl dithiocarbamate were fluorinated.

The solid sample is placed in the preheated extraction thimble, and the modified CO_2 is added to the desired extraction pressure. The system is held static at the extraction temperature and pressure. At the end of the period, the system is vented into a collection vial containing chloroform. This is followed by a dynamic flush of the system with the CO_2 solvent at the same temperature and pressure. Mercury complexed with fluorinated diethyl-dithiocarbamate was extracted from dried aquatic plant material with 95% efficiency by methanol-modified CO_2.

5.3.3. Ultrasonic Sample Preparation

Some sample matrices are inherently difficult to ash. Foodstuffs with high sugar content are an example. Dry ashing must be done slowly and requires over 30 hours. Soluble samples such as sugar can be aspirated in solution directly into the AAS, but the solution must be quite dilute. This leads to high detection limits, and the recovery of analytes tends to be low.

An extraction method that uses an ultrasonic probe has been developed [97]. The sugar is mixed with water and is ultrasonicated for a period of time to ensure thorough solution. Then the pH is adjusted to 9, and aqueous sodium diethyl dithiocarbamate is added. Then the solution is extracted twice with chloroform. The extract is evaporated and the residue taken up in dilute acid for analysis by AAS.

For soil samples ultrasonication in 1:1 diluted aqua regia was found to give excellent recovery of As, Cd, Pb, and Ag from reference samples. The results were comparable to those obtained by microwave digestion, and the speed of extraction and sample throughput were better with the ultra-sonication. The samples were ultrasonicated for three periods of 3 minutes each and shaken by hand between untrasound treatments. A batch of 50 samples could be sonicated simultaneously [23].

5.4. SOLID-PHASE EXTRACTION FOR PRECONCENTRATION

When the sample is a liquid and contains concentrations of analyte below the detection range of the analytical instrument used for the determination, a concentration step is often required. Metals can be concentrated from solution by solid-phase extraction (SPE). This technique has been discussed in detail in Chapter 2. SPE usually involves passing the solution through a

From syringe

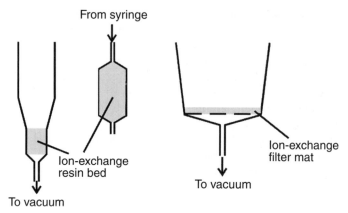

Ion-exchange
resin bed

To vacuum

To vacuum

Ion-exchange
filter mat

Figure 5.7. Solid-phase extraction devices come in a variety of forms, with different sorbents for different applications. Shown are packed bed with a built-in reservoir to hold the sample solution, a syringe tip packed cartridge, and a filter disk in a holder for rapid extraction from large volumes.

column, cartridge, or disk containing a solid material that more or less specifically binds the metal ions present in solution. Some common formats are shown in Figure 5.7. The solid-phase extractant may be an ion-exchange resin, a chelating resin, or other material designed to bind all cations, or anions, or to bind specific groups of ions. In addition, this method may be used in conjunction with a dissolved chelating agent, using a organic-binding solid-phase extractant to separate the chelated metal from the solution. Dithizone complexed metals, for example, can readily be sorbed onto a silica-supported C_{18} phase, commonly used for high-performance liquid chromatographic (HPLC) separations. The metals can then be desorbed by elution with acidified acetonitrile. Packing materials with amino $(-NH_2)$ functionality will bind some cations directly. Since the amino group on the sorbent can be protonated, the pH of the sample solution will have a major effect on the efficiency of sorption. Table 5.4 shows several SPE systems especially designed for concentration of metals from aqueous solution.

Recent development of self-assembled monolayers on mesoporous ceramic supports have led to very efficient, rapid, and highly selective materials for sequestration of metals [112]. The ceramic support material incorporating copper ferrocyanide ethylene diamine was found to remove 99% of the cesium from a 2-ppm solution within 1 minute. The sorption was not hampered by acid or high concentrations of sodium or potassium. The sorbed metal can be removed by eluting with an oxidizing agent. The sorbent can then be regenerated by using a reducing eluent. A similar material treated with alkyl thiol can be used to sequester mercury, lead, and silver, with high efficiency. The metals can be recovered using an acid eluent.

Table 5.4. SPE Materials Used to Extract Metals from Water

SPE Material	Metals Sorbed	References
Amberlite XAD-2, functionalized by coupling to quinalizarin [1,2,5,8-tetrahydroxyanthraquinone] by an $-N=N-$ spacer	Cu(II), Cd(II), Co(II), Pb(II), Zn(II), and Mn(II)	98
Lignin derivatized with methyl thio ether functional groups	Hg, Pb, Cd, and Cu from water; Cr(III) and Fe(III) also strongly adsorbed; Na not adsorbed; Ca only moderately	99
Nanoparticles of TiO_2	Cu, Cr, Mn, and Ni	100
TLC-grade silica gel, functionalized with 8-hydroxyquinoline by catalyzed Mannich amminomethylation reaction	Cu(II), Cd(II), Zn(II), Pb(II), and Fe(III)	101
Dimethylglyoxime (DMG)-doped silica	Ni	102
Cellex P, cellulose sorbent with phosphonic acid groups	Recoveries for Al, Be, Cd, Ni, Pb, and Zn are >90%; also suitable for enrichment of Co and Mn	103
Chelex 100, chelating resin	Recoveries for Al, Be, Cd, Ni, Pb, and Zn are >90%; Pb in natural water	103, 95
SIO_2-TPP sorbent (contains porphyrin ligand covalently attached to aminopropyl silica gel)	Selective sorption of Mo(VI) and V(IV)	103
5-Amino-1,3,4-thiadiazole-2-thiol groups attached to silica gel	Cd(II), Co(II), Cu(II), Fe(III), Ni(II), Pb(II), and Zn(II)	104
Ammonium pyrrolidine dithiocarbamates sorbed on quartz microfiber filter	Co^{2+}, Cr^{6+}, Cu^{2+}, Fe^{3+}, Ni^{2+}, Pb^{2+}, and Zn^{2+}	105
Silica gel-immobilized Eriochrome black-T	Zn^{2+}, Mg^{2+} from Ca^{2+}	106
CeO_2	Cd^{2+}, Co^{2+}, Cu^{2+}, Mn^{2+}, Pb^{2+}, and Zn^{2+} at pH $\geq$ 7	107
3-Hydroxy-2-methyl-1,4-naphthoquinone-immobilized on silica gel	Copper, cobalt, iron, and zinc	108

(Continued)

Table 5.4. *(Continued)*

SPE Material	Metals Sorbed	References
TiO$_2$ (Anatase)	At pH 8 quantitative sorption was detected for Bi, Cd, Co, Cr, Cu, Fe, Ge, In, Mn, Ni, Pb, Sb, Sn, Te, Tl, V, and Zn	109
A tetrameric calixarene with hydroxamic acid functional groups, supported on octadecyl-silica and XAD-4 resin	Quantitative enrichment of Cu(II), Zn(II), and Mn(II)	110
AnaLig sorbents, with predetermined molecular recognition chemistry for specific ions, using immobilized macrocycles	Pb from fresh and seawater, Cu, Ni from drug extract, Fe, NI from petroleum, Hg from water	111

The bulk extractant material may be placed in a column and the sample passed through using gravity or a vacuum. However, there are commercially available disposable cartridges that can be used to pass the sample through, either in a vacuum manifold or on the tip of a syringe. Disks composed of the sorbent trapped in a fiber mesh material are also available. Bonded silica sorbent particles held in a stable inert matrix of PTFE fibrils are used for the solid-phase extraction of analytes from complex sample matrices. A variety of functional groups, such as crown ethers, can be bonded to the silica surface to provide selective interactions. These filter disks have the advantage of rapid filtration of fairly large volumes of sample in a vacuum filtration apparatus. The analyte is then desorbed from the sorbent with an appropriate wash, usually acid, and is ready for analysis.

Commercially available devices for extraction utilize a variety of sorbent types. These include ion-exchange resins for both anions and cations, chelating resins, and organic-coated silica particles as used in HPLC columns. Functional groups in the coatings, such as methylpurazole, benzimidazole, and imidazole, give specificity for different heavy metals (e.g., Polyorgs, AnaLig).

5.5. SAMPLE PREPARATION FOR WATER SAMPLES

Water samples can be acid digested to determine total metal content, using procedures as described above. Trace metals can be determined in this way because the concentrations are brought to a sufficiently high level when the sample matrix is evaporated. Contamination is a constant problem, as it is difficult to evaporate large volumes of acidified water without obtaining high blanks. Very clean surroundings are necessary.

The separation of waterborne metals into filterable and nonfilterable categories may be done if desired. This requires filtration of the water sample as soon after collection as possible, and certainly before any acid is added to the sample. The metals in natural water samples are often sorbed on particulate matter in a larger quantity than is present in solution. The particles may be filtered out and analyzed separately by digesting the filter in acid. On the other hand, if the total metal content of the water sample is required, the entire unfiltered sample is acidified and digested.

A less time consuming method for the soluble fraction of the metal is to extract and concentrate the analytes from the water sample without evaporation. This process can be carried out using solid-phase extraction by exposing the sample to an ion-exchange material and sorbing the free metal ions from the sample. It can also be done by adding a soluble organic chelating agent to the sample and extracting the complexed analyte with an organic solvent.

An example of a method suitable for the determination of cadmium, cobalt, copper, iron, manganese, nickel, and zinc in water, using chelation and sample extraction, is as follows [113]. The sample is filtered through an acid-washed membrane filter as soon as possible after collection. It is then acidified with nitric acid for preservation until analysis. This will give the *soluble metal fraction*. If the total metal content is to be found, the sample is acidified and allowed to stand for 4 days with occasional shaking. Then it is filtered.

The filtered sample is neutralized with ammonia, and then buffered sodium diethyldithiocarbamate (SDDC) is added. The pH is adjusted to approximately 6, and the sample, in a separatory funnel, is shaken thoroughly. The analyte is then extracted twice with organic solvent. Nitric acid is added to the solvent, and it is evaporated to dryness on a hotplate. The residue is taken up in nitric and hydrochloric acids, and the dissolved residue is analyzed by AAS. It should be noted that the "soluble" metals are those that pass through the 0.45-μm filter, while "total metals" do not include those that are so tightly bound into the particles filtered out that they were not solubilized in the slow, mild acid leaching process to which the sample was exposed. For a true total metal analysis, an acid digestion would be required.

These methods must all be tested carefully, as the presence of a chelating agent, solid or dissolved, can shift the equilibrium between sorbed, complexed, and free ions in the sample. Metals in water samples can exist in several different forms. They can be sorbed on filterable particles, complexed with soluble humic materials or other soluble or colloidal materials or they can be free ions in solutions. The metals in each form are in equilibrium with those in the other phases. Depending on the kinetics of the system, the formation and extraction of a complex may change the distribution of metals in the various forms. However, in many filtered natural water samples, the determination of free ions and the total dissolved metal analyses give almost identical results.

The ion exchange or chelating resins may be packed into a column and the water passed through it slowly. The column is then eluted with an acid solution to recover the analytes. The ion-exchange properties of these resins varies widely with pH and the sample should be buffered to the correct pH before passing it through the column. The same process may be carried out in a batch mode, by adding a measured amount of fine grain resin to the sample and shaking or stirring for the requisite amount of time. The resin is then filtered out of the sample and analyzed as a slurry in a nitric acid solution. The slurry may also be allowed to settle and the clear supernatant solution analyzed.

Sorbent materials for solid-phase extraction (SPE) are available as powdered resins, but more convenient forms are in prepared disposable cartridges or filter mats. Some of these are shown in Figure 5.6. Cartridges are available that fit on the tip of a syringe, allowing a measured volume of water to be forced through. Cartridges with the packing at the bottom of an open container allow filling with sample and then application of vacuum for drawing the sample through. Filter mats have the ion exchange or chelating resin bound into a fibrous mat which can be used in a vacuum filtration apparatus. These sample preparation devices have the advantage of rapid throughput, but also provide less sample–resin contact, and breakthrough or saturation of the ion-exchange medium may be a problem. With all these methods, control of flow, pH, and total volume of sample and total amount of analyte loaded are all important. The capacity of the cartridge or filter mat must be determined and the breakthrough characteristics of the system understood to ensure that analyte is not lost in the concentration step.

After the sample is passed through the solid-phase extraction device, the analytes are removed with a small amount of acid and collected for analysis. The advantage of these systems is that the analyte is both separated from a large volume of matrix and concentrated into a small volume of acid, ready for analysis.

Several materials are used in the solid-phase extraction of metal ions from samples. Some have a silica base [114], which may be prepared by doping sol-gel glasses with appropriate complex-forming reagents or by coating these reagents on organic-coated silica beads which are available for reversed-phase HPLC column packing. Others are based on polymeric resins. Macrocyclic ligands are also used to obtain high selectivities for the desired analytes over interferences. Commercially available SPE membrane disks have been tested for removal of cesium, cadmium, and lead from acidic solutions containing substantially higher concentrations of aluminum, sodium, and potassium [115]. It was found that the analyte metals can be separated from these difficult solutions rapidly and efficiently.

5.6. PRECIPITATION METHODS

In some cases it is possible to perform a preseparation by selective precipitation of some components of the solution, either the matrix or the analytes. A different application of precipitation phenomena uses coprecipitation to concentrate an analyte by coprecipitating it with a more abundant species. An example of the application of both selective precipitation and coprecipitation is found in the preparation of high-purity silver samples for determination of trace impurities, including gold, cobalt, iron, mercury, zinc, and copper [116]. In this case, the silver matrix caused a great deal of interference in the analysis. To reduce this interference, the sample was dissolved in nitric acid and 3 M HCl was added. The precipitate of AgCl formed was filtered out, and the filtrate was evaporated to near dryness. More dilute HCl was added and a second filtration was carried out. This process reduced the silver to less than 0.12%. The trace metals in the filtrate could then be preconcentrated by any of several methods: ion exchange, sorption on activated carbon, sorption on an immobilized chelating agent, or coprecipitation. Because in this case the sample was to be analyzed by neutron activation, a small solid sample was desirable. The analytes were therefore coprecipitated at pH 4, with $Pb(NO_3)_2$ and ammonium pyrrolidine dithiocarbamate (APDC). APDC is an excellent chelating agent which forms stable chelates with more than 30 metals. The precipitate containing the analytes was filtered out of the solution, and the entire filter was subjected to NAA.

5.7. PREPARATION OF SAMPLE SLURRIES FOR DIRECT AAS ANALYSIS

Slurries, distribution of fine particles in a liquid, may be analyzed rather than clear solutions. Graphite furnace atomic absorbance analysis is particularly suited to this method. Slurries have also been introduced into ICP-AES and ICP-MS instruments. There are both advantages and concerns when slurries are used. The preparation is simple, so contamination can be lowered. No aggressive reagents are needed. It is relatively quick, and calibration can be done using aqueous standards. However, the particle size of the sample is critical. Particles should be 2 μm or smaller, and a proper high-solids inlet for the instrument should be used if samples are to be aspirated [117]. In addition, high analytical background signals are often found when high-solids samples are analyzed by AAS. A good background correction method should be employed.

Samples are prepared by weighing into a plastic bottle, with zirconia beads and the dispersant solution added. This is placed in a flask shaker for a few hours. Other laboratory mills and grinders are also suitable, but the hardness of both the grinder and the sample as well as the composition of the grinding surfaces must be considered, to be sure that the sample is not contaminated with the analytes of interest by particles ground off the surface of the mill. This is especially important in trace work. The dispersant solution usually includes a nonfoaming surfactant, to assist in keeping the slurry well dispersed. The maximum amount of solid sample in the slurry is usually kept to about 1%, to ensure that the slurry remains free flowing and nonviscous.

Keeping the slurry well homogenized while the sample is being taken for analysis is important. It is relatively easier to keep a sample homogenized if the particle density is similar to that of the solution. Particle size is also a major consideration. Finer particles are more easily suspended and kept in suspension than larger ones. If the sample consists of a matrix that contains several different types of material, slurrying can lead to significant error if the particles settle at different rates and the different types of material present contain substantially different concentrations of analyte. Vigorous shaking just prior to sampling may be sufficient for homogenizing readily slurried materials, and an ultrasonic probe can help with less easily mixed samples. For injection of samples into a graphite furnace AAS, autosamplers with built-in ultrasonic agitation are available. These keep the slurry well homogenized until the sample is aspirated into the syringe and injected into the graphite furnace.

5.8. HYDRIDE GENERATION METHODS

Some metals, for example, arsenic and selenium, are difficult to analyze by atomic absorption because their analytical wavelengths are subject to considerable interference. These metals, however, are readily converted to gaseous hydrides by treatment with strong reducing reagents such as sodium borohydride. Since the hydrides can be readily separated from the sample matrix, interferences are much reduced. A typical hydride generation AAS is

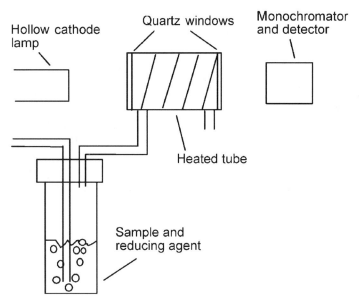

Figure 5.8. The analyte is converted into a gaseous hydride (e.g., As → AsH$_3$), which is purged into the heated furnace. There it decomposes into free As atoms for analysis.

shown in Figure 5.8. The hydrides are formed in a reaction chamber. They are purged into a heated cell in the AAS and are decomposed to free atoms for measurement.

The kinetics of the borohydride reduction of the various arsenic and selenium species differ and must be taken into account. The different oxidation states give different analytical sensitivities, and different interferences are found for each. The optimal pH for reduction to hydride of selenium and arsenic in different oxidation states is also different. Therefore, unless speciation is to be done, it is best to bring all the analyte to a common oxidation state before reaction with borohydride. For example, arsenic acid, containing As(V), is considerably slower to be reduced than is As(III). The As(III) is instantaneously reduced, giving a rapid injection of hydride into the instrument. Therefore, it is best to ensure that all the arsenic is in the As(III) state before adding the borohydride. This is accomplished by digesting the original sample with acid, yielding As(V). This is quantitatively reduced to As(III) with sodium or potassium iodide. The sample is then ready for reaction with the sodium borohydride.

Se(VI) is not readily reduced by sodium borohydride, and samples containing it must be prereduced. Samples containing organic selenium compounds or complexes may require digestion with an oxidizing agent, either alkaline hydrogen peroxide or potassium permanganate. Excess permanganate is removed by reaction with hydroxylamine, and any chlorine formed is removed by boiling in an open container. These digestions leave the selenium in the Se(VI) oxidation state. It is then necessary to reduce this to Se(IV) by boiling with HCl. Se(IV) is rapidly reduced, giving a sharp injection of hydride into the instrument.

Samples of organic matter such as foods may be dry ashed before analysis. Magnesium oxide can be added as an ashing aid. The ashed sample is taken up in HCl solution, and the oxidation state of the analyte is adjusted. For example, KI would be added to convert As(V) to As(III). Then a 3% $NaBH_4$ solution in 0.5% NaOH is added and the hydride flushed into the instrument, AAS or ICP, for analysis.

5.9. COLORIMETRIC METHODS

Fairly rapid and simple analyses can be performed on solutions using a variety of colorimetric reagents. These are reagents that are more or less specific for certain metals and will produce a solution, usually colored, whose absorbance at a particular wavelength is related to the concentration of the analyte. Preparation of samples for colorimetric analysis often requires buffering or pH adjustment of the sample solution and sometimes a treatment to oxidize or reduce the analyte to bring it to the correct oxidation state to react with the reagent. The color-forming reagent is added and the solution diluted to known volume. Specific conditions of temperature and time are usually specified to ensure complete reaction. Some reagents for these determinations are listed in Table 5.5.

Table 5.5. Some Colorimetric Reagents for Metals

Metal	Color Development Reagent	Wavelength (nm)
Cr(VI)	1,5-Diphenylcarbazide	540
Pb	Dicyclohexyl-18-crown-6-dithizone	512
Fe(III)	Thiocyanate	460
Fe(II)	Pyrocatecol violet	570
Cd	Iodide and malachite green	685
Mn	Oxidize to permanganate with KIO_4	525
Mg, Al	Precipitate with 8-hydroxyquinoline, dissolve in acid for determination of hydroxyquinoline	590
Cu	Dithizone	510
Co, Ni, Cu, Zn	4-(2-Pyridylazo)resorcinol	

5.10. METAL SPECIATION

In natural waters, soils, and sediments, trace metals are present in a wide range of chemical forms, in both the solid and dissolved phases. The dissolved phase comprises the hydrated ions, inorganic and organic complexes, together with species associated with heterogeneous colloidal dispersions and organometallic compounds. In soils and sediments, metals may be sorbed to clay particles, bound up in iron or manganese hydroxy compounds, or in calcium oxide minerals, as well as being sorbed to organic solids. Metals may be present in more than one valence state. The solid phase contains elements in a range of chemical associations, ranging from weak adsorption to binding within the mineral matrix. These species are able to coexist, although they are not necessarily in thermodynamic equilibrium with one another. Some common species of selected metals are listed in Table 5.6.

Speciation of an element is the identification and determination of individual physical–chemical forms of that element in the environment, which together make up its total concentration in a sample [118]. Knowledge of the forms that an element can have is of primary importance because their toxicity, mobility, bioavailability, and bioaccumulation depend on the chemical species [119–121]. Speciation studies are thus of interest to chemists doing research on the toxicity and chemical treatment of waters, soils, and sediments, to biologists inquiring about the influence of species on animals and plants, and to geochemists investigating the transport of elements in the environment. It has been noted that there is a difference in the way in which determination of organic compounds and metals is commonly perceived. An analysis of a sample for organics normally entails the determination of

Table 5.6. Selected Metals and Some of Their Chemical Species

Metal	Chemical Forms
Aluminum	Al_2O_3, $Al(OH)_3$, $Al_2Si_2O_5(OH)_4$, $KAlSi_3O_8$, $Al_2Si_2O_5(OH)_4$
Arsenic	AsH_2, AsO_2^-, AsO_4^{3-}, $H_2AsO_3^-$
Cadmium	Cd^{2+}, $Cd(Cl)^+$, and other Cl complexes up to $CdCl_4^{2-}$, CdS
Calcium	$CaCO_3$, Ca^{2+}, CaO, $Ca(OH)_2$
Chromium	$Cr(OH)^{2+}$, CrO_4^{2-}, CrO_3^{3-}
Cobalt	Co^{2+}, Co^{3+}, $Co(OH)_3$, CoAsS, $CoAs_2^-$
Iron	Fe_3O_4, Fe_2O_3, Fe^{2+}, Fe^{2+}, FeS_2, $Fe(OH)_3$
Lead	Pb^{+2}, $PbOH^+$, $Pb_4(OH)_4^{4+}$, Pb-organic complexes
Mercury	Hg_2^+, Hg^{2+}, $HgOH^+$, CH_3Hg, $HgCl_4^{2-}$, $HgCl^-$
Selenium	Se(IV), Se(VI)
Uranium	U_3O_8, $K_2(UO_2)_2(VO_4)_2 \cdot 8H_2O$, UO_2^{2+}

specific compounds, while in most metal determinations, the compounds are destroyed and only the elements are measured [122]. The situation has probably evolved because of the availability of sensitive instrumentation for total metal analyses, and its ease of use. Applicable regulations, in addition, have tended to cast these methods in stone. New methods that examine the chemical species in which metals exist are coming into use only slowly. The equilibrium and kinetic instability of many of these species lends an additional level of difficulty to the actual speciation of metal-bearing chemical species [123].

Measurements of the total concentration of microelements in environmental samples provide little information on their bioavailability. In water, most studies of the susceptibility of fish to heavy metal poisoning have shown that the free hydrated metals ions are the most toxic [118–120]. Ions that are strongly complexed or associated with colloidal particles are usually considered to be less toxic.

Unpolluted fresh water or seawater usually contains low concentrations of toxic metal species, since most of the toxic ions are adsorbed on inorganic or organic particles. Anthropogenic pollution, however, may add metal to water in a toxic form or may cause metals already present to be converted into more toxic forms. For example, acidification of natural waters may release previously bound ions, increasing their toxicity. Changes in the oxidation state of metal ions may also have a profound effect on their bioavailability and toxicity. In soils, metals are present naturally, but dredge spoils or mine tailing waste dump areas bring metal-bearing materials into contact with the biosphere.

The most important reason for speciation measurements is to identify the metal species that are likely to have adverse effects on living organisms, including bacteria, algae, fish, and mammals. The interaction of metals with biological organisms is highly dependent on chemical form. Some species may be able to bind chemically directly with proteins and enzymes, others may adsorb on cell walls, and still others may diffuse through cell membranes and exert a toxic effect. Toxicity is organism dependent and occurs when an organism is unable to cope either by direct use, storage, or excretion with additional metal concentration.

The impact of some metals is strongly related to their chemical form rather than to their total concentration. For instance, arsenic is generally toxic in both its As(III) arsenite and As(V) arsenate forms, but is nontoxic in its organic forms, such as arsenocholine. Mercury, on the other hand, is toxic in all forms but is substantially more toxic as methyl mercury than it is in the elemental state. Chromium in the Cr(III) oxidation state is less toxic and less soluble than it is in the Cr(VI) state.

Therefore, the total metal concentration is inadequate to describe a sample fully. Speciation of the metals present is sometimes required. This is a developing field and presents difficulties to the analyst. The metal concentration present may be near the level of detection for the analysis. If this is further subdivided into several different species, greater analytical sensitivity is required. Further, the different species are usually in equilibrium with each other in the sample. This requires less aggressive extraction processes, as the overall equilibrium should be disturbed as little as possible.

5.10.1. Types of Speciation

Speciation can be defined functionally, operationally, or chemically. A *functional definition* is one which specifies the type of role that the element may play in the system from which the sample was taken. For instance, a functional definition might be "that mercury which can be taken up by plants" or "iron that can be absorbed from a pharmaceutical." This definition is probably closest to what the end user of the information really wants to know but is the most difficult for the analytical chemist to determine. Other than growing the plant in the contaminated water or soil sample and analyzing the plant tissue, or doing feeding studies on the pharmaceutical, it is nearly impossible to obtain this information experimentally.

An *operational definition* is considerably more practical. Operationally determined species are defined by the methods used to separate them from other forms of the same element that may be present. The physical or chemical procedure that isolates the particular set of metal species is used to define the set. "Metals extracted from soil with an acetate buffer" is an operational definition of a certain class. "Lead present in airborne particles of less than 10 μm" is another. In water analyses, simply filtering the sample before acidification can speciate the analytes into dissolved and insoluble fractions. These procedures are sometimes referred to as *fractionation*, which is probably a more properly descriptive term than *speciation*, as *speciation* might imply that a particular chemical species or compound is being determined. When such operational speciation is done, careful documentation of the protocol is required, since small changes in procedure can lead to substantial changes in the results. Standardized methods are recommended, as results cannot be compared from one laboratory to another unless a standard protocol is followed [124]. Improvements in methodology must be documented and compared with the currently used standard methods to produce useful, readily interpretable information.

Finally, particular chemical species can be determined in some cases, as when arsenic content is separated into As(III), As(V), monomethyl arsonic acid, and dimethyl arsinic acid using ion-exchange chromatography. Chemical speciation is sometimes possible but is often very difficult. If the metals present in a sample are to be separated into their different forms, the initial separations are often carried out during the sample preparation.

5.10.2. Speciation for Soils and Sediments

Sieving a soil or sediment will allow determination of metals in each particle size range so that the distribution of the element can be determined. Species defined as biologically active, such as free hydrated ions, may be separated from the bulk of a water sample by exposing the sample to an ion-exchange resin or a chelating resin that will sorb only the species of interest. Then the sorbed species may be removed from the resin by elution with acid or may be determined by analysis of the resin. Even the distinction between the soluble and insoluble forms of an element in a water sample can be considered a type of speciation. The separation of these species is carried out by passing the sample through a membrane filter, usually of 0.45 µm pore size. Both the filtered sample and the material retained on the filter can be analyzed, giving the soluble metal and that present in, or bound to, particles.

Speciation of metal content in solids can be accomplished during the extraction process by subjecting the sample to successive extractions with progressively more aggressive solvents, or by extracting different subportions of sample with the different solvents. It has been shown that it is more difficult to obtain comparable results when using sequential extractions rather than individual extractions of subsamples with different extractants [124]. Some applications of extractions with different solvents are extraction with:

- Aqua regia for a pseudo total metal content, used to determine suitability of sludge for soil application
- Acetic acid or chelating agents such as EDTA to determine trace metal mobility and availability of metals for plant uptake
- Weak extractants such as calcium chloride or nitrate for plant uptake studies, soil fertility studies, and risk assessment
- Ammonium chloride or acid oxalate for differentiation of lithogenic and anthropogenic origins of some soil constituents

The analysis of samples extracted with various solvents will provide information on the most easily removed metal species, the less available, and the most refractory metal content, which is dissolved only by the strongest acid extractants. There are at least a dozen different published speciation schemes for metals in soils and sediments. Many are based on the pioneering work by Tessier et al. [125]. Most include releasing metals from carbonates and hydrous oxides with acids, and an oxidation step to destroy organic materials and sulfides. However, some schemes put the oxidation step early in the scheme, on the theory that there may be an organic coating on the surface of the sample particles. A three-step method, which is being developed in Europe, attempts to divide metals into an easily mobilized fraction, extracted with water or neutral electrolyte, a slowly mobilized fraction extracted by ethylenediaminetetraacidic acid (EDTA) or other chelating agent, and an immobile fraction found using digestion in hydrofluoric acid.

More elaborate methods involve more steps. Sequential extraction schemes attempt to remove metals from soil or sediment in classes, depending on which component of the sample they are bound to, and how readily solubilized or mobilized they are. Sequential extraction procedures use the least aggressive reagent for the first extraction. Solutions of ammonium acetate or magnesium chloride at pH 7 are useful to remove the metals bound to clay particles by simple ionic attraction. Dissolution of carbonates present in the sample by treatment with weak acids releases metals contained in the carbonate minerals. A reducing agent, hydroxylamine hydrochloride, for example, will solubilize iron and manganese oxides and hydroxides, releasing metals bound in, or coprecipitated with, these species. An oxidizing agent such as hydrogen peroxide will destroy organic material, recovering metals complexed with humic substances. Finally, the residue is extracted with strong acid to recover most of the remaining metals in the sample.

The availability of the analytes for uptake by plants, for transport through the soil, and for dissolution into water can be estimated from a well-studied speciation scheme. Risk assessment for disposal of wastes in landfills or for land disposal of dredge spoils or sewage sludges requires knowledge not only of the total metal content but also of the content in each separate fraction to begin to understand how the metals will act in the environment. Table 5.7 summarizes the methods available for speciation of metals in samples.

5.10.3. Sequential Schemes for Metals in Soil or Sediment

One of the classic methods for speciation of metals in soils was developed by Tessier et al. [125], and this method is still substantially in use, although several modifications of this method have also been published. Again, it is important to stress that even small modifications of the methods used can have substantial effects on the data obtained.

The first extraction of easily exchangeable metal ions is done at room temperature with a 1 M solution of $MgCl_2$, at pH 7 for 1 hour with continuous stirring. Extraction 2 in the sequence removes the metals bound to carbonate minerals by extraction with acetate buffer at pH 5. The extraction is complete within 5 hours for fine sediments but might take longer for

Table 5.7. Methods for Pretreatment of Samples for Speciation of Metals

Physical techniques: based on size, density, or charge of the species	Centrifugation Ultrafitration Dialysis Gel filtration chromatography
Chemical techniques: based on redox, complexation, and/or adsorption properties	Oxidative destruction of organics Liquid extraction Ion exchange and adsorbent resins Voltammetry
Species-specific techniques: applicable to particular species	Potentiometry with specific electrodes GC and/or hydride generation, HPLC
Bioassays: influence of the metal ion on the growth or inhibition of organisms	
Comprehensive speciation schemes: combinations of different methods of speciation	

samples of large grain size or those that contain much carbonate. In that case, adjustment of the pH during extraction might be necessary.

The third fraction, that bound to iron and manganese oxides and hydroxides, is extracted with 0.04 M hydroxylamine hydrochloride, in 25% v/v acetic acid at 96°C. This extraction takes 6 hours. For removal of metals from the organic matter present, the fourth fraction, the samples are taken up in 0.02 M HNO_3 and an equal volume of 30% hydrogen peroxide is added. The samples are digested at 85°C for 2 hours, a second portion of H_2O_2 is added, and the digestion is continued for 3 hours more. Then NH_4OAc is added to prevent readsorption of the metals onto the oxidized sample particles. The remaining metals, the final fraction, are dissolved in a 5:1 mixture of HF and $HClO_4$. Two sequential digestions are done, with evaporation to near dryness between.

A similar scheme [126] using the same operationally defined fractions determines 15 elements: Be, Ca, Co, Cr, Cu, Fe, K, Li, Mn, Ni, P, Pb, Ti, V, and Zn, with recoveries of 83 to 110%. The extractants used for each fraction are shown in Table 5.8. The Measurements and Testing Programme of the European Commission (EC) has a recommended method for sequential extractions. The method distinguishes four fractions, as shown in Table 5.9.

5.10.4. Speciation for Metals in Plant Materials

The study of mechanisms of metal uptake in plants often requires knowledge of the specific compounds and complexes in which the metals are present in

Table 5.8. Sequential Extraction Scheme

Fraction	Reagent (for an initial 1.0-g sample)
Exchangeable	8 mL $MgCl_2$, pH 7, agitated at room temperature for 20 min
Bound to carbonates	8 mL NaOAc, adjusted to pH 5 with HAc, agitated at room temp for 5 h
Bound to Fe, Mn oxides	20 mL 0.04 M hydroxylamine hydrochloride, in 25% v/v acetic acid, for 6 h at 96°C
Bound to organic materials or sulfides	3 mL 0.02 M HNO_3, 5 mL H_2O_2 (30%) for 2 h at 85°C; additional 3 mL H_2O_2 added and extraction continued for 3 h; after cooling, 5 mL 3.2 M NH_4OAc in 20% H_2O_2 added and agitated for 30 min
Residual metals	4 mL 70% HNO_3, 3 mL 60% $HClO_4$, 15 mL 40% HF at 90°C for 6 h, 120°C for 10 h, 190°C for 6 h; residue taken up in 5 mL 5 M HCl at 70°C for 1 h

the plant. This is a challenging process, requiring a method that selectively destroys the plant matrix without attacking the metal-bearing complex [128]. For example, washing cells with buffered EDTA solution may remove metal ions reversibly bound to cell walls. Fairly stable organometallic species such as organotin, alkyl lead, or methyl mercury may be separated from a proteinaceous matrix by digesting the matrix with tetramethylammonium hydroxide [129]. Hydrolysis with an aqueous 25% solution of tetramethylammonium hydroxide will dissolve polypeptides and proteins, freeing the stable metal-containing species. However, metals bound to or incorporated in the proteins will not be recovered in their original state. Inorganic alkali

Table 5.9. EC Sequential Extraction Method for 0.5-g Initial Sample

	Samples are centrifuged, filtered, and the residue rinsed with 10 mL DI water between extractions
Exchangeable	10 mL 0.11 M acetic acid at room temperature, with constant agitation
Reducible	20 mL 0.1 M hydroxylamine hydrochloride, acidified to pH 2 with HNO_3; agitated at room temp. for 16 h
Oxidizable	5 mL 8.8 M H_2O_2, 1 h with occasional agitation; heat at 85°C 1 h, evaporate to a few mL, add 5 mL more H_2O_2, evaporate to near dryness, cool and add 25 mL 1 M ammonium acetate, acidify to pH 2, and agitate for 16 h
Residual	Digest with HF, HNO_3, $HClO_4$

Source: Ref. 127.

or acid digestions will not preserve even the stable covalently bonded organometallic compounds.

Another approach is to degrade a biological matrix through use of enzymes. Pectolytic enzymes will break down most pectic polysaccharides and may release metal complexes from the solid parts of plant materials [130]. The resulting digests can be filtered and analyzed by chromatographic methods. Gas chromatography is used for volatile or derivatized organometallics, and HPLC is also commonly used. Size exclusion chromatography can be useful for determination of metals bound to macromolecules. In all chromatographic separations, a detector that responds to the metal being determined is of great advantage. In many cases, ICP/MS or ICP/AES are interfaced to the chromatographic system for this purpose [128].

5.10.5. Speciation of Specific Elements

Some metals are of particular interest in environmental samples, and specific methods for these have been developed. Arsenic, chromium, and mercury are all important in this respect, having very differing toxicities in different forms.

Speciation of Arsenic

Arsenic exists in water primarily as arsenious acid, H_3AsO_3 or As(III), as arsenic acid, H_3AsO_4 or As(V), or organic As compounds such as monomethylarsonic acid and dimethylarsinic acid. The organic compounds are generally found at low levels and are not as toxic as the inorganic species. Therefore, speciation for arsenic tends to be directed toward the determination of As(III) and As(V). The samples in aqueous solution are separated using an anion-exchange column chromatographic column, with the separated anions analyzed by GFAAS.

Speciation of Chromium

The two major species in which chromium exists are the Cr(III) cation and the chromate and dichromate anions, Cr(VI). The Cr(VI) is considered to be considerably more dangerous in environmental samples of soil or water because it is more toxic and also considerably more soluble, and therefore more mobile in the environment. A pH 8, 0.05 M ammonia buffer is used with ultrasonication to extract the Cr(VI) from soil samples, and then the Cr(VI) is sorbed from the extract on an anion-exchange column (Dowex 1-X8). The sorbed analyte is eluted with 10 mL of pH 8, 0.5 M ammonia buffer. The Cr(III) is determined by difference from the total Cr measurement [131]. A cationic ion-exchange resin can also be used to sorb the Cr(III) ions.

Speciation of Mercury

Methyl mercury is of much greater concern when health effects are considered, as it is much more toxic than ionic mercury or free mercury. Methyl mercury is also much more likely to be bioaccumulated, leading to serious contaminations, especially of fish. The speciation for mercury can be accomplished by derivatizing the methyl mercury and Hg^{2+} with sodium tetraethylborate, $NaBEt_4$. The volatile MeHgEt, from methyl mercury, and $HgEt_2$, from Hg^{2+}, species formed are purged from the sample solution and separated in a GC column. An atomic emission spectrometer is used as a detector.

Samples of freeze-dried fish tissues are extracted with 25% tetramethylammonium hydroxide using a microwave digester. After extraction, the pH is adjusted to 4 with acetic acid buffer. A 1% solution of $NaBEt_4$ is added, with some hexane. The solution is shaken for 5 minutes. A fresh portion of $NaBEt_4$ is added, the shaking is repeated, and finally, a third portion is added and allowed to react. The sample is centrifuged and an aliquot of the supernatant hexane is taken for injection into the GC [132].

5.11. CONTAMINATION DURING METAL ANALYSIS

Metals are often determined at trace levels. Since there are substantial amounts of many different metals present in airborne particles, contamination of samples by dust fallout from the laboratory atmosphere can be a serious concern. When an analytical sample has been reduced to a few microliters for injection into the analytical instrument, a single dust particle landing in the sample can make a significant and substantial difference in the results. Because initial preparation steps on real-world samples often involve dust-producing steps such as grinding, sieving, sample homogenization, and division, these steps should be carried out in a separate area of the laboratory. The sample preparation for trace-level samples should be done in a clean area, with the samples protected from atmospheric contamination as much as possible. Samples being analyzed for low trace levels of common metallic elements often require the use of clean-room technologies to allow satisfactory blanks to be obtained.

Strong acids used in digestion can also be a source of contamination when substantial quantities of acid are evaporated in the process of digestion. Ultrapure acids are required in wet digestion processes if traces of metals are to be determined. The container in which the process takes place is another possible source of contamination, and proper cleaning must be carried out. The contamination can arise from the material of the container itself or from carryover from previous use. In trace work some recommended methods of preventing container contamination are segregation of apparatus used for trace work, strict cleaning processes, and selection of proper materials for the digestion vessels.

5.12. SAFE HANDLING OF ACIDS

A word should be said about safety. The mixtures of strong acids and oxidizers used in sample digestion are inherently dangerous. They quickly burn skin, and the danger of an explosive reaction is present. Safety goggles, gloves, and protective aprons should always be used. Rapid reaction with the sample can lead to explosive conditions. Although efficiency requires that digestions take little time, too-rapid reaction of a strong acid with a finely powdered sample can cause a violent reaction. Samples should be treated with a small amount of diluted acid before the stronger acid is added, to begin the reaction more slowly. The strongest oxidizers, such as perchloric acid, should be added only after the majority of the oxidizable material has been decomposed with nitric acid. The final solution containing perchloric acid should never be evaporated completely to dryness directly, but should be evaporated and diluted several times. When hydrofluoric acid is used, special precautions should be taken. HF is easily absorbed into skin and cannot be washed off entirely. It will cause serious, continuing, slow-healing burns. A calcium gluconate ointment should always be on hand if HF is used, and it should be applied immediately to any HF burn.

In addition, the practice of adding acid to water with constant stirring should be observed. When acid mixtures are prepared, only the quantity to be used should be prepared, as these may not be safe to store. Finally, pressure relief valves should be provided to any sealed container in which a digestion is to take place. One should be aware, however, that some analyte can be lost as droplets when these valves vent. This is one of the advantages of the pressure-monitored microwave digestion system. In this, the pressure is controlled by modulating the input power, so venting is avoided.

REFERENCES

1. D. Barclay and G. LeBlanc, *Am. Lab. News*, Oct., p. 12 (2000).
2. K. W. Barnes, *At. Spectrosc.*, **19**(2), 31–39 (1998).

3. P. Zbinden and D. Aubry, *At. Spectrosc.*, **19**(6), 214–219 (1998).

4. K. J. Lambie and S. J. Hill, *Analyst*, **123**, 103r–133r (1998).

5. J. Liu, R. E. Sturgeon, and S. N. Willie, *Analyst*, **120**, 1905 (1995).

6. G. Schnitzer, A Soubelet, C. Testu, and C. Chafey, *Mikrochim. Acta*, **119**, 199 (1995).

7. S. Baldwin, M. Deaker, and W. Maher, *Analyst*, **119**, 1701 (1994).

8. B. S. Sheppard, D. T. Heitkemper, and C. M. Gaston, *Analyst*, **119**, 1683 (1994).

9. B. Sures, H. Taraschewski, and C. Haug, *Anal. Chim. Acta*, **311**, 135 (1995).

10. J. E. Tahan, V. A. Granadillo, J. M. Sanchez, H. S. Cubillan, and R. A. Romero, *J. Anal. At. Spectrom.*, **8**, 1005 (1993).

11. A. M. Yusof, N. A. Rahman, and A. K. H. Wood, *Biol. Trace Elem. Res.*, **43/45**, 239 (1994).

12. R. Mizushima, M. Yonezawa, A. Ejima, H. Koyama, and H. Satoh, *Tohoku J. Exp. Med.*, **178**, 75 (1996).

13. Q. Yang, W. Penninckx, and J. Srneyersverbeke, *J. Agric. Food Chem.*, **42**, 1948 (1994).

14. M. Arruda, M. Gallego, and M. Valcarcel, *J. Anal. At. Spectrom.*, **10**, 501 (1995).

15. M. Arruda, M. Gallego, and M. Valcarcel, *J. Anal. At. Spectrom.*, **11**, 169 (1996).

16. M. J. Campbell, G. Vermeir, R. Dams, and P. Quevauviller, *J. Anal. At. Spectrom.*, **7**, 617 (1992).

17. M. G. Heagler, A. G. Lindow, J. N. Beck, C. S. Jackson, and J. Sneddon, *Microchem. J.*, **53**, 472 (1996).

18. R. Jaffe, C. A. Fernandez, and J. Alvarado, *Talanta*, **39**, 113 (1992).

19. M. Navarro, M. Lopez, M. C. Lopez, and M. Sanchez, *Anal. Chem. Acta*, **257**, 155 (1992).

20. S. A. Pergantis, W. R. Cullen, and A. P. Wade, *Talanta*, **41**, 205 (1994).

21. N. Xu, V. Majdi, W. D. Ehmann, and W. R. Markesbery, *J. Anal. At. Spectrom.*, **7**, 749 (1992).

22. M. Navarro, H. Lopez, M. C. Lopez, and M. Sanchez, *J. Anal. Toxicol.*, **16**, 169 (1992).

23. A. Vaisanen, R. Suontamo, J. Silvonen, and J. Rintala, *Anal. Bioanal. Chem.*, **373**, 93–97 (2000).

24. H. Garraud, M. Robert, C. R. Quetel, I. Szpunar, and O. F. X. Donard, *At. Spectrosc.*, **17**, 183 (1996).

25. P. Hocquellet, *Analusis*, **23**, 159 (1995).

26. K. I. Lambie and S. I. Hill, *Anal. Chim. Acta*, **334**, 261 (1996).

27. P. Quevauviller, I. L. Imbert, and M. Olle, *Mikrochim. Acta*, **112**, 147 (1993).

28. G. S. B. Ianuzzi, F. I. Krug, and M. A. Z. Arruda, *J. Anal. At. Spectrom.*, **12**, 375 (1997).

29. R. Sturgeon, S. Willie, B. Methven, and J. Lam, *J. Anal. At. Spectrom.*, 10, 981 (1995).

30. D.-H. Sun, J. K. Waters, and T. P. Mawhinney, *J. Anal. At. Spectrom.*, 12, 675 (1997).

31. L. Aduna de Paz, A. Alegria, R. Barbera, R. Farre, and M. J. Lagarda, *Food Chem.*, 58, 169 (1997).

32. G. Damkroger, M. Grote, and E. Jansen, *Fresenius' J. Anal. Chem.*, 357, 817 (1997).

33. S. C. Edwards, C. L. Macleod, W. T. Corns, T. P. Williams, and J. N. Lester, *Int. J. Environ. Anal. Chem.*, 63, 187 (1996).

34. A. Lasztity, A. Krushevska, M. Kotrebai, and R. M. Barnes, *J. Anal. At. Spectrom.*, 10, 505 (1995).

35. J. Murphy, P. Jones, and S. Hill, *J. Spectrochim. Acta B*, 51, 1867 (1996).

36. J. McLaren, B. Methven, J. Lam, and S. Berman, *Mikrochim. Acta*, 119, 287 (1995).

37. A. Krushevska, R. M. Barnes, and C. Amarasiriwaradena, *Analyst*, 118, 1175 (1993).

38. C. M. Tseng, A. de Diego, F. M. Martin, D. Amouroux, and O. F. X. Donard, *J. Anal. At. Spectrom.*, 12, 743 (1997).

39. T. I. Gluodenis and I. F. Tyson, *J. Anal. At. Spectrom.*, 7, 301 (1992).

40. T. I. Gluodenis and I. F. Tyson, *J. Anal. At. Spectrom.*, 8, 697 (1993).

41. S. I. Haswell and D. Barclay, *Analyst*, 117, 117 (1992).

42. J. C. Schawnloffel and W. F. Siems, *Rev. Sci. Instrum.*, 67, 4321 (1996).

43. P. H. Towler and J. D. Smith, *Anal. Chim. Acta*, 292, 209 (1994).

44. H. Vanhoe, *J. Trace Elem. Electrolytes Health Dist.*, 7, 131 (1993).

45. P. Schramel and S. Hasse, *Fresenius' J. Anal. Chem.*, 346, 794 (1993).

46. P. V. A. Prasad, J. Arunachalam, and S. Gangadharan, *Electroanalysis*, 6, 589 (1994).

47. E. Stryjewska, S. Rubel, and I. Szynkarezuk, *Fresenius' J. Anal. Chem.*, 354, 128 (1996).

48. D. W. Bryce, A. Izquierdo, and M. D. Luque de Castro, *Analyst*, 120, 2171 (1995).

49. V. Ducros, D. Riffieux, N. Belin, and A. Favier, *Analyst*, 119, 1715 (1994).

50. M. Krachler, H. Radner, and K. I. Irgolic, *Fresenius' J. Anal. Chem.*, 355, 12045 (1996).

51. R. M. Sah and R. Miller, *Anal. Chem.*, 64, 230 (1992).

52. M. Feinberg, C. Suard, and J. Ireland-Ripert, *Chem. Int. Lab. Syst.*, 22, 37 (1994).

53. M. A. B. Pougnet, N. G. Schnautz, and A. M. Walker, *S. Afr. J. Chem.*, 25, 86 (1992).

54. J. L. Burguera, M. Burguera, A. Matousek de Abel de la Cruz, N. Anez, and O. M. Alarcon, *At. Spectrosc.*, **13**, 67 (1992).

55. R. Chakraborty, A. K. Das, M. L. Cervera, and M. de la Guardia, *Fresenius' J. Anal. Chem.*, **355**, 43 (1996).

56. S. Evans and U. Krahenbuhl, *Fresenius' J. Anal. Chem.*, **349**, 454 (1994).

57. D. Chakraborti, M. Burguera, and J. L. Burgueru, *Fresenius' J. Anal. Chem.*, **347**, 233 (1993).

58. E. I. Gawalko, T. W. Nowicki, I. Babb, and R. Tkachuk, *J. AOAC Int.*, **80**, 379 (1997).

59. E. S. Beary, P. I. Paulsen, L. B. Iassie, and I. D. Fassett, *Anal. Chem.*, **69**, 758 (1997).

60. S. Fridlund, S. Littlefield, and I. Rivers, *Commun. Soil Sci. Plant Anal.*, **25**, 933 (1994).

61. I. Matejovic and A. Durackova, *Commun. Soil Sci. Plant Anal.*, **25**, 1277 (1994).

62. H. I. Reid, S. Greenfield, T. E. Edmonds, and R. M. Kapdi, *Analyst*, **118**, 1299 (1993).

63. C. B. Rhoades, Jr., *J. Anal. At. Spectrom.*, **11**, 751 (1996).

64. C. J. Park and J. K. Suh, *J. Anal. At. Spectrom.*, **12**, 573 (1997).

65. S. Wu, Y.-H. Zhao, X. Feng, and A. Wittmeier, *J. Anal. At. Spectrom.*, **12**, 797 (1997).

66. D. C. Baxter, R. Nichol, and D. Littlejohn, *Spectrochim. Acta B*, **47**, 1155 (1992).

67. C. Cabrera, Y. Madrid, and C. Camara, *J. Anal. At. Spectrom.*, **9**, 1423 (1994).

68. C. Cabrera, C. Gallego, M. Lopez, and M. L. Lorenzo, *J. AOAC Int.*, **77**, 1249 (1994).

69. V. E. Negretti de Bratter, P. Brotter, A. Reinicke, G. Schulze, W. O. L. Alvarez, and N. Alvarez, *J. Anal. At. Spectrom.*, **10**, 487 (1995).

70. L. Dunemann and M. Meinerling, *Fresenius' J. Anal. Chem.*, **342**, 714 (1992).

71. P. Quevauviller, I. L. Imbert, and M. Olle, *Mikrochim. Acta*, **112**, 147 (1993).

72. G. S. Banuelos and S. Akohoue, *Commun. Soil Sci. Plant Anal.*, **25**, 1655 (1994).

73. H. Lippo and A. Sarkela, *At. Spectrom*, **16**, 154 (1995).

74. I. Matejovic and A. Durackova, *Commun. Soil Sci. Plant Anal.*, **25**, 1277 (1994).

75. J. S. Alvarado, T. J. Neal, L. L. Smith, and M. D. Erickson, *Anal. Chim. Acta*, **322**, 11 (1996).

76. S. Prats-Moya, N. Grane-Teruel, V. Berenguer-Navarro, and M. L. Martin-Carratala, *J. Agric. Food Chem.*, **45**, 2093 (1997).

77. V. Carbonell, A. Morales-Rubio, A. Salvador, M. de La Guardia, J. L. Burguera, and M. Burguera, *J. Anal. At. Spectrom.*, **7**, 1085 (1992).

78. M. de la Guardia, V. Carbonell, A. Morales-Rubio, and A. Salvador, *Talanta*, **40**, 1609 (1993).

79. G. Heltai and K. Percsich, *Talanta*, **41**, 1067 (1994).

80. K. W. Barnes and E. Debrah, *At. Spectrosc.*, **18**, 41 (1997).

81. L. H. I. Lajunen and I. Piispanen, *At. Spectrosc.*, **13**, 127 (1992).

82. E. V. Alonso, A. G. Detorres, and J. M. C. Pavon, *J. Anal. At. Spectrom.*, **8**, 843 (1993).

83. J. L. Burguera and M. Burguera, *J. Anal. At. Spectrom.*, **8**, 235 (1993).

84. T. Yamane and K. Koshino, *Anal. Chim. Acta*, **261**, 205 (1992).

85. K. Lambie and S. I. Hill, *Analyst*, **120**, 413 (1995).

86. Y. Soon, Y. Kalra, and S. A. Abboud, *Commun. Soil Sci. Plant Anal.*, **27**, 809 (1996).

87. P. Schramel and S. Hasse, *Fresenius' J. Anal. Chem.*, **346**, 794 (1993).

88. M. Mateo and S. Sabale, *Anal. Chim. Acta*, **279**, 273 (1993).

89. E. Stryjewska, S. Rubel, and A. Skowron, *Chem. Anal. (Warsaw)*, **39**, 491 (1994).

90. C. Cabrera, M. Lorenzo, and M. Lopez, *J. AOAC Int.*, **78**, 1061 (1995).

91. M. Navarro, M. C. Lopez, and H. Lopez, *J. AOAC Int.*, **75**, 1029 (1992).

92. P. Zbinden and D. Aubry, *At. Spectrosc.*, **19**(6), 214–219 (1998).

93. Z. Kowalewska, E. Bulska, and A. Hulanicki, *Fresenius' J. Anal. Chem.*, **362**, 125–129 (1998).

94. L. Jorhem, *J. AOAC Int.*, **83**(5), 1204–1211 (2000).

95. B. B. Kebbekus and S. Mitra, *Environmental Chemical Analysis*, Stanley Thornes Publishers, Cheltenham, Gloucestershire, England, 1998.

96. C. M. Wai, S. Wang, and J.-J. Yu, *Anal. Chem.*, **68**, 3516–3518 (1996).

97. F. A. Chimilenko and L. V. Baklanova. *J. Anal. Chem.*, **53**(8), 784–786 (1998).

98. M. Kumar, D. P. S. Rathore, and A. K. Singh, *Fresenius' J. Anal. Chem.*, **370**(4), 377–382 (2001).

99. H. F. Koch and D. M. Roundhill, *Sep. Sci. Technol.*, **36**(1), 137–143 (2001).

100. Pei Liang, Yongchao Qin, Bin Hu, Chunxiang Li, Tianyou Peng, and Zucheng Jiang, *Fresenius' J. Anal. Chem.*, **368**(6), 638–640 (2000).

101. J. P. Bernal, E. Rodriguez de San Miguel, J. C. Aguilar, and G. Salazar de Gyves, *J. Sep. Sci. Technol.*, **35**(10), 1661–1679 (2000).

102. J. Seneviratne and J. A. Cox, *Talanta*, **52**(5), 801–806 (2000).

103. K. Pyrzynska and Z. Jonca, *Anal. Lett.*, **33**(7), 1441–1450 (2000).

104. P. de Magalhaes Padilha, L. A. de Melo Gomes, C. C. Federici Padilha, J. C. Moreira, and N. L. Dias Filho, *Anal. Lett.*, **32**(9), 1807–1820 (1999).

105. Z. Aneva, S. Stamov, and I. Kalaydjieva, *Anal. Lab.*, **6**(2), 67–71 (1997).

106. M. E. Mahmoud, *Talanta*, **45**(2), 309–315 (1997).

107. E. Vassileva, B. Varimezova, and K. Hadjiivanov, *Anal. Chim. Acta*, **336**(1/3), 141–150 (1996).

108. B. S. Garg, J. S. Bist, R. K. Sharma, and N. Bhojak, *Talanta*, **43**(12), 2093–2099 (1996).

109. E. Vassileva, I. Proinova, and K. Hadjiivanov, *Analyst*, **121**(5), 607–612 (1996).

110. S. Hutchinson, G. A. Kearney, E. Horne, B. Lynch, J. D. Glennon, M. A. McKervey, and S. J. Harris, *Anal. Chim. Acta*, **291**(3), 269–275 (1994).

111. R. M. Izatt, J. S. Bradshaw, R. L. Bruening, and M. L. Breuning, *Am. Lab.*, Dec., pp. 28C–28M (1994).

112. Y. Lin, G. E. Fryxell, H. Wu, and M. Englehard, *Environ. Sci. Technol.*, **35**, 3962 (2001).

113. S. J. Haswell, ed., *Atomic Absorption Spectroscopy: Theory, Design, and Application*, Elsevier, New York, 1991.

114. E. Morosanova, A. Velikorodny, and Y. Zolotov, *Fresenius' J. Anal. Chem.*, **361**, 305–308 (1998).

115. R. M. Izatt, J. S. Bradshaw, and R. L. Bruening, *Pure Appl. Chem.*, **68**(8), 1237–1241 (1996).

116. M. Y. Shiue, Y. C. Sun, J. J. Yang, and M. H. Yang, *Analyst*, **124**, 15–18 (1999).

117. N. J. Miller-Ihli, *Fresenius' J. Anal. Chem.*, **345**, 482–489 (1993).

118. S. Caroli, ed., *Element Speciation in Bioinorganic Chemistry*, Wiley, New York, 1996.

119. G. F. Batley, ed., *Trace Elements Speciation: Analytical Methods and Problems*, CRC Press, Boca Raton, FL, 1989.

120. A. M. Ure and C. M. Davidson, eds., *Chemical Speciation in the Environment*, Blackie, Glasgow, 1995.

121. S. J. Hill, *Chem. Soc. Rev.*, **26**, 291–298 (1997).

122. O. F. X. Donard and J. A. Caruso, *Spectrochem. Acta B*, **53**, 157–163 (1998).

123. R. Lobinski, *Spectrochem. Acta B*, **53**, 177–185 (1998).

124. P. Quevauviller, *Trends Anal. Chem.*, **17**, 289–298 (1998).

125. A. Tessier, P. G. C. Campbell, and M. Bisson, *Anal. Chem.*, **51**, 844 (1979).

126. X. Li, B. J. Coles, M. H. Ramsey, and I. Thornton, *Chem. Geol.*, **124**, 109–123 (1995).

127. A. M. Ure, C. M. Davidson, and R. P. Thomas, in P. Quevauviller, E. A. Maier, and B. Griepink, eds., *Quality Assurance for Environmental Analysis*, Elsevier, Amsterdam, 1995, pp. 505–523.

128. M. N. V. Prasad and J. Hagemeyer, *Heavy Metal Stress in Plants*, Springer-Verlag, Berlin, 1999, p. 356.

129. R. J. A. Van Cleuvenberg, D. Chakraborti, and F. Adams, *Anal. Chim. Acta,* **228,** 77–84 (1990).

130. D. S. Forsyth and J. R. Iyengar, *J. Assoc. Off. Anal. Chem.,* **72,** 997–1001 (1989).

131. J. Wang, K. Ashley, E. R. Kennedy, and C. Neumeister, *Analyst,* **122,** 1307–1312 (1997).

132. I. R. Pereiro, A. Wasik, and R. Lobinski, *J. Anal. At. Spectrom.,* **13,** 743–747 (1998).

133. D. McGrath, *Talanta.,* **46,** 439–448 (1998).

CHAPTER

6

SAMPLE PREPARATION IN DNA ANALYSIS

SATISH PARIMOO

Aderans Research Institute, Inc., Philadelphia, Pennsylvania

BHAMA PARIMOO

*Department of Pharmaceutical Chemistry, Rutgers University College of Pharmacy,
Piscataway, New Jersey*

6.1. DNA AND ITS STRUCTURE

It is a well-known fact that the genetic information in an organism is stored and passed on from parents to offspring in the form of *deoxyribose nucleic acid* (DNA). DNA was first isolated more than 100 years ago from salmon sperm by the Swiss biochemist Miescher, who called it nucleic acid. Much later it was realized that, in fact, there are two classes of nucleic acids in all cells. After Feulgen introduced a specific stain for DNA more than seven decades ago, DNA was recognized to be located largely in the nucleus of animal and plant cells. In contrast, the other nucleic acid, *ribonucleic acid* (RNA), occurs mostly in the cytoplasm. Bacteria, which lack a nucleus, possess both DNA and RNA within the cytoplasm itself. DNA and RNA are major components of all cells and together make up from 5 to 15% of their dry weight. It was not until early 1944 that Avery, MacLeod, and McCarty established that the genetic material is indeed DNA [1]. Prior to that, proteins were believed to be carriers of genetic material.

Although the chemical nature of single-stranded DNA was well known by 1950, it was Watson and Crick who finally solved the structure of double-stranded DNA in 1953 and proposed a double helix model of DNA based on x-ray diffraction data [2]. This concept eventually earned them a Nobel prize in 1962. They proposed that DNA consists of two independent strands, each having alternate pentose sugar (deoxyribose) and phosphate units linked via ester linkage (phosphodiester) as part of their backbone

Sample Preparation Techniques in Analytical Chemistry, Edited by Somenath Mitra
ISBN 0-471-32845-6   Copyright © 2003 John Wiley & Sons, Inc.

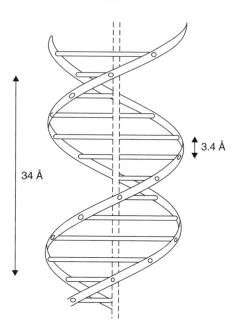

Figure 6.1. Structure of DNA double helix. DNA double helix of 20 Å diameter with its two strands twisted around each other. The nitrogen-containing bases are perpendicular to the helical axis and about 3.4 Å apart from their next base of the same strand. The bases from the two strands opposite each other form hydrogen bonds and help to stabilize the helix. The helix makes a complete turn every 34 Å. (Reproduced from *Textbook of Biochemistry with Clinical Correlations*, T. M. Devlin, ed., Wiley, New York, 1982.)

(Figure 6.1). The two strands are wound around one another every 34 Å and are held together by complementary pairing of nitrogenous bases which are covalently linked to position 1 of the pentose sugar (deoxyribose), these bases being positioned every 3.4 Å (Figures 6.1 and 6.2). A unit of a sugar, a nitrogen heterocyclic base, and phosphoric acid is known as a *nucleotide*. An important feature of the DNA double helix is that its two polymeric strands are *antiparallel*; that is, the orientation of one strand is opposite to the other with respect to 3′–5′ internucleotide phosphodiester bonds (Figure 6.2). The DNA double helix may be visualized as interwinding around a common axis of two right-handed helical polynucleotide strands. Four types of nitrogenous bases exist in DNA: adenine (A), cytosine (C), thymine (T), and guanine (G). DNA's double-helical configuration is governed by the complementary pairing rule of A pairing with T and G pairing with C. Purines comprise A and G, whereas pyrimidines comprise C and T. The pairing of bases occurs via hydrogen bonding, that is, the sharing of two protons between an A–T pair and of three protons between a G–C pair (Figure 6.2). The double-helical structure of DNA is maintained by hydrogen bonding between base pairs as well as stacking interactions between successive bases. The stacking interactions of bases exist as a consequence of the hydrophobic properties of purine or pyrimidine rings. All essential features of DNA, such as its antiparallel nature of strands, specific base pairing, as well as sequence

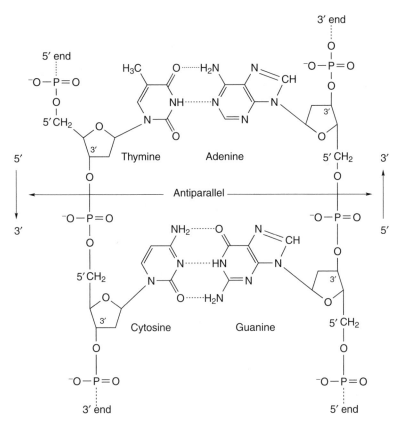

Figure 6.2. Molecular architecture of DNA. Each strand of DNA is composed of alternating pentose sugar (deoxyribose) and phosphate moieties linked to each other via phosphodiester linkage. The first carbon position of the sugar is attached to one of the four nitrogenous bases (A, T, G, or C). The two strands are in opposite orientation to each other with respect to a 5' or 3' phosphate group attached to the sugar moiety. Cytosine (C) pairs with guanine (G) via three hydrogen bonds, and adenine (A) pairs with thymine (T). (Reproduced from *Textbook of Biochemistry with Clinical Correlations*, T. M. Devlin, ed., Wiley, New York, 1982.)

of bases along the strand is maintained through DNA replication when a cell divides. A deviation from a particular sequence due to an error during DNA replication can be fatal to a cell and may even lead to emergence of a cancerous cell.

A *gene* is a segment of DNA whose sequence of four nucleotides (ATCG) along a particular length of DNA ultimately determines which RNA or protein it is going to make. The sequence of nucleotide bases within a gene is transcribed into RNA (ribosomal RNA or transfer RNA) or translated as a triplet code into amino acids of a protein via an RNA intermediate (mRNA) by the cellular protein synthesis machinery. As a result of the human genome sequencing project, it is known that a unicellular yeast cell has about 6000 different genes, whereas a human body has under 40,000 genes, although not all genes are active in all human cells. Together, all genes constitute only about 3% of human genome, and the remaining DNA, although not coding for genes, may have some important functions that are unknown at the present. The sizes of genes are variable within a cell, ranging from a few hundred base pairs of DNA to hundreds of thousands of nucleotide base pairs of DNA.

6.1.1. Physical and Chemical Properties of DNA

Bacteria such as *E. coli*, contain 0.01 pg of DNA per cell and their DNA is about 1 mm in length with about 4 million nucleotide pairs. In contrast, a typical human cell has about 6 pg of DNA, which has a total length of 174 cm. Thus, a human body, which consists of trillions of cells, has anywhere between 10 and 20 billion miles of double-helix DNA. Due to this enormous length of DNA, tremendous compaction of DNA within a cellular nucleus is achieved by interaction of DNA with proteins that form chromosomes in nucleus. The entire human genome in a human cell consists of 23 pairs of chromosomes with over 3 billion base pairs from each parent. The smallest and the largest human chromosome have 50 million and 263 million bases pairs of DNA, respectively. The DNA content and the length of DNA are variable from species to species, as shown in Table 6.1. DNA is also very light—1 μm of DNA weighs only 3.26×10^{-18} g.

DNA is extremely sensitive to mechanical shearing forces because of its huge length. Routine laboratory manipulations such as pipetting can break DNA into shorter fragments. However, once isolated, DNA is a relatively stable macromolecule and can remain so when stored dry or under ethanol

Table 6.1. DNA Content of Various Species

Type of Cell	Organism	DNA/Cell (pg)	Number of Nucleotide Pairs (millions)
Bacteriophage	T4	2.4×10^{-4}	0.17
Bacterium	*E. coli*	4.4×10^{-3}	4.2
Fungi	*N. crassa*	1.7×10^{-2}	20
Erythrocyte (RBC)	Chicken	2.5	2000
Leukocyte (WBC)	Human	3.4	6000

Source: B. Lewis, *Gene Expression*, 2nd ed., Vol. 2, Wiley, New York, 1980, p. 958.

in a freezer. DNA is a polymer of nucleotides and its size and hence its molecular weight can be estimated from a variety of techniques, such as equilibrium centrifugation in a density gradient solution (cesium chloride), electron microscopy, or electrophoresis in agarose gels. Agarose gel electrophoresis is the most commonly used tool for DNA size estimation.

With the exception of a few bacteriophages, which possess single-stranded DNA, bacteria and higher organisms have double-stranded DNA. Viruses contain DNA or RNA as the genetic material but require a host cell for propagation. DNA is soluble in water and can be precipitated with ethanol or isopropanol in the presence of salt (0.1 M NaCl). Precipitated and dried DNA looks like white fibers. Due to its large molecular weight, it can take several hours for DNA to go into solution. DNA is acidic because phosphate groups in the sugar–phosphate backbone are fully ionized at any pH above 4. These phosphate groups are on the outer periphery of the double helix, exposed to the aqueous environment, and impart negative charge to DNA and bind divalent cations such as magnesium and calcium. Bacterial DNA is associated with polycationic amines such as spermine and spermidine, which confer on DNA both stability and flexibility. Even in dilute solutions, DNA is very viscous because of its structure and huge length in relation to its diameter. Hence it displays true solute behavior in only very dilute solutions. However, if DNA strands are separated by physical or chemical means, it leads to a decrease in its viscosity.

Ultracentrifugation in sucrose gradients can be used to determine the molecular weight of DNA by comparing it to a DNA sample of known size and sedimentation coefficient. Equilibrium sedimentation in cesium chloride gradient can distinguish DNA samples based on their densities. For example, single-stranded DNA is denser than double-stranded DNA in CsCl gradients. In addition, the relative abundance of G–C nucleotide base pairs compared to A–T base pairs makes DNA denser, and hence information about G–C content of DNA of different organisms can be obtained by their buoyant density measurements.

Double-helical DNA in solution can undergo strand separation or denaturation as a consequence of extremes of pH, heat, or exposure to chemicals such as urea or amides. Decrease in viscosity, increase in absorbance at 260 nm (hyperchromic effect), decrease in buoyant density, or negative optical rotation indicates denaturation of DNA. The denaturation process disrupts only noncovalent interactions between the two strands of DNA. Since G–C base pairs are held together by three hydrogen bonds in contrast to two for an A–T base pair, A–T rich DNA is easily denatured compared to G–C rich DNA (Figure 6.3). Electron microscopy can detect these A–T-rich regions in a DNA molecule since they form bubblelike structures. Hence the temperature of melting (T_m) of DNA increases in a linear fashion with

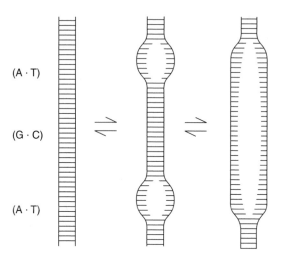

Figure 6.3. Effect of heat on DNA. At high temperature and low ionic strength, the two strands of DNA at A–T-rich regions fall apart, first forming bubble structures along the length of the DNA. As the temperature increases, the size of the bubble increases and the G–C regions also fall apart. Extreme pH ranges also cause DNA denaturation. (Reproduced from *Textbook of Biochemistry with Clinical Correlations*, T. M. Devlin, ed., Wiley, New York, 1982.)

the content of G–C base pairs. In the initial stages of denaturation, DNA strands are not completely separated. The single-stranded regions in this partially denatured molecule assume random conformation. If given favorable conditions for renaturation, the two strands will readily rewind to re-form a complete duplex DNA. However, on complete denaturation, the two strands fall apart and renaturation under favorable circumstances is then a very slow process. Slow cooling of heat-denatured DNA in appropriate ionic strength and temperature is necessary for its renaturation. Strong acidic conditions can cause single-strand breaks within DNA.

6.1.2. Isolation of DNA

Once purified, DNA is a fairly stable polymer if stored appropriately. Since living cells contain many other complex biomolecules besides DNA, methods exist that allow the isolation of DNA in pure form. More details on this topic are presented in Chapter 8. Routine methods of DNA isolation in solution, however, cause some unavoidable shearing of DNA due to hydrodynamic shear forces, and as a result, the average size of isolated DNA is about 100 to 200 kilobases (kb). The basic steps in DNA isolation involve cell disruption and lysis by treatment with detergents, removal of cellular proteins by either enzymatic digestion with a protease or extraction with

Tissue homogenization

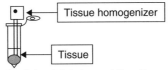

Cell lysis and removal of cellular debris by centrifugation

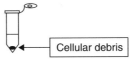

Removal of proteins by phenol-chloroform extraction of cell lysate

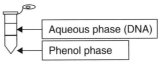

Recovery of DNA by alcohol precipitation

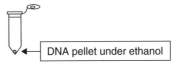

Removal of RNA and polysaccharides if necessary

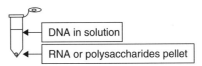

Quality assessment and quantification of DNA

Figure 6.4. Schematic representation of DNA isolation process. After tissue homogenization and cell lysis, the sample is extracted with phenol and the DNA remains in the aqueous phase. DNA is recovered from the aqueous phase by ethanol precipitation.

phenol–chloroform, and precipitation of DNA by a mixture of ethanol and salt. A schematic representation of the isolation process is shown in Figure 6.4. DNA precipitates like a fibrous material that can be collected and further purified by selective enzymatic digestion of any contaminating RNA that may have coprecipitated during the isolation process. The presence of detergents, divalent metal chelating agents, and stable proteases such as proteinase K during the isolation process prevents any hydrolysis of DNA by cellular nucleases and ensures isolation of intact DNA. Selected examples of DNA isolation are presented in this chapter.

6.2. ISOLATION OF DNA FROM BACTERIA

DNA from bacteria such as *Escherichia coli* can be isolated from either small or large volumes of bacterial culture [3]. For small-scale preparation, a 3-mL bacterial culture in LB medium (1% Bacto tryptone, 0.5% Bacto yeast extract, 1% NaCl, pH 7) grown to saturation from a single bacterial colony is chilled in ice and centrifuged in two 1.5-mL Eppendorf tubes in a micro-centrifuge at $14,000 \times g$ for 2 minutes at room temperature. Each bacterial pellet is resuspended in 580 μL of TE [10 mM tris(hydroxymethyl)amino-methane-Cl, pH 8, 1 mM ethylenediaminetetraacetic acid (EDTA)] buffer by vortexing until the pellet is uniformly dispersed. To digest cellular proteins, 6 μL of 10 mg/mL proteinase K solution is added and mixed by gentle brief vortexing. Proteinase K is a sturdy protease and works under harsh con-ditions. Addition of 15 μL of 20% sodium dodecyl sulfate (SDS) solution causes denaturation of proteins and lysis of cellular membranes. Incubation at 50°C for 1 to 2 hours causes proteolytic digestion of the lysate. The sam-ple is processed for phenol and chloroform extraction and ethanol precipi-tation of DNA as described in the next section.

For large-scale preparation of DNA, a single freshly grown bacterial col-ony is inoculated in 5 mL of sterile LB medium and incubated overnight at 37°C on a shaker. This seed culture is in turn inoculated in 500 mL of sterile LB medium and grown at 37°C as described earlier until the late log phase (OD ~ 2 of an aliquot at 600 nm). Some other bacteria may require different culture and incubation conditions. After growing the bulk culture, bacteria are chilled in ice and harvested by centrifugation at $4000 \times g$ for 15 minutes at 4°C. The bacterial pellet is washed by resuspending bacterial pellet in ice-cold 100 mL of TE and centrifuged again as earlier. After discarding the supernatant, the bacterial pellet is processed for cell lysis by resuspending the pellet in 45 mL of TE buffer and addition of 5 mL of 10% SDS and 0.5 mL of 20 mg/mL proteinase K. After mixing thoroughly and incubation at 50°C for 1 to 2 hours, the sample is processed for phenol chloroform extraction and DNA precipitation by ethanol.

6.2.1. Phenol Extraction and Precipitation of DNA

Phenol and chloroform extractions remove other macromolecules, such as proteins and lipids. The phenol should be of good quality and buffered for nucleic acid extraction. It should be free of any oxidized products that can potentially degrade DNA. After phenol extraction, the DNA is recovered by alcohol precipitation [4,5].

Preparation of Buffered Phenol

Although high-grade ready-to-use buffered phenol is available commercially from several vendors, such as Amresco (Solon, OH), Invitrogen (Carlsbad, CA), and others, it is possible to purify phenol by melting solid phenol in a flask and double distislling it in a chemical hood. It is kept in a liquefied state by the addition of water and storing it under an inert gas such as nitrogen or argon in brown bottles. However, phenol is very corrosive and should be handled properly in a chemical fume hood and disposed off appropriately. For use in DNA extraction, liquefied phenol should be adjusted to pH $\sim$ 8 by the addition of 25 mL of 1 M Tris-Cl buffer, pH 8 to 500 mL of phenol, and mixing by stirring for 10 minutes at room temperature. Once the stirring is stopped, the organic and aqueous phase will separate. After removal of the top aqueous phase, the pH of the phenol phase (lower) can be checked with a pH paper. If the pH is still acidic, extraction with the Tris buffer can be repeated. Finally, the buffer layer is replaced with 50 mL of water and phenol and stored in a brown bottle at 4°C. If phenol has been stored for several months in a refrigerator, its color may turn light pink and the pH may be acidic; this phenol should not be used. Some people prefer to use 0.1% of 8-hydroxyquinoline in phenol as an antioxidant. After the addition of 8-hydroxyquinoline, the phenol is yellowish in color and can also be useful in identifying phenol and aqueous phases during DNA extraction. However, the color imparted to phenol (yellow) due to 8-hydroxyquinoline can mask the pink oxidized color in an old sample. Nevertheless, it is good practice to check the pH of the phenol by pH paper before use, especially if it has been stored for a long time.

Phenol–Chloroform Extraction of DNA Sample

The first extraction is carried out with phenol alone by the addition of an equal volume of phenol to the sample containing DNA, mixing by inverting the tube several times in order to mix the phases. Care is taken to be gentle in mixing the samples since vigorous shaking can shear DNA. On the other hand, if phenol mixing is inefficient, the extraction may not be effective. The samples are centrifuged for 10 minutes at 3000 × g for large-volume samples, or in 1.5-mL Eppendorf tubes for small-volume samples in a microcentrifuge for 5 minutes at 12,000 × g in order to separate phases. After centrifugation, the top aqueous phase, containing the DNA, is taken out with a Pasteur pipette in a clean tube and reextracted with an equal volume of phenol–chloroform–isoamyl alcohol mixture (25:24:1) as earlier by gentle inversion and then centrifugation. The top aqueous phase is transferred to a new tube and reextracted until the interphase does not contain any visible precipitate of cellular debris. After the final extraction, the aqueous phase is extracted with an equal volume of chloroform followed by centrifugation. The DNA in the top phase can be recovered after precipitation with ethanol.

An alternative to phenol-based methods for DNA extraction is the use of guanidinium salts and detergents for homogenization of tissues followed by alcohol precipitation of DNA [13]. A commercially available reagent (DNAzol) available from Invitrogen (Carlsbad, CA) uses a proprietary formulation of guanidine–detergent lysis solution for DNA isolation from tissues, including blood.

Recovery of DNA by Ethanol Precipitation

DNA is precipitated from aqueous solutions by ethanol or isopropanol in the presence of salt. The amount of alcohol and salt depends on the type of salt that one wishes to use (Table 6.2). The type of salt used depends largely on downstream applications for which the DNA is to be used. For example, precipitation in the presence of ammonium acetate removes small molecules such as nucleotides, and the DNA can be used for many enzymatic reactions. On the other hand, the phosphorylating enzyme, T_4 kinase, is inhibited by ammonium ions, and unless DNA is reprecipitated in the presence of salts other than ammonium acetate, the phosphorylation reaction may be inhibited. For most routine purposes, alcohol precipitation of DNA with sodium acetate is preferred over sodium chloride because of the higher solubility of the acetate salt in ethanol. Selection of isopropanol or ethanol is more of a convenience than a rule. Although isopropanol precipitation requires an equal volume of isopropanol for the precipitation of DNA, that with ethanol requires 2 volumes and hence can increase the total volume

Table 6.2. Alcohol Precipitation of DNA

Salt	Salt Stock Solution	Final Concentration[a]
Sodium chloride	5.0 M	0.1 M
Sodium acetate	3.0 M (pH 7)	0.3 M
Ammonium acetate	10.0 M	2.0 M

[a] Final salt concentration in DNA solution before the addition of alcohol.

for centrifugation to collect DNA. After addition of alcohol, the method for collection of DNA again depends on the amount of DNA. For large amounts of DNA (milligram quantities), DNA is spooled out on a glass rod or the tip of a Pasteur pipette after the addition of ethanol and inserting a Pasteur pipette/glass rod in the solution and making a swirling motion inside the solution. DNA will spool out onto the Pasteur pipette or glass rod. The advantage of this method is that it does not involve centrifugation, and contamination with RNA is reduced to a minimum. The DNA is rinsed with 70% ethanol and then dried in air. If, however, the amount of DNA is small (micrograms), its is advisable to mix the solution after addition of alcohol by inverting several times so that the alcohol mixes uniformly and then centrifuging the DNA after incubating at least 1 hour at 4°C. Although 80% of DNA will precipitate by incubation for short times in the cold, higher yields are obtained by incubating in the cold (-20°C) overnight. Small sample volumes of DNA can be centrifuged in 1.5-mL Eppendorf tubes in a microcentrifuge at maximum speed ($14,000 \times g$) for 10 minutes at room temperature or at 4°C. For large sample volumes, Sorvall centrifuges can be used and samples centrifuged at $10,000 \times g$ for 15 minutes at 4°C. After centrifugation, the supernatant is carefully removed and the DNA pellet is rinsed with 70% ethanol to remove salts and then recentrifuged as earlier. The DNA pellet is dried in air. Larger quantities may be dried under vacuum. The idea of this drying is to remove ethanol, but excess drying should be avoided, as fully dehydrated DNA is difficult to dissolve. The dried DNA is dissolved in TE and left at 37°C for overnight or until it goes into solution completely. Vortexing should be avoided, as it shears DNA.

Recovery of DNA from dilute solutions (<10 μg DNA/mL) can be enhanced by the addition of an appropriate carrier substance before ethanol precipitation. Molecular biology–grade glycogen, which is available commercially, is added to the sample at a concentration of 20 to 40 μg/mL before the addition of ethanol. The dilute DNA solutions can also be concentrated by repeated extraction with *sec*-butanol (mixing the solution by inverting several times) followed by centrifugation at $3000 \times g$ for 5 minutes at room temperature, discarding the butanol phase each time. Each extraction will extract water out of the solution and hence concentrate the DNA. Finally, DNA can be ethanol precipitated.

Although ethanol precipitation of DNA is recommended for dilute DNA solutions, DNA in an appropriate concentration (ca. 0.5 μg/mL) can be dialyzed against several changes of TE until the $OD_{270 \text{ nm}}$ of the dialysate is less than 0.05. The advantages of dialysis method are that DNA need not be dried and dissolved, which takes 1 to 2 days, and that there is lesser shearing of DNA.

6.2.2. Removal of Contaminants from DNA

Some biological sources of DNA such as bacteria and plants contain a large amount of undesirable biomolecules that coprecipitate with DNA in the presence of salt and ethanol. These include polysaccharides and RNA. These contaminants may make it difficult to dissolve the DNA or interfere in its subsequent use. The amount of RNA contamination is variable, depending on the tissue. For example, when isolating DNA from yeast, a large amount of RNA gets coprecipitated along with DNA. Since RNA also absorbs at 260 nm, it can lead to overestimation of DNA concentration in a sample.

Polysaccharides

DNA from those sources rich in polysaccharides can be purified by the addition of CTAB (hexadecyltrimethylammonium bromide) before chloroform–isoamyl alcohol extraction [6]. After adjusting NaCl concentration to 0.7 M with 5 M NaCl in a DNA solution solution (ca. 0.05 mg/mL in TE), CTAB solution (10% CTAB in 0.7 M NaCl) is added so that the final concentration of CTAB is about 1%. The samples are incubated at 65°C for 10 minutes. It is important to keep the salt at a concentration of greater than 0.5 M so that the DNA does not precipitate as a CTAB–DNA complex. After the addition of an equal volume of chloroform–isoamyl alcohol (24:1 by volume) and gentle but complete mixing, the phases are separated by centrifugation for 10 minutes at 2000 × g. The interphase will appear as a white precipitate of CTAB–polysaccharides/protein complex. The aqueous phase containing DNA is transferred with a wide-bore pipette to a tube, and the CTAB chloroform–isoamyl alcohol extraction can be repeated until no cellular material is visible at the interphase. The DNA from the aqueous phase is precipitated with ethanol as described earlier, and any residual CTAB is washed with 70% ethanol washes.

RNA

Residual RNA in a DNA preparation can be removed by treatment with ribonuclease (RNase). RNase A, which is free of DNase, is available commercially, or the contaminant DNase in the crude RNase A solution can be heat inactivated by heating RNase A solution (10 mg/mL in 10 mM Tris-Cl, pH 7.5, 15 mM NaCl) at 100°C for 15 minutes [4]. DNA solution in TE at a concentration of at least 100 μg/mL is treated with RNase to a final concentration of 1 μg/mL followed by incubation at 37°C for 1 hour [3]. RNase can be removed by phenol and phenol–chloroform extraction. After the RNase treatment, removal of broken-down ribonucleotides can be accomplished by reprecipitation of DNA in the presence of ammonium acetate and isopropanol. Alternatively, the DNA can be dialyzed against TE using the appropriate size cutoff membrane. A combination of Rnase A and Rnase T_1 is preferred, as it can lead to better fragmentation of RNA than can RNase A alone. In addition, if highly purified DNA is desired, DNA can be purified by centrifugation in CsCl–ethidium bromide [3]. DNA solution in TE is adjusted to a concentration of 50 to 100 μg/mL and then for every 4 mL of the DNA solution, 4.3 g of CsCl and 200 μL of 10 mg/mL ethidium bromide is added, mixed, and centrifuged in an ultracentrifuge as described in Section 6.3. Alternatively, commercially available anion-exchange matrix columns can be used that are available as part of genomic DNA isolation kits from various vendors, such as Qiagen (Chatsworth, CA), Promega (Madison, WI), Invitrogen (Carlsbad, CA), Stratagene (La Jolla, CA), and Clontech (Palo Alto, CA), among others. The kits are advantageous since the use of phenol–chloroform is avoided.

6.3. ISOLATION OF PLASMID DNA

Plasmids are autonomously replicating small DNA molecules present in a variety of bacterial species. They are double-stranded, closed circular, and supercoiled DNA molecules that range in size from 1 kb to more than 200 kb in length. Many plasmids contain genes that confer antibiotic resistance to the bacterial host. In nature, plasmid DNA gets transferred from one bacterial host to another, and as a result, may transfer drug resistance to the recipient host. The commercial importance of plasmid DNA lies in the fact that it can readily be propagated, isolated, and manipulated in a test tube (genetically engineered) to accommodate (clone) any foreign DNA, which will then grow as part of the plasmid DNA once introduced back into bacteria. A large number of genetically engineered plasmid DNA vectors are available commercially for inserting and replicating foreign DNA. These recombinant plasmid DNA vectors can be grown in an appropriate host of interest, such as bacteria, yeast, or mammalian cells. Plasmid DNA can be introduced into various host cells by appropriate chemical treatments or with the help of electric current (electroporation). In expression-based vectors, specific proteins can be produced by growing plasmid DNA bearing foreign DNA insert (recombinant plasmid DNA) in appropriate hosts (Figure 6.5). Among the first proteins that were produced commercially by this technology were growth hormone and insulin.

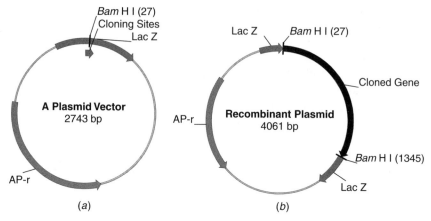

Figure 6.5. Maps of circular double-stranded vector plasmid and its derivative recombinant plasmid. (*a*) Map of a 2743-base pair plasmid DNA vector that carries a gene for antibiotic (ampicillin) resistance. The ampicillin resistance (*Amp-r*) gene allows selective growth of those bacteria that carry the plasmid DNA with the gene. The *Lac Z* gene codes for a β-galctosidase peptide, as a result of which bacterial colonies containing the plasmid are blue when grown in a medium containing appropriate substrates and an inducing agent for the enzyme. Within the *Lac Z* gene there are cloning sites wherein a foreign DNA segment can be cloned at one of the several restriction sites. (*b*) The same vector DNA, in which a foreign gene (1318 base pairs) has been cloned at the Bam HI restriction site using T_4 DNA ligase enzyme. Cloning within the *Lac Z* gene region disrupts the *Lac Z* gene and causes recombinant clones to be distinguished by their white color from those without the foreign DNA, which are blue.

6.3.1. Plasmid DNA Preparation

As mentioned earlier, plasmid DNA is present as small supercoiled circular double-stranded DNA in bacteria. In this conformation, plasmid DNA is more resistant to alkaline denaturation than is host genomic DNA. Hence disruption of cells bearing plasmid DNA followed by the addition of alkali and subsequent neutralization and centrifugation leads to the precipitation of denatured genomic DNA and proteins, whereas plasmid DNA remains in the solution [7]. Plasmid DNA can be recovered from the supernatant by ethanol precipitation and may be further purified [3,4].

In order to grow bacteria for plasmid DNA isolation, typically, a single bacterial colony is inoculated into 5 mL of LB medium. If the plasmid codes for antibiotic resistance, that antibiotic is added to the LB medium. The bacterial culture is grown to saturation overnight at 37°C on a shaker. The cells are centrifuged in 1.5-mL Eppendorf tubes in microcentrifuge for 2 minutes at maximum speed. After removal of the supernatant, the pellet is resuspended in 100 μL of GTE solution (50 mM glucose/25 mM Tris-Cl, pH

8.0/10 mM EDTA). After the addition of 200 μL of NaOH/SDS solution (0.2 M NaOH/1% SDS), the sample is mixed by inversion several times. After 5 minutes, 150 μL of 3 M potassium acetate solution (pH 4.8) is added, mixed by inversion several times, and incubated in ice for 5 minutes. After centrifugation for 5 minutes in microcentrifuge at 14,000 $\times$ g for 5 minutes, the clear supernatant is extracted with phenol, phenol–chloroform, and chloroform. The plasmid DNA is then precipitated after the addition of 2 volumes of ethanol. After incubation at room temperature for 5 minutes, the DNA is centrifuged in microcentrifuge by centrifugation for 10 minutes at room temperature. The pellet is then washed with 70% ethanol and dried under vacuum.

For large quantities, the bacterial culture is grown by transferring 1 mL of active overnight grown culture into 500 mL of LB medium containing the appropriate antibiotic in a 2-L flask and grown until saturation overnight. For low-copy plasmids such as pBR 322, the bulk culture is grown to an OD of 0.6 at 600 nm, and then chloramphenicol (from a stock of 34 mg/mL in 50% ethanol) is added to a concentration of 35 μg/mL for 18 hours with shaking. The bacterial pellet from a 500-mL culture is resuspended in 4 mL of GTE solution. Residual bacterial clumps may cause inefficient lysis and reduce the plasmid yield; hence complete resuspension of bacterial pellet is desirable. Egg white lysozyme in GTE solution is added to a final concentration of 25 mg/mL in order to disrupt the bacterial cell wall. After the addition of 10 mL of NaOH/SDS solution, the lysed material is mixed by inverting the tube a couple of times and incubating in ice for 10 minutes. Then 7.5 mL of 3 M potassium acetate solution (pH 4.8) is added and mixed thoroughly until the viscosity is reduced. The samples are kept in ice for 10 minutes. After centrifugation for 10 minutes in an SS-34 rotor in a Sorvall centrifuge at 13,000 rpm, the clear supernatant is passed through four layers of cheesecloth to remove any flocculent precipitate. To digest RNA, the samples are incubated for 60 minutes at room temperature after the addition of DNase-free RNase to a final concentration of 10 μg/mL. After extracting the samples successively with phenol, phenol–chloroform–isoamyl (25:24:1), and chloroform–isoamyl alcohol (24:1), the plasmid DNA from the aqueous phase is precipitated by the addition of $\frac{1}{4}$ volume of 10 M ammonium acetate and 2 volumes of ethanol. The samples are held for 10 minutes in ice before centrifugation. After washing the pellet with 70% ethanol, the pellet is dissolved in 2 mL of TE buffer.

6.3.2. Purification of Plasmid DNA

The plasmid DNA prepared as described above still has some broken RNA fragments and polysaccharides and may contain some bacterial DNA. Various approaches exist [3,4] for the purification of plasmid DNA.

PEG Precipitation

One of the simpler means to purify plasmid is the selective precipitation of plasmid DNA by PEG (polyethylene glycol). However, the yield of DNA obtained depends on the duration of incubation with cold PEG. The DNA sample (2 mL in TE) is added to 0.8 mL of PEG solution* and incubated at 0°C for at least 1 hour, preferably overnight. The DNA is centrifuged at 12,000 rpm in an SS-34 Sorvall for 15 minutes at 40°C. The pellet is dissolved in 1 mL of TE and after extraction with phenol–chloroform and chloroform, the DNA is precipitated with 2 volumes of ethanol after adjusting concentration to 0.3 M sodium acetate with a 3 M solution. The DNA pellet is washed with 70% ethanol, dried, and dissolved in TE buffer at a concentration of 1 to 3 mg/mL in TE and stored at 4°C, or at −20°C for long-term storage.

CsCl/Ethidium Bromide Equilibrium Centrifugation

The plasmid DNA from a 500-mL culture is taken in 4 mL of TE, and 4.4 g of CsCl is added and allowed to dissolve. After the addition of 0.4 mL of 10 mg/mL ethidium bromide, the samples are centrifuged in 5 mL of quick-sealing ultracentrifuge tubes for 3.5 hours in a VTi 80 rotor at 77,000 rpm, or overnight at 65,000 rpm at 20°C. Alternatively, a Beckman type 50 or 65 rotor can be used and centrifuged at 45,000 rpm for 36 hours at 20°C. The tubes are filled completely with the TE/CsCl solution. After centrifugation, the supercoiled plasmid DNA band is visualized with the help of a long-wave ultraviolet (UV) lamp held sideways and away from the eyes. A UV protective face shield is used while using UV light. The RNA forms a pellet and the distinct plasmid DNA band above in the gradient is taken out from the side of the tube with the help of a syringe and a 20-G needle by piercing the needle below the plasmid band and using syringe suction gently to take out the DNA, the beveled edge of the needle should face upward. The upper minor band, which often may not be visible, is the chromosomal DNA band and should be avoided. The ethidium bromide from the sample is removed by repeated extractions with TE saturated n-butanol until no red color remains in the aqueous phase. The sample is either dialyzed or diluted with 2 volumes of TE to lower the concentration of CsCl before precipitating the plasmid DNA with ethanol. Ethidium bromide is a mutagen and possibly a carcinogen and hence should be handled carefully and disposed off as a hazardous chemical.

---

*PEG solution: 30% PEG (W/V) 8000, 1.6 M NaCl.

Column Chromatography

Plasmid DNA can be purified from degraded RNA pieces and other contaminants by size exclusion chromatography using either Sephacryl-450 or Bio-Gel A-150. The plasmid DNA can also be purified by using kits that include columns containing binding matrix according to instructions provided by various vendors, such as Qiagen (Chatsworth, CA), Promega (Madison, WI), and Invitrogen (Carlsbad, CA), among others. More details are presented in Chapter 8.

6.4. GENOMIC DNA ISOLATION FROM YEAST

Yeast cells have a very rigid outer cell wall, and hence the first step in cell disruption for DNA isolation involves weakening of the cell wall by enzymatic (zymolase) treatment [8]. Yeast cells with their cell wall removed, called *spheroplasts*, are readily susceptible to cell lysis by detergents. For DNA isolation, yeast cells are grown to a density of about 10^8 cells/mL in YPD medium (1% yeast extract, 2% peptone, 2% glucose, pH 5.8) at 30°C with shaking. Yeast cells from 1 L of culture (about 6 to 8 g) are harvested by centrifugation at $3000 \times g$ for 15 minutes and washed in 100 mL of water by resuspension and centrifugation. Another wash is given in 100 mL of 50 mM EDTA, pH 7.5 by resuspending the cell pellet and centrifugation as earlier. The cellular pellet can be either processed further or stored at $-20°C$ until processed further. The first step in DNA isolation is the preparation of spheroplasts. The cellular pellet from 1 L of culture is resuspended in 14 mL of SCEM (1 M sorbitol, 0.1 M sodium citrate, 60 mM EDTA, 50 mM β-mercaptoethanol, pH 7). A small aliquot (100 μL) is taken in 900 μL of SCEM to serve as a negative control for cell lysis. The rest of the 14-mL yeast sample is incubated at 37°C with intermittent shaking with 20 mg of zymolase in 0.5 mL of SCEM. Aliquots (100 μL) are taken out every 15 minutes and the absorbence measured at 660 nm in comparison to the 0 time negative control. If the ratio of the absorbence of the sample to that of the 0 time negative control is 0.1, the spheroplast reaction is complete. This may take 30 minutes or more, depending on the activity of the zymolase. After the addition of 28 mL of lysis buffer (0.5 M Tris-Cl, pH 9, 3% N-lauroylsarcosine, 0.2 M EDTA, 0.5 mg/mL proteinase K) and mixing by gentle inversion several times, the sample is incubated at 50°C for at least 2 hours. After further incubation at 65°C for 30 minutes, 11.5 mL of 3 M potassium acetate, pH 4.8 is added and the tubes mixed by inversion. After leaving the sample tube in ice for 1 hour, the precipitated proteins–salt complexes are removed by centrifugation at $3500 \times g$ for 15 minutes. The supernatant containing the DNA is transferred by decanting into a clean tube and mixed with an equal volume of isopropyl alcohol by inverting the tube several times. After 10 minutes the DNA precipitate is collected by centrifugation at $5000 \times g$ for 15 minutes at room temperature. After draining out excess alcohol by inverting the tubes and wiping the edges with Kim wipes tissue, the DNA is resuspended in 20 mL of TE and incubated with 100 µg (in 200 µL) of DNase-free RNase A at 37°C for 1 hour. The sample is extracted with an equal volume of Tris-Cl buffer (pH 8) equilibrated phenol by inverting tubes several times gently. After centrifugation at room temperature at $5000 \times g$ for 15 minutes, the top aqueous phase containing DNA is transferred to a new tube and extracted with PCI (phenol–chloroform–isoamyl alcohol in the ratio of 25:24:1) two more times and once with chloroform alone. Each time the sample is centrifuged, as described earlier, to separate the phases.

After the final extraction, the DNA is precipitated by the addition of a $\frac{1}{4}$ volume of 10 M ammonium acetate and an equal volume of isopropyl alcohol. DNA is either spooled out on a tip of Pasteur pipette (made blunt-ended in a flame) or centrifuged. After washing with 70% ethanol, the Pasteur pipette is kept in an upright position with the DNA facing up, or the tube containing DNA is allowed to air dry (the pellet is better dissolved if it is not dried excessively). The DNA is dissolved in TE at 37°C overnight so that the final concentration is 0.5 µg/mL or less.

6.5. DNA FROM MAMMALIAN TISSUES

6.5.1. Blood

In mammals, red blood cells do not contain DNA since they are devoid of nucleus. The hemoglobin in them can get adsorbed to DNA if it is present during the isolation procedure. Hence for the isolation of DNA from blood, red blood cells are first removed either by Ficoll/Hypaque gradient centrifugation, or lysed by the detergent Triton X-100 followed by recovery of nuclei of white blood cells, which carry DNA.

The Triton X-100 lysis method [9] is relatively simple and is a cost-effective method. To 10 mL of blood, 90 mL of cold (4°C) RBC lysis buffer (0.32 M sucrose, 10 mM Tris-Cl, pH 7.5, 5 mM MgCl$_2$, 1% Triton X-100) is added and mixed by inverting the tube a few times. This step releases cellular contents, and the crude nuclear fraction containing the DNA is then centrifuged at $3500 \times g$ at 4°C for 30 minutes. The supernatant is decanted and the pellet is resuspended in a small volume of cold RBC buffer by vortexing and brought upto 40 mL with the same buffer and recentrifuged as before. The supernatant is decanted and the pellet is dissolved in 4 mL of proteinase K solution (10 mM Tris-Cl, pH 8.0, 0.75 mM NaCl, 25 mM EDTA, 0.5% SDS, 200 μg/mL proteinase K). At this stage as DNA is released from proteins, the solution becomes viscous. After incubation at 37°C overnight or at 55°C for 2 hours, the proteins gets digested by proteinase K and the DNA is extracted with phenol chloroform and ethanol precipitated as discussed earlier. DNA can be similarly isolated from lymphoblastoid cells, which are immortalized white blood cells, by directly resuspending a cellular pellet from 50 million cells in a 1-mL solution (10 mM Tris-Cl, pH 8.0, 0.75 mM NaCl, 25 mM EDTA), followed by a 10-fold dilution with proteinase K containing buffer, as described above.

6.5.2. Tissues and Tissue Culture Cells

DNA can be isolated from fresh or previously frozen tissue [3,10]. The first step is to "snap freeze" the fresh tissue in liquid nitrogen. Frozen tissues can be stored in a −80°C freezer. An aliquot of frozen tissues (1 g or less) is ground with a prechilled mortar and pestle to a fine powder in liquid nitrogen. Care is taken not to let the tissue thaw while grinding in liquid nitrogen. The frozen tissue powder in liquid nitrogen is transferred to a disposable polypropylene tube, and DNA extraction buffer is added before the tissue thaws such that 100 mg of tissue has 1.5 mL of DNA extraction buffer (100 mM NaCl, 10 mM Tris-Cl, pH 8, 25 mM EDTA, pH 8, 0.5% SDS). After heating the tubes briefly at 37°C, proteinase K is added to a final concentration of 100 μg/mL. The digestion volume is scaled up accordingly for larger amounts of tissue. The samples are incubated overnight at 50°C and then processed for phenol and chloroform extraction and ethanol precipitation as described earlier. Small amounts of tissues can be chopped into small pieces and digested directly in the proteinase K–containing buffer without freezing the tissue first.

In contrast to tissues, tissue culture cells are readily lysed with detergent [11]. Adherent cells from a culture plate are first scraped using a rubber policeman into a small volume of phosphate-buffered saline (137 mM NaCl, 2.7 mM KCl, 4.3 mM Na$_2$HPO$_4$, 1.4 mM KH$_2$PO$_4$, pH 7.3) and harvested by centrifugation at 1500 × g for 10 minutes at 4°C. The cellular pellet is resuspended in ice-cold TE so that 1 mL contains 100 million cells. After the addition of 10 volumes of freshly prepared digestion buffer (10 mM Tris-Cl, pH 8, 0.05 EDTA, pH 8, 0.5% Sarcosyl, and 100 μg/mL proteinase K), the sample is incubated at 50°C for 3 hours. DNA is recovered by ethanol precipitation after extraction with phenol, phenol–chloroform, and chloroform as described earlier.

6.6. DNA FROM PLANT TISSUE

Younger plants stored in dark for 1 to 2 days prior to extraction are preferred for DNA isolation since such tissue is poor in starch. The DNA is isolated essentially as described in Ref. 12. Plant tissue, frozen in liquid nitrogen, is ground to a fine powder with a mortar and pestle in liquid nitrogen. To 10 to 50 g of tissue powder, 5 to 10 mL of freshly prepared extraction buffer (100 mM Tris-Cl, pH 8, 100 mM EDTA, 250 mM NaCl, 100 µg/mL proteinase K, 10% Sarkosyl) is added and mixed by gentle stirring. The sample is incubated at 55°C for 1 to 2 hours. After centrifugation at $5500 \times g$ for 10 minutes at 4°C, the crude DNA is precipitated by 0.6 volume of isopropanol in cold (-20°C) for 30 minutes and centrifugation at $7500 \times g$ for 15 minutes at 4°C. The pellet is dissolved in 9 mL of TE buffer, and 9.7 g of solid CsCl is mixed and the sample is incubated in ice for 30 minutes. After centrifugation at $7500 \times g$ for 10 minutes at 4°C, the supernatant is filtered through two layers of cheesecloth. To the filtered sample, 0.5 mL of 10 mg/mL ethidium bromide is added and incubated in ice for 30 minutes and then centrifuged at $7500 \times g$ for 10 minutes at 4°C. The supernatant is centrifuged in ultracentrifuge at $525,000 \times g$ for 4 hours or $300,000 \times g$ overnight at 20°C. The DNA band is visualized against UV light (using eye protection), and the band is collected using a 15-G needle and syringe. The ethidium bromide is removed from the DNA-containing solution by repeated extraction with isoamyl alcohol or 1-butanol saturated with water. After a few hours of dialysis against TE to remove CsCl, DNA is ethanol-precipitated after addition of $\frac{1}{10}$ volume of 3 M sodium acetate, pH 7.

6.7. ISOLATION OF VERY HIGH MOLECULAR WEIGHT DNA

Routine methods of DNA isolation described above give DNA that can range from 50 to 200 kb. However, it has been possible to separate very high molecular weight DNA (megabase range) by pulsed field gel electrophoresis [14–16]. Cells such as yeast or lymphoblastoid cells are first mixed with melted agarose at an appropriate cell density and then cast into a mold that creates small plugs that can be processed for cell lysis. Since DNA does not experience any shearing forces during the isolation procedure, the DNA inside the agarose plugs is without breaks. The DNA released inside the agarose plugs in then size-separated in special electrophoresis equipment called *contour clamped hexagonal electrophoresis*, where the electric field is applied in a hexagonal shape and the current is pulsed at a preset rate. This equipment is available commercially, and using such gel equipment it is possible to separate intact yeast chromosomes ranging up to a size of a couple of megabase pairs of DNA. For details, the reader is referred to the references cited above.

6.8. DNA AMPLIFICATION BY POLYMERASE CHAIN REACTION

Polymerase chain reaction (PCR) is a method for amplifying DNA from a small amount of DNA catalyzed by thermostable DNA polymerase under appropriate reaction conditions with a pair of primers (oligonucleotides) that are complementary to DNA. K. Mullis, who invented the technique in the 1980s, was awarded a Nobel prize in 1994. Since its invention, various refinements and modifications have been described, and several review articles and books have been written on the subject [17–20].

The aim here is to provide an overall view of and how a typical PCR experiment is set up and how small amounts of DNA can be isolated from various sources for PCR. A schematic representation of PCR is shown in Figure 6.6. A typical PCR reaction mixture consists of a DNA template, a pair of specific primers, four deoxynucleotides (dATP, dGTP, dCTP, dTTP), thermostable DNA polymerase enzyme, and the appropriate buffer. PCR reaction cycles consist of thermal denaturation of template DNA, followed by annealing of specific primers to complementary template DNA strands in opposite directions and DNA chain elongation by the enzyme. These cycles are repeated many times. After each denaturation, DNA strands fall apart to allow small oligonucleotide primers (typically, 21 to 24 nucleotides) to anneal to the two strands of DNA during the annealing step. Once annealed, the primers are each extended in the $5'$-to-$3'$ direction by the thermostable DNA polymerase to yield a complementary replica of their template DNA strands. Each of these newly synthesized strands of DNA contains a site for binding one of the two primers and serves as a template for further amplification of DNA. Each cycle of amplification leads to the exponential amplification of the target DNA sequence flanked by its priming sites for the primers at the two opposite ends on the two strands of each of the amplified DNA molecules. After a certain number of cycles (typically, about 30 to 40 cycles), product DNA accumulation reaches a plateau. The duration of the exponential phase of amplification of the target sequence depends on the initial number of target sequences and the efficiency of the PCR reaction itself.

6.8.1. Starting a PCR Reaction

A typical PCR reaction set up protocol is shown in Table 6.3. The first step in PCR involves the selection of oligonucleotide primer sequences that are optimal for the amplification of a particular target DNA sequence. In addi-

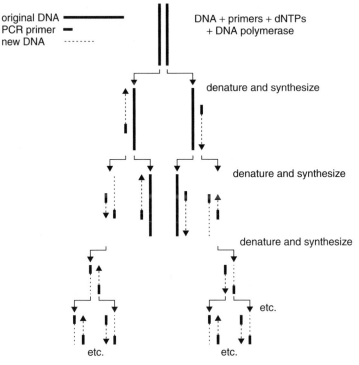

Figure 6.6. Amplification of DNA by PCR. Target DNA sequence from a complex genome can be amplified by heat denaturation, providing appropriate conditions for the enzyme (Taq DNA polymerase) that allow it to cause exponential amplification of a particular DNA segment. Among components besides the enzyme that are essential for amplification process are oligonucleotide primers in opposite orientation to each other, shown by dotted arrows, deoxynucleotide triphosphates (dNTPs), Mg^{2+}, and buffer. A 30-cycle amplification leads to a many-million-fold amplification of the discrete DNA segment, flanked by oligonucleotide primer sequences. (Reproduced from *Short Protocols in Molecular Biology*, 4th ed., F. M. Ausubel, R. Brent, R. E. Kingston, D. D. Moore, J. G. Seidman, J. A. Smith, and K. Struhl, eds., Wiley, New York, 1999, p. 15-1.)

tion to primers, the reaction mix contains template DNA that should contain the target sequence of interest and primer binding sites. In theory, PCR can amplify DNA from a single target molecule. The amount of template DNA added depends on the complexity of the DNA. For example, in the case of mammalian DNA, 60 to 200 ng of DNA in a 15- to 100-μL PCR reaction is used, whereas considerably less DNA needs to be used for less complex genomic DNA, such as yeast or bacterial DNA. The amount of Mg^{2+} depends on the type of Taq polymerase used and the length of oligonucleotide primers used. Typically, the concentration of $MgCl_2$ is 1.5 mM but can be 2.5 mM with other enzymes, such as Amplitaq DNA polymerase

Table 6.3. Components of a Typical PCR Reaction

Components	Final Concentration
10 × PCR buffer	1 × concentration
MgCl$_2$	1.5 mM or more
Forward primer	0.2–1 μM
Reverse primer	0.2–1 μM
20 mM four-dNTP mix	0.2 mM
Template DNA	60–200 ng
Taq polymerase	2.5 U/100 μL
Water	To adjust volume

Gold (ABI-Perkin Elmer, Foster City, CA). Too much DNA can lead to PCR artifacts and hence should be avoided. The four deoxynucleotide triphosphates (dNTP) include dATP, dCTP, dGTP, and dTTP. The mixture can be prepared and stored in aliquots at −80°C. The commercial vendors that sell the Taq DNA polymerase enzyme also provide PCR reaction buffer either with or without Mg^{2+}. The basic constituents of PCR buffer include 100 mM Tris-Cl, pH 8.3 (at room temperature), 500 mM KCl, and other additives in some brands.

Primers can be custom synthesized commercially at a fairly reasonable rate. Primer sequences are selected so that they are typically 21 to 24 nucleotides in length with an average G + C content of 40 to 60%. The primers should not be part of repetitive-sequence DNA, nor should the DNA have palindromic sequences. There are computer programs to help select oligonucleotide primers for PCR, such as PRIMER (from White Head Institute; *www-genome.wi.mit.edu/ftp://distribution/software/primer.0.5/manual.asc*). In addition, commercial sources exist for software (National Biosciences) that includes the OLIGO program.

Various thermostable DNA polymerases are available commercially from various vendors. The first thermostable DNA polymerase enzyme that became available commercially was Taq DNA polymerase, isolated from *T. aquaticus*. This enzyme lacks 3′-to-5′ proofreading exonuclease activity and hence has a higher error rate than those enzymes that possess this proofreading activity, such as pfu enzyme. For most routine purposes any thermostable DNA polymerase should suffice, irrespective of its error rate during PCR.

PCR reaction cycles are carried out in a commercially available programmable thermocycler. A typical PCR cycle consists of initial denaturation of DNA at 94°C for 3 to 10 cycles (depending on the enzyme used), followed by 30 to 40 cycles of brief denaturation (94°C, 30 seconds), annealing (50 to 55°C), and elongation (72°C). Variations of these conditions occur if amplifying larger DNA (several kilobase pairs) segments, and commercial vendors provide specific cycling instructions. Various factors can be controlled to optimize PCR reaction; such factors include cycling conditions (annealing temperature, duration of annealing and elongation cycles), Mg^{2+} concentration, pH of reaction, concentration of dNTPS, and primers. Some hard-to-amplify templates may yield product in the presence of 10% DMSO or formamide [21]. Priming at low temperature allows priming to occur at nonspecific sites in the template DNA, and hence depending on the primers, false positive amplifications could result. This can be reduced considerably by using thermostable DNA polymerases such as Amplitaq Gold (Perkins-Elmer, Foster City, CA), or similar enzymes which are activated only after preheating at high temperature (Hot-Start PCR). Amplitaq Gold requires higher Mg^{2+} (2.5 mM). Previously, all PCR samples were layered with a thin layer of mineral oil to avoid evaporation during a PCR cycle. However, in more recent machines a top-heated lid in contact with the lid of the PCR tubes minimizes condensation during PCR and bypasses the use of oil in the PCR tubes.

Due to its exponential amplification of target sequences, the extraordinary sensitivity of PCR makes it prone to the amplification of irrelevant sequences if a contaminating DNA sequence exits in the reaction mixture and if the primers are able to prime the contaminating DNA template sequence. Hence extreme precautions are taken to avoid false amplifications due to contamination of template DNA. DNA can be also amplified from RNA by first converting RNA into DNA by an enzyme known as *reverse transcriptase*; hence this method can be used to scan for expression of various genes in different tissues starting with RNA from various tissues.

6.8.2. Isolation of DNA from Small Real-World Samples for PCR

Buccal DNA

When blood samples are difficult to obtain, buccal samples can often be used for DNA isolation [22]. A buccal sample can provide enough material for several PCR reactions. The buccal epithelial cell samples are obtained by rotating a sterile buccal brush around the inside of cheeks. The method involves first rinsing a person's mouth with water and rolling the inside of the cheek with a soft nylon brush (Cytobrush Plus, Medscand, Hollywood, FL) for about 30 seconds. Gloves are used to avoid contamination of the brush by hands. The brush can be air-dried and put into an individual container and transported. The brush is dipped into 600 μL of 50 mM NaOH in a 1.5-mL microcentrifuge tube. After cutting off the end of the brush, the sample, along with the brush tip, is vortexed vigorously for 30 seconds twice and then incubated at 95°C for 5 minutes to release DNA. The brush tips are removed with forceps, avoiding contact with the solution. The solution is neutralized by the addition of 60 μL of 1 M Tris-Cl, pH 6.5 and vortexed again. The samples are centrifuged in the microcentrifuge at 12,000 × g for 5 minutes and the supernatant is ready for PCR. About 3 to 4 μL is usually enough for 50 μL of PCR reaction. The yield is variable, and addition of too much material can be inhibitory for PCR. It may be a good idea to test a range of concentrations for PCR for each sample. The samples should be stored in aliquots at −20°C. A commercially available kit for isolation of DNA from buccal samples is also available (Epicenter Technologies).

Small Amounts of Blood

A few drops of blood collected onto FTA paper (Life Technologies or Promega) after drying is stable for shipment at room temperature. The DNA binds to the FTA paper and inactivates bloodborne pathogens. The paper can be stored indefinitely at room temperature. The dried paper can be cut into small pieces and cells are lysed and DNA is immobilized within the paper matrix. Additional washes remove heme and other cellular debris and the paper-bound DNA can be used directly for PCR. The details involved in processing the FTA paper for PCR are given by the manufacturer of the FTA Gene Guard System (Life Technologies or Promega).

DNA for PCR can also be obtained from blood-stained material using the Chelex method [23]. A small piece of blood-stained material (3 mm^2) is taken in a clean microcentrifuge tube and 1 mL of water is added and left as such at room temperature for 15 to 30 minutes. Then 200 μL of 5% Chelex suspension in water (Biorad) is added and incubated for 15 minutes at 56°C on a rotatory shaker. After vortexing the tubes, the tubes are placed in a heating block at 98 to 100°C for 8 minutes. The tubes are vortexed again for 10 seconds and centrifuged for 3 minutes at 12,000 × g at room temperature to pellet the Chelex resin. The supernatant can be quantitated for DNA or concentrated by ethanol precipitation.

Hair Root

A hair shaft contains little or no DNA. The major source of DNA from hair is the hair root pulled from the scalp. The hair that sheds by itself or comes out easily on pulling from the scalp is most likely from the resting phase of the hair follicle and is not a good source of DNA, as it contains mostly cellular debris at its root. However, it is possible to isolate DNA from pulled hair roots [24].

The hair is washed in a petri dish to remove any surface debris. About a 1-cm portion of the hair root and shaft is cut with a scalpel and put into a microcentrifuge. The samples are incubated with 200 μL of 5% Chelex and 2 μL of 10 mg/mL proteinase K. After incubation at 56°C for 6 hours to overnight, the tubes are vortexed for 15 seconds at maximum speed and then placed in a heating block at 98 to 100°C for 8 minutes. The tubes are again vortexed as before and centrifuged for 3 minutes at 12,000 × g at room temperature to pellet Chelex resin. The DNA in the supernatant can be used directly for PCR or extracted and concentrated if it is too dilute.

6.9. ASSESSMENT OF QUALITY AND QUANTITATION OF DNA

6.9.1. Precautions for Preparing DNA

Properly isolated, DNA should be >50 kb in size without low-molecular-weight streaks as judged by agarose gel electrophoresis. DNA shearing can happen because of improper cell lysis and consequent degradation by cellular nucleases or by mechanical shearing during the isolation process itself. The former can be controlled by following the procedure appropriately so that cellular lysis in the lysis buffer is carried out as quick as possible. Mechanical shearing can be minimized by using wide-bore pipettes and tips while transferring the DNA. DNA should be devoid of any proteins, as indicated by a ratio of >1.8 between absorbence at 260 and 280 nm. If the ratio is less than 1.8, an additional extraction with phenol chloroform and ethanol precipitation may be necessary. DNA should be without contamination such as that by RNA, which contributes to the 260-nm absorbance and hence can cause overestimation of DNA. As described earlier, a simple RNase treatment can remove RNA from a DNA preparation.

6.9.2. Assessment of Concentration and Quality

A simple method of estimating DNA concentration is to measure its absorbance at 260 nm (UV range) in a spectrophotometer using a quartz cuvette. Absorbence measurements should take into account any contribution due to buffer components of the DNA solution. Absorbance is also measured at 280 nm to assess the purity of the DNA sample. A pure DNA solution should have a 260/280 nm ratio of >1.8. Contaminants that absorb at 280 nm, such as proteins, will lower this ratio. A 260-nm reading of 1 in a 1-mL cuvette with a 1-cm path length indicates a concentration of 50 μg/mL, whereas the same OD reading given by a single-stranded DNA solution will have a concentration of 36 μg/mL.

Although direct absorbance measurement in the UV range gives very reliable measurements of DNA concentration, its sensitivity is in the range

1 to 50 μg/mL. For dilute solutions of DNA, more sensitive dye-based fluorescence methods are preferred. Various commercially available kits (Biorad, CA; Molecular Probes, Inc, Eugene, OR) use dyes such as Hoechst 33258 or PicoGreen. Hoechst 33258 can measure 10 ng/mL DNA concentration in a cuvette, but picogreen is several times more sensitive and hence can be used to estimate as little as 50 pg of DNA per milliliter. The fluorescence, which is a direct measure of DNA content, can be measured in a fluorometer. The exact concentration can be estimated if a set of known DNA standards are included in the assay. The respective excitation and emission wavelength is 365 and 448 nm for Hoechst 33258 and 480 and 520 nm for PicoGreen-containing samples. It is important to include control DNA of the same type as the test sample. For example, double-stranded circular plasmid DNA is used as control DNA if the test sample is plasmid DNA, and linear double-stranded DNA is used as control for linear double-stranded DNA. These dyes are not only advantageous for small-quantity DNA estimations, but also do not allow any contaminating RNAs to interfere in the estimation. These dyes can be used to determine the approximate DNA concentration of the sample by comparing the fluorescence intensity with a range of concentrations (nanogram quantities in a volume of 5 to 10 μL) of a standard DNA after taking Polaroid pictures of DNA samples (5 to 10 μL) on top of a UV light source. Care should be taken in handling these dyes, and proper procedures for disposal should be followed, as they are known to be mutagenic.

Although the spectrophotometric measurements at different wavelengths can determine the extent of contaminants, the overall quality of DNA can be determined by analyzing samples on a horizontal 0.7% agarose gel

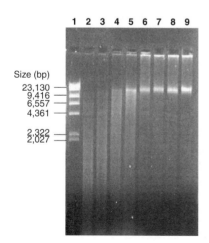

Size (bp)
23,130
9,416
6,557
4,361

2,322
2,027

Figure 6.7. Agarose gel electrophoresis of yeast genomic DNA. Agraose (1%) gel electrophoresis of yeast DNA before (lane 9) or after digestion with increasing quantities of a restriction endonuclease, Sau 3A. Lane 8 had DNA with minimum enzyme digestion, and lane 2 had DNA with maximum digestion. Formation of smear as a result of smaller DNA fragments of varying sizes is evident upon complete digestion with the restriction endonuclease. Lane 1 has DNA size markers in base pairs created by digesting bacteriophage λ DNA with Hind III restriction enzyme.

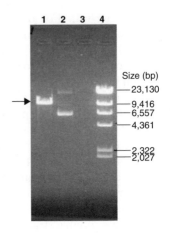

Figure 6.8. Agarose gel electrophoresis of plasmid DNA. Agarose (1%) gel electrophoresis of a plasmid DNA (ca. 0.2 µg) after restriction enzyme digestion (lane 1) or before restriction enzyme digestion (lane 2). Lane 3 is blank. Lane 4 has bactriophage λ DNA cut with restriction enzyme Hind III as DNA size markers. Note the uncut plasmid (lane 2 has two bands [the faster-moving band is a supercoiled form of the plasmid DNA, and the slower-moving band is the relaxed form (single-strand nick)]. The linear form (lane 1) results from a cut introduced at a single site in both strands of DNA by the restriction endonuclease. This plasmid has only a single site for the enzyme, and hence enzyme treatment creates only a single band, as shown in lane 1.

Size (bp)
— 23,130
— 9,416
— 6,557
— 4,361
— 2,322
— 2,027

electrophoresis and staining with 0.5 µg/mL ethidium bromide in the gel buffer (40 mM Tris-acetate, pH 8.3, 1 mM EDTA). A good preparation of genomic DNA should have DNA of the size >50 kb, as determined by running appropriate commercially available DNA size standards such as that derived from bacteriophage λ DNA (Figure 6.7). The quality of DNA is also reflected by digestion of DNA by various restriction enzymes, which indicates the absence of any inhibitors for enzymatic activity that might originate during the isolation process. Plasmid DNA on agarose gels displays more than one band because the mobility of various conformations (supercoiled DNA, open-circle, and linear DNA) is different. After restriction enzyme digestion that cuts the plasmid DNA at a single site, a single band of plasmid DNA should be visible, as shown in Figure 6.8. The yield of DNA depends not only on the type and the amount of tissue, but also on

Table 6.4. Yield of DNA Isolated from Various Types of Tissues

Source of DNA	Yield
Bacteria	0.5–2 mg/100 mL culture (10^8–10^9 cells/mL)
Plasmid DNA	2–3 mg/L bacterial culture
Yeast	2–8 mg/L yeast culture (ca. 8-g cells)
Human blood	10–20 µg/mL[a]
Lymphoblastoid cells	300 µg from 50 million cells
Mammalian tissue	About 2 mg/g tissue
Tissue culture cells	120 µg from 20 million cells
Plant tissue	10–40 µg/g fresh tissue

[a] The yield can be variable depending on the WBC count in the blood.

the extent of tissue homogenization. The typical yield of DNA from various sources is given in Table 6.4.

6.9.3. Storage of DNA

Pure DNA can be stored in TE (pH 8) at 4°C for several months. However, traditionally, storage of DNA under ethanol at −20°C has been preferred for long-term storage (years). If subjected to repeated freeze and thaw, an aqueous solution of DNA can lead to single- and double-strand breaks. Hence storing aliquots in a freezer at −20°C (frost-free) or −80°C is another preferred method for long-term storage.

REFERENCES

1. O. T. Avery, C. M. MacLeod, and M. McCarty, *J. Exp. Med.*, **79**, 137–158 (1944).
2. J. D. Watson and F. H. C. Crick, *Nature*, **171**, 737–738 (1953).
3. F. M. Ausubel, R. Brent, R. E. Kingston, D. D. Moore, J. G. Seidman, J. A. Smith, and K. Struhl, eds., *Short Protocols in Molecular Biology*, 4th ed., Wiley, New York, 1999.
4. J. Sambrook, E. F. Fritsch, and T. Maniatis, *Molecular Cloning: A Laboratory Manual*, 2nd ed., Cold Spring Harbor Laboratory Press, Cold Spring Harbor, NY, 1989.
5. D. M. Wallace, *Methods Enzymol.*, **152**, 41–48 (1987).
6. M. G. Murray and W. F. Thomas, *Nucleic Acids Res.*, **8**, 4321–4325 (1980).
7. H. C. Birnboim, *Methods Enzymol.*, **100**, 243–255 (1983).
8. G. A. Silverman, Purification of YAC-containing total yeast DNA, in D. Markie, ed., *YAC Protocols: Methods in Molecular Biology*, Vol. 54, Humana Press, Totowa, NJ, 1996.
9. G. I. Bell, J. H. Karam, and W. J. Rutter, *Proc. Natl. Acad. Sci. USA*, **78**, 5759–5763 (1981).
10. M. Gross-Bellard, P. Oudet, and P. Chambon, *Eur. J. Biochem.*, **36**, 32 (1972).
11. N. Blin and D. W. Stafford, *Nucleic Acids Res.*, **3**, 2303–2308 (1976).
12. S. L. Dellapotra, J. Wood, and J. B. Hicks, *Plant Mol. Biol. Rep.*, **1**, 19 (1983).
13. R. A. Cox, in L. Grossman and E. Moldave, eds., *Methods in Enzymology*, Vol. 12, Part B, Academic Press, New York, 1968, p. 120.
14. D. C. Schwartz and C. R. Cantor, *Cell*, **37**, 67–75 (1984).
15. G. Chu, D. Vollrath, and R. W. Davis, *Science*, **234**, 1582–1585 (1986).
16. H. Rietman, B. Birren, and A. Gnirke, Preparation, manipulation and mapping of HMW DNA, in B. Birren, E. D. Green, S. Klapholz, R. M. Myers, and

J. Roskams, eds., *Genome Analysis: A Laboratory Manual*, Vol. 1, Cold Spring Harbor Press, Cold Spring Harbor, NY, 1997, pp. 83–248.

17. R. K. Saiki, D. H. Gelfand, S. Stoffel, S. J. Scharf, R. Higuchi, G. T. Horn, K. B. Mullis, and H. A. Erlich, *Science*, **239**, 487–491 (1988).

18. T. J. White, N. Arnheim, and H. A. Erlich, *Trends Genet.*, **5**, 185–189 (1989).

19. C. W. Dieffenbach and G. S. Dveksler, eds., *PCR Primer: A Laboratory Manual*, Cold Spring Harbor Laboratory Press, Cold Spring Harbor, NY, 1995.

20. H. A. Erlich, ed., *PCR Technology: Principles and Amplifications for DNA Amplification*, Stockton Press, New York, 1989.

21. D. Pomp and J. F. Medrano, *Biotechniques*, **10**, 58–59 (1991).

22. B. Richards, J. Skoletsky, A. Shuber, R. Balfour, R. Stern, H. Dorkin, R. Parad, D. Witt, and K. Klinger, *Hum. Mol. Genet.*, **2**, 159–163 (1994).

23. P. S. Walsh, D. A. Metzger, and R. Higuchi, *Biotechniques*, **10**, 506–513 (1991).

24. R. E. Bisbing, The forensic identification and association of human hair, in R. Saferstein, ed., *Forensic Science Handbook*, Prentice Hall, Englewood Cliffs, NJ, 1982.

CHAPTER

7

SAMPLE PREPARATION IN RNA ANALYSIS

BHAMA PARIMOO

Department of Pharmaceutical Chemistry, Rutgers University College of Pharmacy, Piscataway, New Jersey

SATISH PARIMOO

Aderans Research Institute, Inc., Philadelphia, Pennsylvania

7.1. RNA: STRUCTURE AND PROPERTIES

The genetic information present in DNA is expressed in a cell via ribonucleic acid (RNA). Although all cells in an organism contain the same DNA, tissues differ with respect to the quantitative and qualitative profile of their RNA. The timing and the regulated level of expression of RNA in a cell are crucial for the proper development of a tissue in an organism. RNA is a long polymer made up of a linear array of ribonucleoside monophosphate monomers joined to each other via phosphodiester linkages (Figure 7.1). These monomer units consist of a five-carbon sugar (ribose), a phosphate group, and a heterocyclic nitrogenous base. There are four nitrogenous bases in RNA: cytosine (C), uracil (U), guanine (G), and adenine (A). A nitrogenous base linked to a pentose sugar is known as *ribonucleoside*. Hence, RNA is similar to DNA, but there are two major differences. First, the sugar of RNA is ribose, which is identical to the deoxyribose sugar of DNA except that it contains an additional OH group. The second difference is that RNA contains no thymine, but contains the closely related pyrimidine base uracil. During enzymatic synthesis of RNA, two phosphate groups from a ribonucleoside triphosphate are released and the monophosphate form (with the phosphate group nearest to sugar residue) is incorporated in the growing RNA chain. The sequence of RNA is dependent on DNA, which acts as a template during the enzymatic synthesis of RNA. Alternating moieties of

Sample Preparation Techniques in Analytical Chemistry, Edited by Somenath Mitra
ISBN 0-471-32845-6   Copyright © 2003 John Wiley & Sons, Inc.

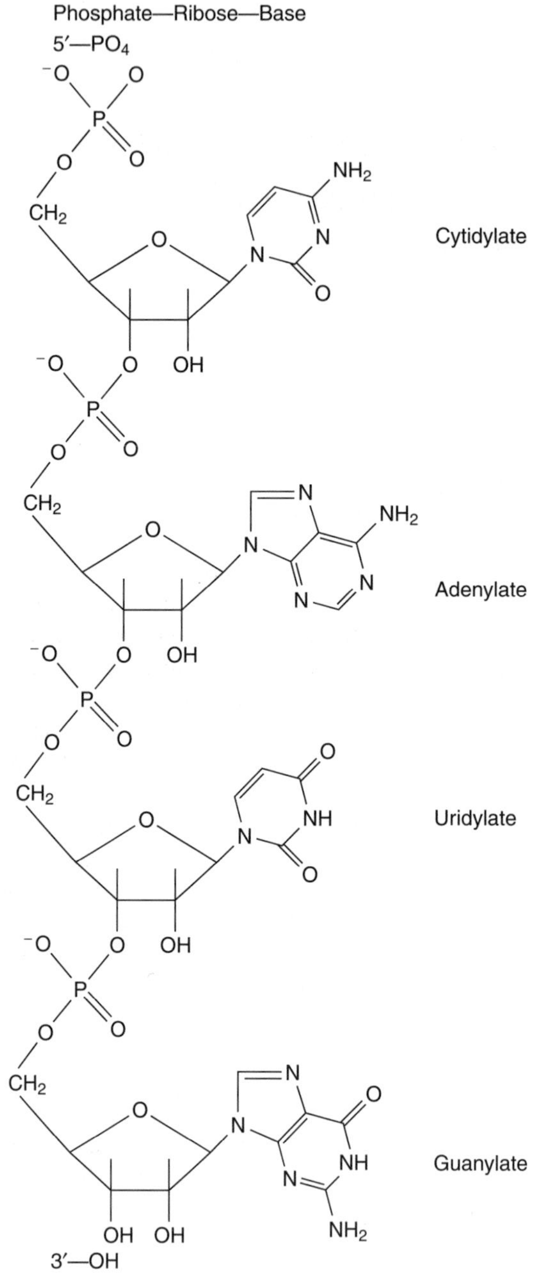

Phosphate—Ribose—Base

5′—PO₄

Cytidylate

Adenylate

Uridylate

Guanylate

3′—OH

sugar and phosphate groups, constituting the phosphodiester backbone of RNA molecules, imparts a net negative charge to the molecule. Although RNA is a single-stranded molecule, it can form a hybrid with another complementary RNA or DNA molecule via hydrogen bonding, where guanine binds to cytosine and adenine binds to uracil (RNA) or thymine (DNA). This base pairing of RNA with a complementary strand of nucleotides occurs under optimal conditions of ionic strength, temperature, and pH. For example, the formation of double-stranded molecules (complementary in sequence) is favored under high salt and low temperature, while their dissociation into single strands is favored at low salt and high temperature. The temperature at which RNA and its complementary sequence fall apart depends on both the length of the complementary sequence and its GC content. Organic solvents such as formamide and formaldehyde can lower the temperature at which RNA and its complementary RNA or DNA strand hybridize to each other. In other words, the two strands of nucleic acids can dissociate into single strands at a lower temperature. These organic solvents interfere with hydrogen bonding of the nitrogenous bases. Thermodynamically, RNA:RNA hybrids are most stable and are followed by the RNA:DNA hybrids.

7.1.1. Types and Location of Various RNAs

There are three major classes of RNA in cells: messenger RNA (mRNA), transfer RNA (tRNA), and ribosomal RNA (rRNA). Of these, the latter two are termed *stable RNAs*, as they have a longer half-life than that of mRNA [1]. Ribosomal RNA is the most abundant class of RNA in a cell. In a typical eukaryotic cell (yeast, plant, and animal), there are other RNAs, such as organelle RNA and small RNAs in nuclei (snRNAs) or in the cytoplasm (7S RNA). In eukaryotic cells, most RNAs are synthesized as larger precursor molecules and are then processed into smaller mature RNAs. Total RNA in a human cell may range from 10 to 30 pg, with most of it in the cytoplasm (about 85%), while the rest is in the nucleus.

mRNA molecules account for 1 to 5% of the total cellular RNA and are synthesized on a DNA template by the transcription machinery of a cell that

◄───

Figure 7.1. Molecular structure of RNA. The single-stranded RNA molecule consists of ribonucleoside residues linked to each other via phosphodiester bonds. The four nitrogenous bases in RNA are shown with their linkage at the C_1 position of ribose. The RNA chain elongates from the 5′ to the 3′ direction as the new nucleotide residues are added at the 3′-OH end of the chain during RNA synthesis in a cell. (Adapted from *Textbook of Biochemistry with Clinical Correlations*, T. M. Devlin, ed., Wiley, New York, 1982.)

includes one of the RNA polymerase enzymes. The average size of mRNA in a human cell is about 2 kilobases (kb). Most mRNAs are associated with ribosomes in the form of polyribosomes, also known as *polysomes*. Unlike in bacteria, mRNA is synthesized in the nucleus of eukaryotic cells such as yeast, plants, and animals as a large precursor molecule called *heterogeneous nuclear RNA* (hnRNA). This precursor RNA is processed, resulting in the deletion of several segments of varying lengths (introns), and the spliced RNA is modified before it leaves the nucleus and enters the cytoplasm. In the case of mRNA, modifications occur at both the 5' and 3' ends before the mature mRNA is transported from the nucleus to the cytoplasm, where it is used by the cellular machinery as a template to make proteins (Figure 7.2). A distinguishing feature of mRNA from a eukaryotic cell, as a result of 3'-end enzymatic processing, is the presence of a characteristic long poly-A^+ tail at the 3' end that facilitates its isolation by affinity methods using oligo(dT) or poly-U bound to solid matrix [2,3]. At both the 5' end (beginning of RNA) and the 3' end (preceding the poly-A tail) of mRNA

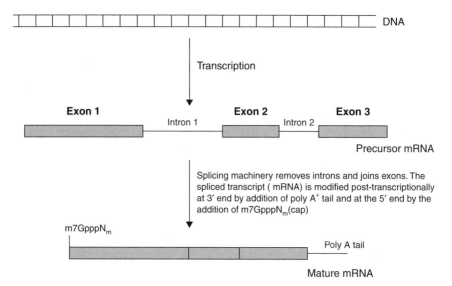

Figure 7.2. Relationship of the mature RNA to its precursor RNA. RNA molecules are synthesized as larger precursor molecules in the nucleus. As they enter cytoplasm, they are proceeded by the cellular splicing machinery to generate mature RNAs by eliminating intervening sequences and joining segments of RNA, the exons. As shown, additional modifications in the case of mRNA include addition of a stretch of poly-A residues at the 3' end and a cap at the 5' end. The 5' cap consists of a terminal 7-methylated guanosine in an unusual 5'-5' linkage via triphosphates, rather than the usual 3'-5' phosphodiester linkage, to the adjacent nucleotide that has methylated ribose sugar at its 2'-OH position. The 5'-cap is believed to help in the attachment of ribosomes to mRNA during protein synthesis.

Table 7.1. Abundance, Number, and Copies of Various Classes of mRNA

Abundance	Number of Different RNA Species/Cell	Copies/Cell
Low	11,000	5–15
Intermediate	500	200–400
High	<10	12,000

are sequences that do not code for proteins and are hence known as *noncoding sequences*. The length of noncoding sequences vary from one gene to another. Only the coding portion of the mRNA sequence codes for a protein. In general, the length of the coding sequence in mRNA is proportional to the size of the protein. In a typical eucaryotic cell, the number of mRNAs exceeds several thousand, and they are heterogeneous in terms of length, sequence, and relative abundance. The total number of RNA molecules in a human cell may range from 0.2 to 1 million and fall in three abundance classes. Table 7.1 shows the abundance of various classes and copies of molecules of mRNA. Out of the total number of genes in a typical cell, only a third of the genes are active. Each cell has its characteristic RNA expression profile, which may change during its growth and differentiation.

tRNAs constitute about 15 to 20% of the total RNA in a cell. An average tRNA is 75 nucleotides long, with the range being 73 to 93 nucleotides. The primary role of tRNA is to transport amino acids to a growing protein chain in the cellular protein synthesis machinery. Although there are only 20 amino acids, there are at least 56 different tRNAs in a typical cell because of degeneracy of the genetic code. Due to the ability of tRNAs to fold into a specific secondary and tertiary structure, it resembles a clover leaf with four loops and four stems. Each of the 20 amino acids is recognized by a specific aminoacyl-tRNA synthetase, which also recognizes multiple tRNAs for a specific amino acid and catalyzes the attachment of amino acids to tRNAs to form aminoacyl-tRNAs. It is these aminoacyl-tRNAs, carrying specific amino acids at their ends, that along with mRNA and ribosomes participate in the protein synthesis by recognizing nucleotide triplet code of mRNA via their anticodon loop. Aminoacyl-tRNAs can be isolated selectively from nonaminoacylated tRNAs in the laboratory by the biotin-avidin affinity method [4]. This is also described in Chapter 8.

Ribosomal RNAs (rRNAs) play an active structural role in ribosomes that are essential components of the cellular protein synthesis machinery. rRNAs are also believed to participate in tRNA binding, ribosomal subunit association, and antibiotic interactions. In a typical eukaryotic cell, there are four types of rRNA (28S, 18S, 5.8S, and 5S) that vary in size and sequence, and together constitute more than 80 to 85% of the total RNA in a cell. Ribosomal RNAs vary in size between species. In humans the size of 18S and 28S RNA are 1868 and 5025 nucleotide residues, respectively, whereas in a mouse, their sizes are 1869 and 4712 nucleotide residues. The smaller 5.8S and 5S RNAs are 158 and 120 nucleotides in length.

7.2. RNA ISOLATION: BASIC CONSIDERATIONS

Unlike DNA, RNA is very susceptible to rapid degradation due to ribonucleases (RNases), which are highly stable RNA degrading enzymes. In addition, RNA is more labile than DNA, especially at higher temperatures (>65°C) and at alkaline pH (>9). The sensitivity of RNA toward alkaline hydrolysis can be used for selective hydrolysis of RNA in a mixture of RNA and DNA [5]. Isolation of intact RNA is crucial to the success of many applications, such as the measurement of qualitative and quantitative changes in gene expression, preparation of cDNA or cDNA libraries, and in the synthesis of a probe for various molecular hybridization experiments.

Several methods exist for RNA isolation and have been described in detail in the literature [6–10]. Details of some of the techniques that can be used in the extraction and isolation of RNA are also discussed in Chapter 8. Some methods that may work for tissues poor in RNases may not yield good quality RNA from tissues that are rich in RNases. Moreover, the success of isolation of a good-quality RNA depends not only on a particular isolation method and reagents, but also on how the tissue is handled (storage condition and the time from dissection) and how rapidly the tissue is homogenized for RNA isolation. Although the biggest source of RNases is the tissue itself, there are additional exogenous sources, such as, hands, skin, hair, contaminated solutions, and laboratory supplies. Certain tissues, such as pancreas and spleen, are particularly abundant in RNases that rapidly degrade RNA. Due to the high activity of RNases and the fact that they are very stable not requiring any cofactors to function, extreme caution needs to be exercised in the extraction procedures to ensure that good-quality RNA is obtained. It is possible to curtail endogenous tissue RNase activity by rapid disruption using a tissue homogenizer in the presence of a strong chaotropic agent (a biologically disruptive agent) such as the guanidinium salts, phenol, and a detergent [e.g., sodium dodecyl sulfate (SDS)].

Although RNA is stable during the extraction procedure when strong protein denaturing agents are present, it is susceptible to degradation if the RNases get introduced from exogenous sources at a postextraction stage. To eliminate RNases contamination from external sources, the use of sterile disposable plasticware for reagents is preferred. The use of hand gloves and keeping all solutions covered with lids or aluminum foil is a good laboratory practice. If used at all, glassware should be baked at 200°C for 4 to 12 hours. The water used for preparing reagents should be treated with diethylpyro-carbonate (DEPC) to inactivate RNase by stirring with the water to a final concentration of under 0.1%. This should be carried out in a chemical hood for 1 hour, and the treated water should be autoclaved to destroy excess DEPC. For those solutions that cannot be treated with DEPC (Tris buffer) or autoclaved (heat-labile biochemicals), DEPC-treated water should be used to make solution from high-quality molecular-biology-grade chemicals that are certified to be RNase-free. Because DEPC is a suspected carcino-gen, caution is required in its handling. Many RNase-free reagents, includ-ing water, are commercially available.

The major steps in RNA isolation include rapid cell or tissue disruption, RNase inactivation by denaturants, which also dissociate RNA and protein complexes, and the recovery of RNA after removal of macromolecules (Figure 7.3). The severity of the treatment of cells for their lysis depends on whether a cell wall is present and its nature. Rapid disruption of tissues and mixing with denaturants is one of the most important steps in RNA isola-tion as it quickly inactivates RNases. The separation of RNA from proteins is achieved by extraction with a chaotropic agent in the presence of a deter-gent. This is followed by the separation of RNA either by gradient cen-trifugation or by partitioning into an aqueous phase where proteins go into the organic phase (phenol/chloroform). RNA from the aqueous phase is precipitated by the addition of alcohol (2.5 volumes of ethanol or equal volume of isopropanol) in the presence of a salt (Table 7.2). Sodium or ammonium acetate salts are preferred over sodium chloride because of the higher solubility of the acetate salts. It is a good practice to reprecipitate RNA with sodium acetate and ethanol if the RNA was precipitated pre-viously in the presence of lithium chloride. Some enzymatic reactions, such as reverse transcriptase, are inhibited by lithium ions. RNA from dilute solutions (<100 ng/mL) can be precipitated efficiently by the addition of glycogen (20 to 50 μg/mL) as a carrier before the addition of alcohol.

7.2.1. Methods of Extraction and Isolation of RNA

The methods of RNA isolation depends on the tissue and type of RNA to be extracted. Procedures to isolate total cellular RNA include chemical extrac-tions and centrifugation. mRNA is isolated from total RNA using affinity chromatography or magnetic beads, while high-pressure liquid chromatog-raphy methods are used for small RNA molecules. Phenol extraction was one of the first techniques to isolate RNA successfully from many sources

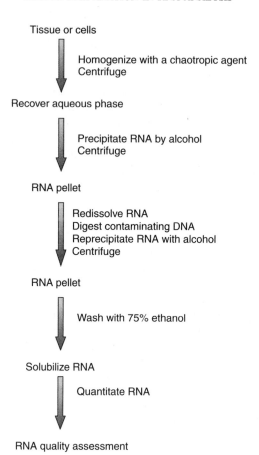

Figure 7.3. Flowchart of the basic steps in RNA isolation process. Important to recovery of intact RNA is the rapid disruption of cells or tissues in the presence of a chaotropic agent. Extent of DNA contamination is variable depending on the isolation procedure and the tissue. DNA can be removed by treatment with DNase I.

Table 7.2. Concentration of Various Salts Used for Alcohol Precipitation of RNA

Salt and Stock Solution Concentration	Salt Concentration in RNA Solution before Addition of Alcohol
Sodium acetate, pH 5.2; 3 M	0.3 M
Ammonium acetate; 10 M	2 M
Potassium acetate; 2.5 M	0.25 M
Sodium chloride; 5 M	0.1 M
Lithium chloride; 8 M	0.8 M

[11–13]. However, guanidinium salts have been found to be a better option, even for those tissues that are rich in RNA-degrading enzymes [14–17].

7.3. PHENOL EXTRACTION AND RNA RECOVERY: BASIC PRINCIPLES

The principle of this method is based on the ability of organic phenol to denature and precipitate proteins without altering the solubility of RNA. The method involves thorough mixing of the sample with an equal volume of a mixture of phenol–chloroform and isoamyl alcohol in the ratio 25:24:1. This is followed by centrifugation to separate the organic phenol phase from the inorganic aqueous layer containing the RNA. Being denser, the phenol layer remains below the aqueous layer, and the proteins are trapped in the phenol phase or at the interface. Extraction of the top aqueous phase with the phenol–chloroform mixture is repeated until the interface is no longer visible after centrifugation. Chloroform is used with phenol to improve the deproteinization efficiency of phenol. Isoamyl alcohol in the extraction mixture ensures that a well-defined interface is produced, thus improving the success of aspirating the supernatant containing the RNA. RNA in the aqueous phase is aspirated into a sterile tube and precipitated by the addition of salt and alcohol (Table 7.2). After incubation at $-20°C$ for 1 hour or overnight (if the yield of RNA is low), RNA is harvested by centrifugation at $10,000 \times g$ for 20 minutes at 4°C, and the RNA pellet is washed with ice-cold 75% v/v ethanol to remove excess salt. Small quantities of RNA, after alcohol precipitation, are harvested at higher speeds of centrifugation ($60,000 \times g$ for 1 hour) or by the addition of RNase-free glycogen ($25 \ \mu g/mL$) as carrier before mixing with alcohol. The RNA pellet is dissolved in sterile water (1 to 10 mg/mL). Difficulty in dissolving RNA in water suggests that the sample may be contaminated with macromolecules such as polysaccharides and DNA, and the RNA should be purified further.

   The phenol used for RNA extraction should be of the highest purity (double distilled), as oxidation products of phenol can cause degradation of RNA. Before use, the phenol is saturated with DEPC-treated water, and the phases are allowed to separate. The pH of the top aqueous layer is tested with pH paper. If phenol is acidic, it is equilibrated with Tris buffer by mixing it with a $\frac{1}{40}$ volume of 1 M Tris-Cl, pH 7.0. After phase separation, the top buffer layer is removed and the phenol is equilibrated twice with water. Additives such as 8-hydroxyquinoline (0.1%) are used in phenol to inhibit the activity of nucleases. Caution should be exercised in the use of phenol, as it is corrosive and a suspected carcinogen. Use in a chemical hood is recommended. Phenol–chloroform extraction should be carried out in containers that can withstand these compounds during processing and centrifugation.

7.3.1. Examples of RNA Isolation Using Phenol Extraction

RNA from Bacteria

Like plants, bacteria have a rigid cell wall. Hence appropriate measures are taken to weaken or break the cell wall before the cells are lysed for the extraction of RNA. Separate methods [10] exist for gram-negative or gram-positive bacteria because of differences in their cell compositions.

Gram-Negative Bacteria. A 100-mL culture of bacteria such as *Escherichia coli*, grown to log phase, is placed on ice and chilled for 10 minutes. Bacteria are centrifuged for 5 minutes at $5500 \times g$ at 4°C. The bacterial pellet is dissolved in 2 mL of STET (8% sucrose, 5% Triton X-100, 50 mM EDTA, 50 mM Tris-Cl, pH 7) lysing solution, and 0.1 mL of 0.2 M vanadylribonucleoside complex (VRC) is added as an RNase inhibitor. After the addition of 1 mL of phenol and vortexing for 1 minute, 1 mL of chloroform is added, and the solution is vortexed again as earlier. The top aqueous phase, containing the RNA, is separated from the organic phase by centrifugation at $10,000 \times g$ at 4°C. RNA is precipitated by the addition of a $\frac{1}{10}$ volume of 3 M sodium acetate, pH 5.2 and 2.5 volumes of cold ethanol to the aqueous phase. After incubation in ice for 1 hour, RNA is collected by centrifugation at $10,000 \times g$ for 10 minutes at 4°C and dissolved in 2 mL of 10 mM VRC. After extracting with 1:1 mixture of phenol and chloroform twice, RNA is precipitated with ethanol in the presence of sodium acetate as described earlier. The RNA pellet is dissolved in 6 mL of DEPC-treated water and purified on cesium chloride gradient. To 6 mL of RNA solution, 4.5 g of solid CsCl is added and the volume made to 9 mL with DEPC-treated water. It is layered over a 3-mL cushion of 5.7 M CsCl made in 100 mM ethylenediaminetetraacetic acid (EDTA) (pH 7.0) in an SW-41 rotor and centrifuged in a Beckman (Palo Alto, CA) ultracentrifuge for 14 hours at $150,000 \times g$ at 20°C. DNA at the interphase as well as the CsCl above the DNA is removed and the rest of the CsCl is poured out. After rinsing the RNA pellet with 70% ethanol, RNA in the pellet is dissolved in 0.36 mL of DEPC-treated water. RNA is precipitated with a $\frac{1}{10}$ volume of 3 M sodium acetate, pH 5.2 and a 2.5 volume of chilled ethanol. The RNA pellet obtained after centrifugation at $12,000 \times g$ for 10 minutes at 4°C is washed with ice-cold 75% ethanol by vortexing and recentrifugation as earlier. After air drying the pellet, the RNA is dissolved in 200 μL of DEPC-treated water.

Gram-Positive Bacteria. The bacteria from a 100 mL are centrifuged as described earlier, and resuspended in 5 mL of lysis buffer (30 mM Tris-Cl, pH 7.4, 100 mM NaCl, 5 mM EDTA, 1% SDS) to which 100 μg/mL proteinase K is added just before use. After freezing on dry ice and thawing, the culture is sonicated three times for 10 seconds each without foaming. After addition of an equal volume of 25:24:1 phenol–chloroform–isoamyl alcohol and vigorous mixing, the aqueous phase containing the RNA is separated by centrifugation in phenol-resistant tubes at 10,000 × g for 10 minutes at 4°C. The aqueous phase is reextracted twice with an equal volume of 25:24:1 phenol–chloroform–isoamyl alcohol, as described earlier. RNA in the aqueous phase is precipitated with a $\frac{1}{10}$ volume of sodium acetate and 2.5 volumes of ethanol. After incubation in ice for 15 to 30 minutes, the RNA is centrifuged at 12,000 × g for 10 minutes, washed with 75% ethanol as described earlier, and air-dried.

RNA from Plants

Plant cells have a rigid cell wall surrounding their cell membrane, hence, like bacteria, their cell wall must be broken by strong mechanical action such as grinding, or weakened by cell wall degrading enzymes (protoplasting) before the extraction of RNA. Since the plants are rich in polysaccharides, repeated precipitation of RNA by LiCl becomes necessary. The phenol–SDS method has been used for a variety of eukaryotic tissues. The method described has been reported earlier [10]. Typically, plant tissue (15 g) is snap frozen in liquid nitrogen and ground in the same in a precooled mortar and pestle. The slurry in liquid nitrogen is transferred quickly to a beaker containing 150 mL of grinding buffer (0.18 M Tris-Cl, pH 8.2, 0.09 M LiCl, 4.5 mM EDTA, 1% SDS) and 50 mL of Tris-Cl buffer (pH 8) equilibrated phenol. The tissue is homogenized with Polytron (Beckman) at about 80% of its maximum speed for 2 minutes. After the addition of 50 mL of chloroform and further homogenization for 30 seconds at low speed, the mixture is incubated at 50°C for 20 minutes. After centrifugation at 17,700 × g at 4°C for 20 minutes the aqueous layer is removed and saved. The interphase is reextracted with an equal volume of phenol–chloroform, and centrifuged at 12,000 × g for 20 minutes at 4°C. The aqueous layers are pooled and extracted repeatedly with phenol–chloroform followed by centrifugation at 17,7000 × g for 15 minutes at 4°C until no interphase is visible (usually, three extractions). Finally, the aqueous phase is extracted with chloroform, and centrifuged as earlier. The aqueous phase is made 2 M with respect to LiCl by the addition of 0.33 volume of 8 M LiCl, and after overnight incubation at 4°C, the precipitated RNA is collected by centrifugation for 20 minutes at 15,000 × g at 4°C. The pellet is rinsed with 3 mL of 2 M LiCl and then dissolved in 5 mL of DEPC-treated water, and the RNA is reprecipitated by bringing LiCl concentration to 2 M and incubation at 4°C for at least 2 hours. The RNA is centrifuged at 12,000 × g for 20 minutes at 4°C and rinsed with 2 M LiCl. The RNA in the pellet is dissolved in 2 mL of DEPC-treated water and precipitated after the addition of a $\frac{1}{10}$ volume of 3 M sodium acetate and 2.5 volumes of ethanol. After overnight incubation at −20°C or 30 minutes in dry ice, the RNA is collected by centrifugation at 15,000 × g at 4°C. RNA pellet is rinsed with chilled 75% ethanol, air dried, and dissolved in 1 mL of DEPC-treated water.

Yeast RNA

Yeast cells also have rigid cell walls that need to be broken by enzymatic or mechanical means before extracting RNA by the phenol-based method [10]. A 10- to 20-mL yeast culture grown to a midlog phase ($OD_{600\ nm}$ = 0.5 to 1) is centrifuged so as to harvest 2 × 10^8 cells by centrifugation at 4°C and 2000 × g. The pellet is resuspended in 1 mL of RNA buffer (0.5 M NaCl, 10 mM EDTA, 200 mM Tris-Cl, pH 7.5) and transferred to a 1.5-mL microcentrifuge tube and centrifuged at 4°C for 30 seconds. The pellet is resuspended in 300 µL of RNA buffer. An equal volume of chilled acid-washed glass beads equivalent to a 200-µL volume is added. The glass beads are pretreated with concentrated nitric acid for 1 hour and washed extensively with deionized water and DEPC-treated water, and baked in an oven at 300°C overnight. To the mixture of yeast cells and glass beads, 300 µL of 25:24:1 phenol–chloroform–isoamyl alcohol equilibrated with RNA buffer is added. The tubes are centrifuged at room temperature for 1 minute. The aqueous (top) layer, without the interphase, is transferred to a clean tube. The aqueous phase is extracted twice with an equal volume of 25:24:1 phenol–chloroform–isoamyl alcohol with intervening centrifugations. Finally, the aqueous layer is extracted with 24:1 chloroform–isoamyl and centrifuged. RNA is precipitated by the addition of a $\frac{1}{10}$ volume of 3 M sodium acetate and 2.5 volumes of ethanol. It is recovered by centrifugation after an overnight incubation at −20°C. The RNA pellet is rinsed with chilled 75% ethanol, air-dried, and dissolved in 50 µL of DEPC-treated water. For larger cultures of yeast, volumes can be scaled up appropriately.

The major disadvantage of the glass bead–based shearing method described above is that the yield of RNA can be poor if cell disruption is not complete. Another method using phenol and SDS takes advantage of freezing and thawing to enhance cell disruption [18]. A 20-mL early–midlog phase culture of yeast is centrifuged and the cells are collected by centrifugation. The cell pellet is resuspended in 400 µL of acetate–EDTA buffer (50 mM sodium acetate, pH 5.2, 10 mM EDTA, 10 mM VRC) followed by the addition of 40 μL of 10% SDS. After brief vortexing, the sample is mixed with an equal volume of buffer-equilibrated phenol and heated at 65°C for 4 minutes. After freezing in a dry-ice ethanol bath, the sample is centrifuged in a microfuge for 2 minutes to separate the phases. The aqueous phase is extracted once with phenol–chloroform–isoamyl alcohol (25:24:1) and once with chloroform–isoamyl alcohol (24:1). Each time it is followed by centrifugation to recover the aqueous phase. RNA is recovered from the aqueous phase by ethanol precipitation as described earlier.

7.4. GUANIDINIUM SALT METHOD

This method is widely applicable to several types of tissues, and RNA has been recovered successfully from animals, plants, and bacteria. Guanidinium hydrochloride and guanidinium thiocyanates are very powerful chaotropic agents. The guanidinium thiocyanate–based method has become the method of choice for the isolation of good-quality RNA from a variety of tissues. Cells (or tissues) are homogenized directly in a solution containing guanidium salt and reducing agents such as 2-mercaptoethanol (2-ME) or dithiothreitol (DTT) to break intramolecular protein disulfide bonds. These conditions rapidly inactivate RNases by distorting the secondary and tertiary folding of the enzymes when the cells are disrupted. Using these reagents, it is possible to isolate intact RNA even from RNase-rich tissues such as pancreas and spleen.

Homogenization in guanidinium salt–containing solution releases RNA as well as DNA into the homogenate. RNA can be separated from DNA due to differential buoyant densities of DNA (1.5 to 1.7 g/mL) versus RNA (1.7 to 2 g/mL) on a CsCl or cesium trifluoroacetate (CsTFA) gradient (Pharmacia LKB, Piscataway, NJ) by ultracentrifugation. Nucleic acids have lower buoyant densities in CsTFA and dissociate more readily from proteins than in CsCl; this is due to the salting-in effect of TFA ions. Unlike CsCl, CsTFA is an excellent inhibitor of RNases and is hence preferred over CsCl. Centrifugation through a CsCl gradient or precipitation with LiCl does not recover efficiently small RNAs such as tRNA and 5S ribosomal RNA.

7.4.1. Examples of RNA Isolation Using Guanidinium Salts

RNA from Animal Tissues and Cells Using Guanidinium Thiocyanate

Freshly dissected soft tissues are cut into small pieces and processed immediately for RNA isolation. Alternatively, the tissue can be flash frozen in liquid nitrogen and stored at −80°C until RNA isolation. Freezing in liquid nitrogen is the method of choice for RNA isolation from hard tissues. Cells grown in tissue culture can be recovered by centrifugation for 10 minutes at $1000 \times g$ (for suspension cultures) or in the case of adherent cells by scraping with a rubber policeman in the presence of phosphate-buffered saline (PBS) and centrifugation at $1000 \times g$ for 10 minutes. Although more efficient recovery of adherent cells is possible by detaching cells from plates by treatment with trypsin, the procedure can cause cell lysis and compromise the quality of RNA. The cell pellet recovered by centrifugation is then resuspended in PBS, and after recentrifugation the cell pellet alone can be either frozen in liquid nitrogen and stored at −80°C or processed immediately for RNA isolation.

The frozen tissues are powdered in liquid nitrogen with a prechilled mortar and pestle. The tissue slurry is transferred to a container and homogenized immediately with the RNA homogenization solution (4 M guanidinium thiocyanate, 25 mM sodium citrate, pH 7, 0.1 M 2-ME or 0.01 M DTT) in a homogenizer such as the Polytron (Brinkman) at its near-maximum speed for about 1 minute, until it disperses completely and uniformly. Reducing agents (2-mercaptoethanol or DTT) are added to the solution just immediately before the use. For every gram of tissue, 10 mL of RNA homogenization solution is used. After homogenization, Sarkosyl is added from a 20% stock solution so that the final concentration is 0.5%, and the sample is heated for 2 minutes at 65°C. The tissue homogenate is centrifuged for 10 minutes at $12,000 \times g$ and at room temperature. After low-speed centrifugation, the supernatant is subjected to gradient centrifugation through CsCl or CsTFA. This yields good-quality RNA even from small amounts of tissue. To purify RNA by CsCl gradient centrifugation, the homogenate (3.5 mL) is layered onto a cushion of 9.7 mL of 5.7 M CsCl in DEPC-treated water, 10 mM EDTA, pH 7.5, and centrifuged in an ultracentrifuge at 32,000 rpm for 24 hours at 22°C in an SW41 Beckman rotor. For larger volumes, a Beckman SW 28 rotor can be used with 12 mL of homogenate layered over a 26.5-mL cushion of CsCl and centrifuged at 25,000 rpm for 24 hours. According to the method of Okayama et al. [19], the concentration of guanidinium thiocyanate is higher in the homogenization solution (5.5 M), and 18 mL of the homogenization solution is used for every gram of tissue or cells. This 18-mL homogenate is layered over 19 mL of CsTFA solution (density, 1.51 g/mL in 100 mM EDTA) and centrifuged at 15°C for 20 hours at 25,000 rpm in a SW 28 Beckman rotor. Smaller amounts of tissue (25 to 140 mg) homogenate in a 2.5-mL volume can be layered over 2.7 mL of CsTFA and centrifuged for 20 hours in a Beckman SW 50.1 rotor at 31,000 rpm. In both CsCl and CsTFA gradient centrifugation methods, RNA forms a pellet at the bottom of the tube. DNA and proteins remain above in the solution. The supernatant is taken out carefully without disturbing the RNA pellet at the bottom. Using a sterile blade, the top portion of the tube is cut just above the RNA pellet. This avoids contamination with protein or DNA that may be sticking to the sides of the tube when the RNA pellet is subsequently dissolved. The RNA pellet is dissolved in a $\frac{1}{3}$ volume (with respect to tissue homogenate) of a solution consisting of 10 mM Tris-Cl, pH 7.5, 1 mM EDTA, and 0.1% SDS. The RNA is then allowed to dissolve for several minutes by drawing the fluid repeatedly through the disposable tip of an automatic pipettor. It is advisable to use the RNA dissolving buffer in two aliquots to recover all RNA efficiently from the pellet. Following successive extractions with equal volumes of phenol–chloroform and chloroform, RNA is precipitated with 2.5 volumes of ethanol after the addition of a $\frac{1}{10}$ volume of 3 M sodium acetate, pH 5.2. When precipitating microgram quantities of RNA, it is advisable to use siliconized tubes to avoid losses due to nonspecific binding of RNA to glass or plastic surface of tubes.

An alternative to gradient centrifugation for removal of DNA is to extract the tissue homogenate in guanidinium thiocyanate with water-saturated phenol under acidic pH [17]. Under these conditions DNA goes into the phenol phase, whereas RNA remains in the aqueous phase. The acidic pH is achieved by the addition of a $\frac{1}{10}$ volume of 2 M sodium acetate, pH 4, to the tissue homogenate in guanidinium thiocyanate solution. The tissue homogenate is then extracted with an equal volume of water-saturated phenol and 0.2 volume of chloroform by thorough mixing of phases and incubation at 4°C for 15 minutes. After centrifugation at 10,000 × g at 4°C for 20 minutes, the proteins and DNA go into the organic phase, and the RNA is recovered from the top aqueous phase by mixing it with an equal volume of isopropanol, incubation at −20°C for 60 minutes, and centrifugation at 10,000 × g for 20 minutes at 4°C. The RNA pellet is dissolved in a $\frac{1}{3}$ volume (with respect to original tissue homogenate) of guanidine thiocyanate tissue homogenization solution containing sarcosyl. The RNA is reprecipitated by the addition of an equal volume of isopropanol and incubation in cold for 1 hour followed by centrifugation. After rinsing the RNA pellet with 75% ethanol and air drying, the RNA in the pellet is dissolved in DEPC-treated water or deionized formamide for long-term storage at −80°C. If the pellet is hard to dissolve, heating at 55°C for about 10 minutes can help to dissolve RNA. RNA is usually dissolved easily at a concentration of 1 to 5 μg/μL. Whenever needed, RNA can be recovered from formamide by ethanol precipitation followed by centrifugation. During the past several years, the use of a single-phase solution of guanidine–thiocyanate and phenol has become popular. Many commercial vendors sell such reagents as TRIzol (Invitrogen, Carlsbad, CA), Tri-Reagent (Sigma-Aldrich,

St.Louis, MO), and RNA STAT-60 (TEL-TEST B, Inc., Friendswood, TX), or similar ready-to-use RNA extraction reagents. It is also possible to isolate RNA, DNA, and proteins simultaneously from the tissue homogenate using guanidine–thiocyanate and phenol [20,21]. Since DNA is trapped in the interphase and the phenol phase under acidic pH during extraction, the addition of an equal volume of 1 M Tris solution (pH 10.5) to the phenol phase raises the pH and thereby increases the solubility of the DNA in the aqueous phase. This DNA can be recovered by ethanol precipitation.

To isolate RNA from tissue culture cells, fresh or frozen cell pellet is either homogenized in a Polytron homogenizer or is ground in the RNA homogenization buffer with a mortar and pestle. For tissue culture cells, 3.5 mL of RNA homogenization buffer (4 M guanidinium thiocyanate containing solution) is added for 100 million cells. For small volumes where homogenization may not be possible, the tissue homogenate, along with Sarkosyl, is sheared by passing through a syringe with 20-G needle several times. The rest of the extraction and purification procedure is similar to that followed for animal tissues.

RNA from Animal Tissues and Cells Using Guanidinium Hydrochloride

Although guanidine hydrochloride is a potent chaotropic agent, its use for RNA isolation has not been as popular as that of guanidine thiocyanate. The reasons may be that it needs to be used at a much higher concentration to be an effective protein denaturant. The method described here is a modification of methods described in Refs. 15 and 16.

Cells or tissues are prepared as described above. Tissues or cell pellets are homogenized with Polytron homogenizer (Brinkman) for 1 minute in 10 volumes of homogenization buffer (8 M guanidine HCL, 0.1 M sodium acetate, pH 5.2, 5 mM dithiothreitol, 0.5% sodium lauryl sarcosinate). After centrifugation of the homogenate at $10,000 \times g$ for 10 minutes at room temperature, the RNA in the supernatant may be either purified by CsCl gradient centrifugation as described earlier or by alcohol precipitation followed by the removal of DNA. To precipitate nucleic acids by ethanol, 0.1 volume of 3 M sodium acetate, pH 5.2 and 0.5 volume of chilled ethanol are added to tissue homogenate after centrifugation and the nucleic acids are centrifuged at $10,000 \times g$ for 10 minutes after incubation at 0°C for at least 2 hours. The pellet is dissolved again in the homogenization buffer and nucleic acids are reprecipitated using ethanol as earlier. This process of nucleic acid solubilization and ethanol precipitation is repeated once more, and finally, the RNA is rinsed with 75% ethanol and air dried. The nucleic acid pellet is dissolved at 37°C for 30 to 60 minutes in a minimal volume of Tris-SDS containing buffer with freshly added proteinase K (10 mM Tris-Cl, pH 7, 0.1% SDS, 100 µg/mL proteinase K). After phenol–chloroform extraction, RNA is precipitated with ethanol and sodium acetate as before. Contaminating DNA is removed by either DNase treatment or by lithium chloride precipitation, as described later.

7.5. ISOLATION OF RNA FROM NUCLEAR AND CYTOPLASMIC CELLULAR FRACTIONS

To isolate RNA from cellular fractions such as nuclear or cytoplasm, the first step is to isolate that particular cell fraction. During the fractionation process, caution is exercised not to contaminate one fraction with another. The crude nuclear fraction is often contaminated with mitochondria and endoplasmic reticulum, both of which carry their RNA components. Hence it is highly desirable to further purify the nuclear fraction before isolating the RNA.

The initial tissue homogenization is carried out in a nondisruptive buffer similar to that as described in Ref. 22, except that it contains VRC RNase inhibitor (10 mM Tris-Cl, pH 8.6, 0.14 M NaCl, 1.5 mM MgCl$_2$ 1 mM DTT, 0.5% NP 40, and 20 mM VRC). The required concentration of NP 40 maintains nuclear membrane integrity but disrupts the outer cell membrane, and the VRC prevents RNA degradation from lysosomal RNases. The tissue homogenate is centrifuged at 500 × g for 3 minutes at 4°C to remove cellular debris and unbroken tissues. The supernatant is centrifuged at 2500 × g for 10 minutes at 4°C to separate the crude nuclear pellet from the cytosolic fraction. The crude nuclear pellet is further purified by centrifugation through sucrose cushion to remove contaminating mitochondria and endoplasmic reticulum [23]. The purity of nuclei is assessed by phase-contrast microscopy and purification steps may be repeated until pure nuclei are obtained. RNA is then extracted from the nuclei and cytosol using phenol–SDS extraction described earlier. Some tissues form bulky precipitates at the interphase during phenol extraction and may lead to poor RNA recovery. In such instances, the treatment of samples with proteinase K–containing buffer (an equal volume of the buffer containing 0.02 M Tris-Cl, pH 8, 25 mM EDTA, 0.3 M NaCl, 2% SDS, and 100 µg/mL proteinase K) prior to phenol extraction improves recovery of RNA [6,24]. It is important to add proteinase K to the buffer just before use.

7.6. REMOVAL OF DNA CONTAMINATION FROM RNA

Most RNA preparations are contaminated with varying amounts of DNA, depending on the tissue and the method of RNA isolation. Although small amounts of DNA may not interfere in some experiments, it can certainly be problematic in procedures such as RT-PCR, which involves reverse transcriptase–mediated synthesis of cDNA from RNA followed by polymerase chain reaction (PCR). Since PCR can use both cDNA and contaminating DNA as a template during the amplification process, the contaminating products can lead to serious artifacts.

The amount of DNase used essentially depends on the amount of DNA present in the total RNA preparation. For example, during bacterial RNA preparation, where the level of contaminating DNA may be high, the RNA is dissolved in 950 μL of DNase digestion buffer, and 40 μL of 2.5 mg/mL RNase-free DNase I is added and incubated for 60 minutes at 37°C to degrade the DNA. Other samples where the contamination of DNA may be low, the RNase-free DNase I can be used at a lower concentration (2 μg/mL) at 37°C for 1 hour to degrade the DNA. The proteins are removed by treating the sample with proteinase K (100 μg/mL proteinase K in 10 mM Tris-Cl, pH 7, 0.1% SDS, and 5 mM EDTA) for 1 hour at 37°C. After extracting the sample once with an equal volume of phenol–chloroform–isoamyl alcohol (25:24:1) and centrifugation, the separated aqueous phase is reextracted with an equal volume of chloroform–isoamyl alcohol (24:1). The RNA from the aqueous phase is precipitated by the addition of sodium acetate and ethanol as described previously. In lieu of DNase treatment, RNA can be purified from DNA by selective precipitation of RNA by lithium chloride. The RNA in the aqueous phase is precipitated with 1.4 volumes of 6 M LiCl at 4°C for at least 15 hours before centrifugation at 10,000 × g for 30 minutes at room temperature.

7.7. FRACTIONATION OF RNA USING CHROMATOGRAPHY METHODS

7.7.1. Fractionation of Small RNA by HPLC

Small RNA such as tRNA, 5S rRNA, and snRNAs cannot easily be isolated by the methods described previously. These can be fractionated from total RNA by high-performance liquid chromatography (HPLC). The sample determines the choice of the stationary and the mobile phases. RNA separation by HPLC depends on their polyanionic nature (anion exchange), lipophillic nucleobases (reversed phase), or their chain length. Samples are collected in fractions, and the RNA-containing fractions are identified by their ultraviolet absorbance. One approach to the separation of small RNAs is based on anion-exchange chromatography [25]. The negatively charged RNA is bound to a support matrix that is positively charged. The bound RNAs are eluted with increasing ionic strength of an eluant such as ammonium phosphate or ammonium sulfate. Larger RNA species with high negative charges are more tightly bound and therefore elute at higher ionic strengths. RNAs are detected by absorbance at 260 nm. Elution profile of specific tRNAs can be monitored by using labeled tRNA.

7.7.2. mRNA Isolation by Affinity Chromatography

For all functional analyses, it is necessary to purify mRNA from the other types of RNA. mRNA constitutes only a small fraction (a few percent) of the total RNA. For separation of mRNA from the rest of RNA, advantage is taken of the fact that most mRNA species have a long poly-A$^+$ tail at their 3' end (2-3). Oligo(dT) or poly-U affinity matrix is used to bind poly A$^+$-containing mRNA that can then be eluted from the column. Several such methods exist with variations in the type of oligo(dT) used or the matrix to which it is attached.

Isolation of mRNA Using Oligo(dT)–Cellulose Matrix

The poly-A$^+$ tail of mammalian mRNAs is 200 to 250 nucleotides long, although it can range from 50 to 300 nucleotides. By the virtue of hydrogen bonding between poly-A$^+$ stretches of RNA and the oligo-T (or oligo-U) bound to a solid matrix, the bound fraction representing mRNA can easily be isolated by elution with water, or a low-ionic-strength buffer. Total cellular RNA, extracted by any one of the methods described above is allowed to bind to oligo(dT) in high salt buffer (HSB) (0.5 M NaCl, 0.1% SDS, 1 mM EDTA, 10 mM Tris-HCl, pH 7.5). These conditions favor binding of mRNA poly-A$^+$ tails to the oligo(dT)–cellulose. Although the binding is greater with KCl, there is an increase in nonspecific hybridization to the matrix when it is used. Generally, 1 g of oligo(dT) can bind to 20–50 OD$_{260 \text{ nm}}$ units of poly A. In other words, 1 mg of total RNA requires 25 mg of oligo(dT)–cellulose, or the amount of oligo(dT)–cellulose is $\frac{1}{20}$ of the original tissue weight.

Binding of mRNA to the matrix can be carried out by batch adsorption in a small microfuge tube or by passing RNA through a column of binding matrix. Oligo(dT)–cellulose is first washed with 0.1 M NaOH and then equilibrated by washing with several bed volumes of HSB. Some commercial vendors supply oligo(dT)–cellulose that is recommended to be used directly without washing with alkali. Prior to binding, RNA in water is heated for 5 minutes at 65°C to disrupt any secondary structure in RNA, chilled in ice, and then diluted with an equal volume of double-strength HSB. RNA (1 mg/mL) in the high salt buffer is then adsorbed to the matrix at room temperature. In the batch adsorption method, RNA and the oligo(dT)–cellulose matrix are shaken on a rotatory shaker for 30 minutes at room temperature, followed by centrifugation at $750 \times g$ for 1 minute to collect the oligo(dT)–cellulose. The supernatant containing unbound RNA is discarded, and the pellet is washed four times by resuspending the cellulose matrix in high salt buffer and centrifugation. The washes are repeated four more times with low salt buffer (LSB) (0.15 M NaCl, 1 mM EDTA, 10 mM Tris-HCl, pH 7.5) before eluting bound mRNA.

In the column method, the oligo(dT)–cellulose is packed in a column. The heat-denatured RNA is then loaded in HSB directly onto the column. The column is then drained under gravity and washed with 5 bed volumes of HSB and 5 bed volumes of LSB buffer. The poly-A$^+$ RNA, in either batch or column method, is eluted with 4 bed volumes of water preheated to 65°C. The eluted RNA, after heat denaturation and addition of HSB, is passed again through a second round of poly-A$^+$ RNA selection on oligo(dT)–cellulose as before to enhance the enrichment of poly A$^+$–containing RNA. Although the column method is slow, the advantage is that fractions (50 to 100 μL) can be collected during elution of RNA. Those fractions with RNA in them (detected by UV absorbance or ethidium bromide fluorescence of an aliquot) are pooled in a siliconized RNase-free tube. Poly-A$^+$ RNA is concentrated by ethanol precipitation in the presence of 0.3 M sodium acetate at −20°C overnight, followed by centrifugation at $10,000 \times g$ for 20 minutes.

Oligo(dT)–cellulose can be regenerated for future reuse by washing with 10 bed volumes of a regeneration solution containing 0.1 M NaOH, 5 mM EDTA. After two washes with 3 bed volumes of HSB, the oligo(dT)–cellulose is resuspended in 1 volume of HSB and stored at 4°C for future use. Alternatively, for long-term storage, it can be washed with ethanol, dried, and stored at −20°C.

Isolation of mRNA by Biotin-Streptavidin Affinity Method

mRNA can be separated from the rest of the RNA by biotinylated oligonucleotide-mediated affinity chromatography (Figure 7.4). The amount of biotinylated probe used depends on the amount of mRNA in a tissue, which may range from 1 to 5% of the total RNA. Typically, 0.4 μg of biotinylated probe [biotinylated oligo(dT)] is hybridized to 0.25 μg of mRNA [7]. It is carried out in pH 7 buffer containing 1 M NaCl, 50 mM PIPES, and 2 mM EDTA at 85°C for 2 minutes, then at 55°C for 1 hour. An oligonucleotide can be conjugated to biotin moiety chemically [26,27]. Reagents for photobiotinylation and thiol-reactive labeling with biotin maleimide are available commercially (Vector Laboratories, Burlingame, CA; Sigma-Aldrich, St. Louis, MO). Moreover, biotinylated oligonucletides can also be custom synthesized by many commercial oligonucleotide-synthesizing vendors.

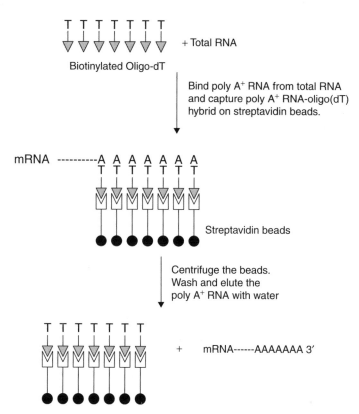

Figure 7.4. Isolation of poly-A$^+$ RNA by biotin–streptavidin affinity matrix. Poly-A$^+$ RNA is captured as a hybrid between poly-A$^+$ RNA and biotinylated oligo(dT) by streptavidin matrix. Most mRNAs carry poly-A$^+$ stretch at their 3′ end, and hence poly A–containing RNA can be enriched substantially by this affinity capture method. Poly-A$^+$ RNA can be eluted from the beads by low salt or water. The eluted RNA can be ethanol precipitated.

The mRNA that is bound to biotinylated oligonucletide can be recovered by separation of hybrids using affinity chromatography on streptavidin agarose beads. Biotin–avidin interaction offers one of the tightest-binding (K_d 10 to 15 M) systems. Streptavidin agarose is one such affinity support materials. The agarose, carrying mRNA bound to a biotinylated oligo(dT), is collected by centrifugation, batch washed as described for oligo(dT)–cellulose, and the RNA eluted with 10 mM Tris-HCl (pH 7.8), 30% formamide by incubating at 60°C for 10 minutes. The RNA is precipitated with ethanol as described before. Another variation of this approach is to use commercially available streptavidin-coated magnetic beads [28]. The advantage is that the streptavidin matrix is collected with the help of a magnetized

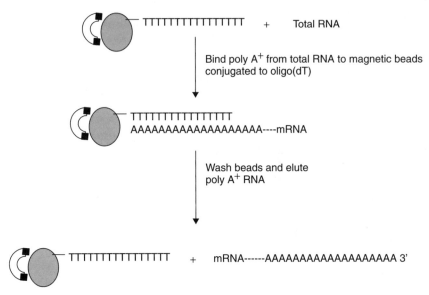

Figure 7.5. Isolation of poly-A$^+$ RNA by magnetic oligo(dT) capture. The principle of the method is again based on hydrogen bonding between poly-A$^+$ stretch at the 3' end of mRNA and the oligo(dT) that is bound to magnetic beads under high salt conditions. The beads are rapidly harvested after binding to RNA with help of a magnet. The isolated mRNA hybrid and oligo(dT) hybrid on beads can also be used for the in vitro synthesis of reusable magnetized cDNA that can be recovered after simple alkaline hydrolysis of RNA.

test-tube rack rather than by centrifugation, as in the case of streptavidin agarose. This is described in detail in Chapter 8. It is also possible to isolate total RNA or mRNA from a large number of samples using affinity-binding matrices and a robotic workstation such as MagNA Pure LC (Roche Molecular Biochemicals, Indianapolis, IN).

Isolation of mRNA Using Oligo(dT)-Coated Magnetic Beads

Yet another approach for mRNA isolation is the use of oligo(dT)-coated magnetic beads for mRNA isolation (Figure 7.5). Such beads with attached oligo(dT) are available commercially (Dynal, Lake Success, NY). Alternatively, the oligonucleotide can be chemically attached to magnetic beads [29].

The magnetic beads with attached oligo(dT) probes can be used for purifying mRNA either from a total RNA preparation or directly from the tissue [30–32]. The advantage of the latter is that mRNA can be isolated from very small amounts of tissue or cells. The tissue homogenate obtained after homogenization of the tissue with guanidinium thiocyanate buffer is diluted 1.6-fold with water to lower the molarity of guanidine thiocyanate to 2.5 M. An aliquot of the diluted tissue homogenate (0.25 mL) is vortexed for 10 seconds with 0.5 mL of oligo(dT)-coated beads [1 mg of beads, carrying 62.5 pmol of oligo(dT)] that represents a five- to sixfold molar excess over mRNA. Following a 5-minute hybridization of the mRNA to the oligo(dT) at 37°C in the bead hybridization buffer (0.1 M Tris-HCl, pH, 7.5; 0.01 M EDTA; 4% BSA; 0.5% sodium lauroyl sarcosine), the beads are harvested using a test-tube-rack permanent magnet. The clear supernatant is discarded and the beads resuspended in water for the elution of bound poly-A$^+$ RNA.

There are distinct advantages of using magnetic beads or streptavidin-coated magnetic beads over the conventional column methods for isolating RNA. These methods are rapid and offer ease of direct isolation from tissues per se, particularly from extremely limited amounts of tissue cells. In contrast, oligo(dT)-based columns tend to get clogged, and the loss of RNA on them is a concern, especially if starting samples are small.

7.8. ISOLATION OF RNA FROM SMALL NUMBERS OF CELLS

At times there are very small amounts of tissue available for experimentation. Nevertheless, it is possible to isolate RNA from limited amounts of samples such as tumor biopsies or laser capture microdissected material [33]. Although one can expect only small amounts of RNA from such tissues or cells, yields of RNA are enough for use in various applications, such as gene expression–related qualitative and quantitation studies using RT-PCR or even cDNA libraries [31,34]. Small-scale RNA isolation is possible using commercially available kits that are available to isolate either total RNA or mRNA (Table 7.3). These kits can be used to isolate RNA from a few cells, or less than 1 mg of tissue.

Most of these kits use standard guanidine thiocyanate lysis solution, and some have been standardized for isolation of RNA form a variety of tissues, blood, and cells. Some kits use silica-based membrane in a spin cartridge that sits in a microcentrifuge and binds RNA under high salt conditions. After washing out contaminants, RNA is eluted with water or low salt buffer using a tabletop microfuge. Others kits are designed to isolate mRNA by the selection of poly-A$^+$ RNA either from total RNA, or directly from the tissue homogenate. These kits use oligo(dT) bound to cellulose, or some synthetic beads to isolate mRNA directly from the tissue homogenate using a centrifugation method. Some kits use biotinylated oligo(dT) and strepta-vidin-coated magnetic beads (Promega, Madison, WI), microfuge tubes (Roche Molecular Biochemicals, Indianapolis, IN), or oligo(dT)-coated

Table 7.3. Commercial Kits for the Isolation of RNA from Small Tissue or
Small Number of Cells

Supplier	Name of the Kit
Ambion (Austin, TX)	RNAqueous
	MicroPoly(A)Pure
Amersham-Pharmacia (Piscataway, NJ)	QuickPrep Micro mRNA purification
Dynal (Lake Success, NY)	Dynabeads mRNA DIRECT Micro
Invitrogen (Carlsbad, CA)	S.N.A.P. Total RNA Isolation
	Micro-FastTrack 2 mRNA
Promega (Madison, WI)	SV Total RNA Isolation system
	PloyATtract system 1000
Qiagen (Valencia, CA)	Rneasy Mini
	Oligotex Direct mRNA Micro
Roche Molecular Biochemicals	High Pure RNA
(Indianapolis, IN)	mRNA Capture
Sigma-Aldrich (St.Louis, MO)	GenElute mammalian Total RNA
	GenElute mRNA Miniprep
Stratagene (La Jolla, CA)	StrataPrep Total RNA Miniprep and
	Microprep Kits

magnetic beads (Dynal, Lake Success, NY) to capture mRNA using magnetic separation of mRNA-carrying beads.

7.9. IN VITRO SYNTHESIS OF RNA

Sometimes, a single species of RNA may be needed, and isolation of the target RNA from a complex RNA mixture from a tissue can be quite difficult. However, a pure single species of RNA can be made in a laboratory test tube, provided that a recombinant plasmid DNA carrying the complementary DNA sequence is available. Plasmid vectors with a cloned DNA fragment flanked by promoters such as T3, T7, or SP6 are useful for generating RNA in vitro. The cloned DNA can be transcribed into RNA by enzymatic means using one of the RNA polymerases (T3 or T7 or SP6), depending on the promoter element present in the plasmid vector (Figure 7.6). Several commercial suppliers, such as Ambion (Austin, TX), Promega (Madison, WI), and Invitrogen (Carlsbad, CA), sell kits or reagents that allow synthesis of RNA in a test tube. Before transcription, the double-stranded circular plasmid DNA is made linear with a suitable restriction enzyme such that only the foreign DNA segment of interest gets transcribed into RNA, not the plasmid vector DNA (Figure 7.6). The reaction conditions differ when small amounts of high-specific-activity radioactively

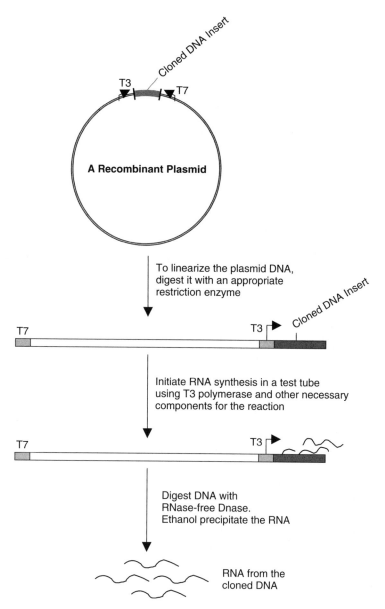

Figure 7.6. Schematic representation of in vitro synthesis of RNA. Shown is a plasmid molecule containing a cloned DNA that is flanked by T3 and T7 promoter sequences. The recombinant plasmid DNA is linearized in such a way that the transcription from one of the promoter elements generates RNA molecules corresponding to the cloned insert DNA and not the plasmid vector DNA. At the end of the reaction plasmid DNA is removed after enzymatic digestion with DNase I, and the pure RNA species is ethanol precipitated.

labeled RNA are generated is contrast to the large amounts of unlabeled RNA. A typical reaction, besides buffer components and the enzyme, contains four ribonucleotides (CTP, GTP, UTP, ATP), placental RNase inhibitor (RNasin), and linearized plasmid DNA. The DNA template, after the completion of the reaction is removed by DNase I treatment followed by extraction with phenol–chloroform to remove the enzymes. The RNA is then precipitated by ethanol, centrifuged, and dissolved in water. The advantage of this system is that large quantities of pure RNA can be made from the template.

7.10. ASSESSMENT OF QUALITY AND QUANTITATION OF RNA

Before proceeding with any experimentation involving RNA, it is essential to test the integrity of RNA. This is usually tested by agarose gel electrophoresis. When total cellular RNA or cytoplasmic RNA is subjected to electrophoreses under denaturing conditions, such as formaldehyde-containing agarose gels, two distinct bands of rRNA (28S and 18S) that constitute the majority of cellular RNA should be clearly visible after removal of formaldehyde by soaking the gel in water followed by staining with 0.5 μg/mL ethidium bromide in water (Figure 7.7). Minimal smearing in addition to the two distinct bands of ribosomal RNA is normal. However, if the 28S and 18S bands appear smaller than their expected sizes, or if a smear of these bands is observed, degradation of RNA has occurred and the RNA is of poor quality. Small RNAs such as transfer RNA or 5.8S ribosomal RNA, all comigrate at the leading edge of the gel. Caution is exercised since formaldehyde is a suspected carcinogen, and the gel is handled under a chemical hood and disposed of appropriately. Usually, to visualize two ribosomal RNA bands, 1 to 2 μg of RNA/lane in an agaorse gel is required.

In cases where very small quantities of RNA are isolated, quality assessment can be made by probing Northern blots prepared with small quantities of RNA with probes such as ribosomal RNA, β-actin, or oligo(dT). Such blots can be prepared by size fractionation of nanogram quantities of RNA in formaldeyde–agarose gels, followed by transfer to nylon membrane under high salt conditions [10,35].

If RNA degradation is noticed on an agarose gel, it is important to determine whether it occurred during the isolation procedure, during the running of gel electrophoresis, or if the RNA was degraded in the tissue prior to RNA isolation. Running an RNA sample in a nondenaturing agarose gel may indicate a smear because of RNA secondary structure and does not necessarily indicate RNA degradation. Hence, denaturing gels are preferred over nondenaturing gels for quality assessment of RNA. DNA con-

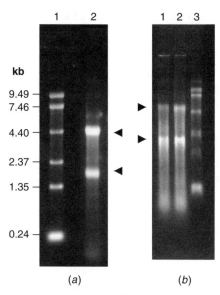

(a)   (b)

Figure 7.7. Agarose gel electrophoresis of total RNA. Total RNA from mouse skin (panel *a*, lane 2) and two human cadaver skin samples (panel *b*, lanes 1 and 2) were isolated by guanidine thiocyanate method and size fractionated on denaturing formaldehyde containing 1% agarose gel and stained with 0.5 µg/mL ethidium bromide. Note that in case of mouse skin RNA, two distinct ribosomal RNA bands (upper 28S and lower 18S bands) are clearly visible. In contrast, in case of human skin samples, which were collected several hours postmortem, there is partial RNA degradation as is evident by fuzzy 28S and 18S ribosomal RNA bands. RNA degradation is more pronounced in one of the samples than the other (panel *b*, compare lane 1 and lane 2). Ribosomal RNA bands are indicated by arrowheads. RNA size markers (Invitrogen, Carlsbad, CA) in the range 0.24 to 9.5 kb are in lane 1 (panel *a*) and lane 3 (panel *b*).

tamination appears mostly toward the top of the gel well above the 28S-RNA band. The integrity of RNA can also be judged by functional assay using in vitro cell-free translation system, such as commercially available reticulocyte lysate.

Although gel electrophoresis of RNA indicates the integrity of an RNA sample, contamination of RNA with proteins, salts, or organic reagents such as phenol or chloroform is detected spectrophotometically by measuring absorbance ratios at 260 and 280 nm. The ratio for pure RNA is typically between 1.8 and 2. Lower ratios indicate contamination. Strong absorbance at 280 nm indicates contamination of RNA with proteins, and strong absorbance at 270 and 275 indicates contaminating phenol. Reprecipitation of RNA followed by washing the RNA pellet with 75% ethanol should improve 260/280 ratios after removal of salts and organic solvents. Protein contamination can be eliminated by phenol–chloroform extraction. Some-

Table 7.4. Typical Yields of RNA from Various Tissues

Cells or Tissue (1 g)	Yield of Total RNA (µg)
Epithelial cells (1 million cells)	8–15
Fibroblast cells (1 million cells)	5–7
Kidney	3–4
Liver	6–10
Spleen	6–10
Skeletal muscle	~1
Placenta	1–4
Pea seedling	466

times, lower ratios could be due to acidity of the water in which absorbance measurements are taken. Absorbance measurements in dilute buffer solutions should avoid this problem.

Absorbance measurement at 260 nm also provides information about the quantity of RNA. An RNA solution of 44 µg/mL concentration will give an absorbance of one A_{260} unit when using a 1-mL quartz cuvette and a 1-cm path. However, most laboratories use a value of 40 µg OD rather than 44. It is important to remember that both DNA and RNA absorb at 260 nm. Hence, before making RNA measurements, it is advisable to remove DNA contamination. Fluorescence at 530 nm using RiboGreen (Molecular Probes, Eugene, OR) is another method of quantitation of RNA. Yield of RNA varies depending on the tissue and the method of isolation. Typical yields of RNA from various tissues are given in Table 7.4.

7.11. STORAGE OF RNA

Once purified of proteins, RNA is fairly stable. However, it is less stable than DNA. For long-term storage, it is advisable to store RNA in aliquots. Although RNA can be stored in RNase-free water at $-80°C$ for extended periods of time, it has been observed that at $-80°C$, it is more stable in formamide than in RNAse-free water [36]. Formamide is believed to protect RNA against degradation by RNase. It is equally important to know that the formamide must be deionized so as to remove oxidizing products that can degrade RNA. RNA can be recovered from formamide by addition of four volumes of ethanol in the presence of 0.2 M NaCl. After incubation for 5 to 10 minutes, RNA is recovered by centrifugation at 10,000 × g for 10 minutes.

REFERENCES

1. R. L. P. Adams, *The Biochemistry of Nucleic Acids*, Chapman & Hall, London, 1992.

2. M. Edmonds, M. H. Vaughan, Jr., and H. Nakazato, *Proc. Natl. Acad. Sci. USA*, **68**, 1336–1340 (1971).

3. H. Aviv and P. Leder, *Proc. Natl. Acad. Sci. USA*, **69**, 1408–1414 (1972).

4. J. Putz, J. Wientges, M. Sissler, R. Giege, C. Florentz, and A. Schwienhorst, *Nucleic Acids Res.*, **25**, 1862–1863 (1997).

5. R. M. Bock, *Methods Enzymol.*, **12A**, 224–228 (1968).

6. J. Sambrook, E. F. Fritsch, and T. Maniatis, *Molecular Cloning: A Laboratory Manual*, 2nd ed., Cold Spring Harbor Laboratory Press, Cold Spring Harbor, NY, 1989.

7. P. Jones, J. Qiu, and D. Rickwood, *RNA Isolation and Analysis*, Bios Scientific Publishers, Oxford, 1994.

8. R. E. Farrell, Jr., *RNA Methodologies: A Laboratory Guide for Isolation and Characterization*, Academic Press, San Diego, CA, 1993.

9. J. Adamovicz and W. C. Gause, in C. W. Dieffenbach and G. S. Dveksler, eds., *PCR Primer: A Laboratory Manual*, Cold Spring Harbor Laboratory Press, Cold Spring Harbor, NY, 1995.

10. F. M. Ausubel, R. Brent, R. E. Kingston, D. D. Moore, J. G. Seidman, J. A. Smith, and K. Struhl, eds., *Short Protocols in Molecular Biology*, 4th ed., Wiley, New York, 1999.

11. M. Girard, *Methods Enzymol.*, **12A**, 581–588 (1968).

12. K. S. Kirby, *Methods Enzymol.*, **12B**, 87–99 (1968).

13. R. D. Palmiter, *Biochemistry*, **13**, 3606–3615 (1974).

14. J. M. Chirgwin, A. E. Przybyla, R. J. MacDonald, and W. J. Rutter, *Biochemistry*, **18**, 5294–5299 (1979).

15. R. A. Cox, *Methods Enzymol.*, **12B**, 120–129 (1968).

16. R. J. MacDonald, G. H. Swift, A. E. Przyyla, and J. M. Chirgwin, *Methods Enzymol.*, **152**, 219–227 (1987).

17. P. Chomczynski and N. Sacchi, *Anal. Biochem.*, **162**, 156–159 (1987).

18. M. E. Schmitt, T. A. Brown, and B. L. Trumpower, *Nucleic Acids Res.*, **18**, 3091 (1990).

19. H. Okayama, M. Kawaichi, M. Brownstein, F. Lee, T. Yokota, and K. Arai, *Methods Enzymol.*, **154**, 3–28 (1987).

20. P. Chomczynski, *Biotechniques*, **15**, 532–537 (1993).

21. D. Majumdar, Y. J. Avissar, and J. H. Wych, *Biotechniques*, **11**, 94–101 (1991).

22. J. Favaloro, R. Treisman, and R. Kamen, *Methods Enzymol.*, **65**, 718–749 (1980).

23. H. Busch, *Methods Enzymol.*, **12A**, 421–448 (1968).

24. M. L. Frazier, W. Mars, D. L. Florine, R. A. Montagna, and G. F. Saunders, *Mol. Cell. Biochem.*, **56**, 113–122 (1983).

25. B. S. Dudock, in M. Inouye and B. Dudock, eds., *Molecular Biology of RNA: New Perspectives*, Academic Press, New York, 1987.

26. T. Kempe, W. I. Sundquist, F. Chow, and S. L. Hu, *Nucleic Acids Res.*, **13**, 45–57 (1985).

27. D. Y. Kwoh, G. R. Davis, K. M. Whitfield, H. L. Chappelle, L. J. DiMichele, and T. R. Gingeras, *Proc. Natl. Acad. Sci. USA*, **86**, 1173–1177 (1989).

28. B. H. Thorp, D. G. Armstrong, C. O. Hogg, and I. Alexander, *Clin. Exp. Rheumatol.*, **12**, 169–173 (1994).

29. V. Lund, R. Schmid, D. Rickwood, and E. Hornes, *Nucleic Acids Res.*, **16**, 10861–10880 (1988).

30. E. Hornes and L. Korsnes, *Genet. Anal. Tech. Appl.*, **7**, 145–150 (1990).

31. E. E. Karrer, J. E. Lincoln, S. Hogenhout, A. B. Bennett, R. M. Bostock, B. Martineau, W. J. Lucas, D. G. Gilchrist, and D. Alexander, *Proc. Natl. Acad. Sci. USA*, **92**, 3814–3818 (1995).

32. D. V. Morrissey, M. Lombardo, J. K. Eldredge, K. R. Kearney, E. P. Groody, and M. L. Collins, *Anal. Biochem.*, **181**, 345–359 (1989).

33. F. Fend, M. R. Emmert-Buck, R. Chuaqui, K. Cole, J. Lee, L. A. Liotta, and M. Raffeld, *Am. J. Pathol.*, **154**, 61–66 (1999).

34. K. Schutze and G. Lahr, *Nat. Biotechnol.*, **16**, 737–742 (1998).

35. P. S. Thomas, *Proc. Natl. Acad. Sci. USA*, **77**, 5201–5205 (1980).

36. P. Chomczynski, *Nucleic Acids Res.*, **20**, 3791–3792 (1992).

CHAPTER

8

TECHNIQUES FOR THE EXTRACTION, ISOLATION, AND PURIFICATION OF NUCLEIC ACIDS

MAHESH KARWA AND SOMENATH MITRA

Department of Chemistry and Environmental Science, New Jersey Institute of Technology, Newark, New Jersey

8.1. INTRODUCTION

The quality of isolated nucleic acids is critical in obtaining accurate and meaningful analytical data. To obtain high-purity nucleic acids from a complex matrix, such as a cell lysate requires well-designed sample preparation procedures. Typical impurities to be removed include cell debris, small molecules, proteins, lipids, carbohydrates, inactivation of cellular nucleases, and unwanted nucleic acids. In the early years of chromatography, attempts were made to inject the lysate directly into a column right after centrifugation. It was quickly realized that this resulted in deterioration of the column's performance, introduced background interferences, and increased column backpressure. In addition, the irreversible adsorption of background species such as proteins altered the chromatographic selectivity.

Since then much progress has been made in sample preparation techniques that reduce sample complexity. An overview of the sequence of extraction, isolation, and purification of nucleic acids is presented in Figure 8.1. It can be categorized in several unit steps beginning with the extraction of DNA until its sizing and sequencing. The different options within each step are listed in Table 8.1 and are described in this chapter. The technique best suited in a given application depends on:

- The starting material (whole organ, tissue, cell culture, blood, etc.)
- The source organism (mammalian, lower eukaryotes, plants, prokaryotes, and viruses)
- Target nucleic acid (ssDNA, dsDNA, total RNA, mRNA, etc.)

Sample Preparation Techniques in Analytical Chemistry, Edited by Somenath Mitra
ISBN 0-471-32845-6   Copyright © 2003 John Wiley & Sons, Inc.

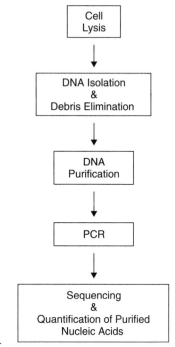

Figure 8.1. Steps in sample preparation.

Table 8.1. Techniques in Extraction, Isolation, and Purification of Nucleic Acids

Unit Steps	Techniques Available
Cell lysis	Mechanical methods: pressure shearing, ultrasonic disintegration, bead-mill homogenizers Nonmechanical methods: enzymatic lysis, osmotic lysis, freezing and thawing, detergent-based lysis and electroporation
Solids and debris removal	Centrifugation, filtration, membrane separation and precipitation
DNA purification	Solvent extraction and precipitation, gel electrophoresis, chromatography: size exclusion, ion exchange, solid-phase extraction, SPRI, affinity purification
Isolation of purified DNA	Washing, elution, precipitation, and centrifugation
DNA amplification	PCR
DNA analysis (size sequencing and quantification)	Capillary gel electrophoresis, CE and microchip-based CE

- Desired results (yield, purity, and purification time available)
- Downstream application [PCR (polymerase chain reaction), cloning, labeling, blotting, RT (reverse transcriptase)-PCR, cDNA synthesis, RNAse protection assays, gene therapy, etc.]

8.2. METHODS OF CELL LYSIS

The first step in the extraction of nucleic acids may require the lysis of cells and the inactivation of cellular nucleases. These two steps may be combined into one. For instance, a single solution may contain detergents to solubilize the cell membrane and strong chaotropic salts to inactivate the intracellular enzymes.

The choice of lysis procedures depends on the properties of the cell wall. So it is important to know the components/structure of the cell wall under investigation. For example, there are two major types of eubacterial cell walls. They can be identified by their reaction to certain dyes (characterized by Christian Gram in the 1880s). Gram-positive cells are stained purple and gram-negative cells are stained red in the presence of Gram stain (crystal violet dye along with iodine). The gram-positive cell wall is thicker (30 to 100 nm) than the gram-negative cell wall (20 to 30 nm thick). Approximately 40 to 80% of the gram-positive cell wall is made of a tough, complex polymer called *peptidoglycan* (Figure 8.2), which is highly cross-linked (linear heteropolysaccharide chains cross-linked by short peptides). As a result, the gram-positive cell wall (e.g., *Streptococcus pyogenes*) is very sensitive to the action of penicillin (or its derivatives) and lysozyme (an enzyme found in tears and saliva), which have the ability to hydrolyze the peptide linkage. The gram-negative cell wall (e.g., *Escherichia coli*) has a distinctively layered appearance, as shown in Figure 8.3. Its inner region consists of a monolayer of peptidoglycan, while the outer region is essentially a protein-containing lipid bilayer. In gram-negative bacteria, 15 to 20% of the cell wall is made up of peptidoglycan and is cross-linked only intermittently. The extent of cross-linking determines the toughness of the cell wall. In general, gram-positive cells are relatively harder to lyse than gram-negative cells. Older cells may be more easily lysed than younger ones, and the larger cells may be lysed more easily than smaller ones.

There are several methods of cell lysis (Table 8.2) [1,2], but there is none that works with cells of all biological origins. Each technique has its advantages and disadvantages, and the specific method of choice depends on the cell characteristics, the cell type, and the final application. A combination of more than one method may also be used. For example, enzymatic lysis uses specific enzymes to target the cell wall. However, to disrupt the cytoplasmic

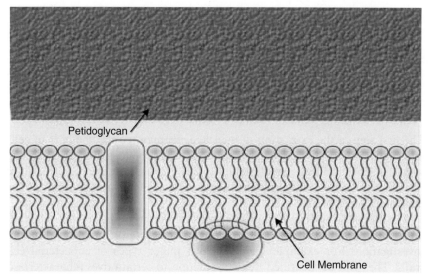

Figure 8.2. Gram-positive type cell wall. (Reprinted with permission from *http://www.bact.wisc.edu/MicrotextBook/BacterialStructure/CellWall.html*)

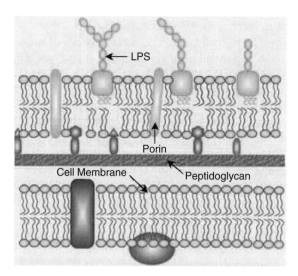

Figure 8.3. Gram-negative type cell wall. (Reprinted with permission from *http://www.bact.wisc.edu/MicrotextBook/BacterialStructure/MoreCellWall.html*)

Table 8.2. Comparison of Various Cell Lysis Methods

Lysis Method	Instrument Requirements	Mode of Lysis	Mechanism
Pressure shearing	Required, moderate cost	Harsh	Shear forces
Ultrasonic disintegration	Required, moderate cost	Moderate	Shear forces
Bead milling	Required, inexpensive	Harsh	Shear forces
Enzymatic	Not required, cost of enzyme (moderate)	Gentle	Breaks covalent bonds— specific to cells
Osmotic lysis	Not required, inexpensive	Gentle	Osmotic shock
Freezing and thawing	Not required, inexpensive	Gentle	Shock
Detergents	Not required, inexpensive	Gentle	Solubilization of the lipid bilayer
Electroporation	Required, moderate cost	Moderate	Irreversible permeation of the membrane

cell membrane made of lipid bilayer, detergents that solubilize the lipid are needed. The prerequisite of every lysis method is that it should be rigorous enough to lyse the cells and at the same time be gentle enough to preserve the integrity of the target nucleic acids. Complete lysis of all microorganisms in a target sample is required if the recovered nucleic acids truly represent the sample. Cell lysis procedures could be classified into two broad categories: mechanical methods and nonmechanical methods. A brief discussion of these follows.

8.2.1. Mechanical Methods of Cell Lysis

There are several methods of lysing a cell mechanically.

Pressure Shearing

Pressure shearing is a widely used mechanical means of cell breakage. A sample of bacterial suspension (5 to 40 mL, up to 30% cells by volume) is placed in a steel cylinder fitted with a piston and a small relief valve connected to an outlet tube. The entire assembly is placed in a 10-ton hydraulic press. When the piston is forced down, high shear forces are generated as the

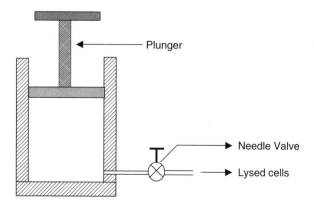

Figure 8.4. French pressure cell.

suspension passes through the small orifice of the relief valve. This breaks the cells. The French Pressure Cell Assembly [American Instrument Co. (Aminco), Rockville, MD] is a popular pressure shear device. Figure 8.4 shows the schematic of the French pressure cell. Pressures as high as 20,000 psi are applied during lysis. It is effective in the lysing of gram-negative bacteria and some gram-positive bacilli.

Ultrasonic Disintegration

Cell disintegration by ultrasound is due to the rapid vibration of an ultrasonic probe tip, which causes *cavitation*. The cavitiation results in the formation of microscopic gas bubbles streaming at high velocity in the vicinity of the tip. The high-shear forces generated by the fast-moving bubbles result in cell breakage. It is not instantaneous, and a cell suspension may need to be treated for several minutes to lyse a reasonable fraction of the cells. Although it is not a good method for primary cell breakage, it is useful for the lysis of spheroplasts and for the separation of inner and outer membranes in gram-negative bacteria. Ultrasonic disintegrators generate considerable heat during processing. For this reason the sample should be kept ice cold if possible. The addition of 0.1- to 0.5-mm-diameter glass beads in a ratio of one volume of beads to two volumes of liquid is recommended for microorganisms. Free radicals can be generated during sonication, which can react with most biomolecules. However, the damage caused by these oxidative free radicals can be minimized by including scavengers such as cysteine, dithiothreitol, or other −SH compounds in the media.

The ultrasonic probe consists of an electronic oscillator and an amplifier whose ac output is converted to mechanical waves. The transducer output is

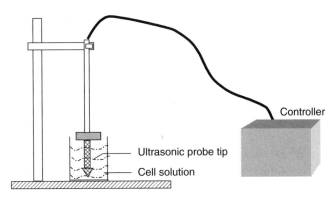

Figure 8.5. Schematic diagram of an ultrasonic disruptor.

coupled to the suspension undergoing treatment by a half-wave metal probe, which oscillates at the circuit frequency. Most ultrasonic disintegrators work in the frequency range 15 to 25 kHz. Typical power densities should be on the order of 100 W/cm^2. Figure 8.5 shows the schematic diagram of an ultrasonic probe. Some manufacturers of ultrasonic disintegrators are Artek Systems (Farmington, NY), BioSpec Products (Bartlesville, OK), Branson Sonic Power Company (Danbury, CT), B. Braun Biotech (Bethlehem, PA), RIA Research Corp. (Hauppauge, NY), Sonic Systems (Newton, PA), and VirTis Company (Gardiner, NY). A new development is a cordless disrupter from BioSpec Products (Bartlesville, OK). With a $\frac{1}{8}$-in. tip diameter, it easily fits into microtubes and 96-well titer plates.

Bead Mill Homogenizers

This is the most widely used mechanical method of cell lysis. A large number of minute glass beads are vigorously agitated by shaking or stirring in a bead mill. Disruption occurs by the crushing action of the glass beads as they collide with the cells. The schematic of a shaker bead mill is shown in Figure 8.6. It is possible to shake a mixture of cell suspension with glass beads manually and bring about cell disruption. The common approach is to use a vortex mixture, where the process can be completed in a few minutes. The handheld approach is slow and tedious, so mechanical devices that use either shaking or stirring actions are more common. After treatment, the beads settle down by gravity, and the cell extract is easily removed.

The size of the glass beads is important. The optimal size for bacteria and spores is 0.1 mm; it is 0.5 mm for yeast, mycelia, microalgae, and unicellular animal cells such as leucocytes or tissue culture cells. The speed of disruption

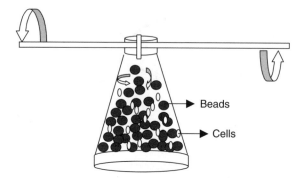

Figure 8.6. Schematic diagram of a shaking bead mill.

is increased by about 50% when higher-density ceramic and zirconia–silica beads are used in place of glass. The loading of the beads should be at least 50% of the total liquid-biomass volume, but can be as high as 90%, provided that adequate agitation is still possible. Generally, a larger fraction of beads in the cell suspension leads to faster cell disruption.

Shaking Bead Mills. The simplest example of shaking device is the Mickle shaker. This device had no provision for cooling the sample during shaking, and it is necessary to interrupt the shaking frequently to cool the sample container. This device has been replaced by the Braun MSK tissue disintegrator (B. Braun Melsungen Apparatebau, Melsungen, Germany). It disintegrates most samples within 3 to 5 minutes at a temperature below 4°C. Cooling is provided by a stream of liquid CO_2 delivered to the sample container. The sample container can be shaken horizontally at a speed of 2000 to 4000 oscillations per minute and can hold sample sizes up to 40 mL. Smaller volume bead mills have smaller breakage chambers with high surface/volume ratios for adequate heat dissipation without requiring external cooling. Some commercial products include Mini-BeadBeater (BioSpec Products, Bartlesville, OK), the Micro-Dismembrator II (B. Braun Biotech, Bethlehem, PA), Retsch Mixer (Brinkmann, Westbury, NY), and the Fast-Prep (Bio 101, Vista, CA). BioSpec Products manufactures two versions, one that holds a single 2-mL screw-cap microvial, and the other capable of handling eight vials at a time. Disruption of microorganisms takes about 1 to 5 minutes.

Rotor Bead Mills. Larger-capacity laboratory bead mill cell disrupters agitate the beads with a rotor rather than by shaking. Equipped with efficient cooling jackets, larger sample volumes can be processed without overheating. A popular model is the Bead Beater (BioSpec Products, Bartlesville, OK). This unit can handle sample volumes as large as 250 mL and takes 3 to 5 minutes to lyse cells. Cell concentrations as high as 40% can be used.

Although the foregoing cell disrupters are used primarily for the microorganisms, they can also be used to homogenize and extract plant and animal tissue. They are suitable for both soft tissues and tough/fibrous samples such as skin, tendon, or leaves. Extraction yields are often higher than those by other methods. For nucleic acid isolation, the lysis can be carried out directly in the extraction solution (e.g., phenol or guanidinium SCN), where nuclease concerns are eliminated and yields are enhanced. For PCR applications, the use of disposable microvials eliminates cross-contamination between samples. Selective homogenization is sometimes possible using different bead sizes or by manipulation of agitation speed.

8.2.2. Nonmechanical Methods of Cell Lysis

Enzymatic Lysis

Enzymes target specific bonds in the cell wall and provide the most gentle cell lysis with minimum mechanical damage. This method is often limited to releasing periplasmic or surface enzymes. Some common enzymes involved in cell lysis are $\beta(1, 6)$ and $\beta(1, 3)$ glycanases, proteases, and mannase.

Egg white lysozyme has been used extensively for the preparation of spheroplasts from gram-positive microorganisms such as *Bacillus megaterium*, *Micrococcus lysodeikticus*, *Sarcina lutea*, and *Streptococcus faecalis*. Egg white lysozyme in the presence of chelating agents is also used for the partial dissolution of the cell wall of certain gram-negative bacterial species, such as *Escherichia coli*, *Proteus*, *Aerobacter*, *Pseudomonas*, and *Rhodospirillum rubrum*. Lysozyme digestion is normally carried out in a suitable osmotic buffer such as the hypotonic sucrose dilution (0.3 to 0.5 M) at neutral or slightly alkaline pH. Thirty-minute treatment at lysozyme concentrations of 0.1 to 1.0 mg/mL generally results in complete protoplast formation. Low levels of Mg^{2+} or Ca^{2+} (in the range 0.5 to 5 mM) may be required to stabilize the protoplasts. Muramidases, which have broader substrate specificity than egg white lysozyme, are commercially available. They are effective in digesting the cell wall of the gram-positive organisms that are resistant to egg white lysozyme.

Unlike gram-positive bacteria, virtually all gram-negative bacteria have a peptidoglycan structure, which is sensitive to lysozyme. However, the outer membrane of gram-negative bacteria is impermeable to lysozyme and must be destroyed before lysozyme will act. This can be done by freezing and thawing, by pretreatment with ethylenediaminetetraacetic acid (EDTA), or

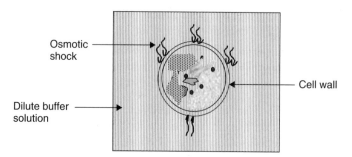

Figure 8.7. Cell undergoing osmotic shock.

in some cases by membrane-active antibiotics such as polymyxin B. Digestion of the peptidoglycan of gram-negative bacteria does not result in the removal of the outer membrane, and the osmotically sensitive structures thus formed are referred to as *spheroplasts* instead of protoplasts.

Osmotic Lysis

The solute concentration inside the cell and the pressure on the membrane are relatively high (on the order of 75 psi). When the solute concentration outside the cell is low, the concentration gradient makes the water flow in while the solute tends to flow out. The membrane holds the solute in and is only permeable to water. Without something supporting the membrane, the cell can swell and burst due to this osmotic shock caused by the hydrostatic pressure on the cell membrane (Figure 8.7). Most microorganisms cannot be disrupted by osmotic shock unless their cell walls are first weakened by enzymatic attack or by the growth under conditions that inhibit cell wall synthesis. Rapid dilution (10- to 20-fold) with dilute buffer (or distilled water) effectively lyses spheroplasts and protoplasts. The size of the vesicles produced may be determined by the speed and the extent of the dilution. A commonly used procedure is to pellet the protoplasts (or spheroplasts) by centrifugation, followed by suspension in the lysing buffer.

Freezing and Thawing

Freezing and thawing may render gram-negative cells sensitive to lysozyme and detergents. This procedure can be applied to the large-scale isolation of membranes or subcellular organelles. The following procedure is effective for *E. coli*. A thick suspension of washed cells (about 30% cells by volume) in 0.02 M Tris buffer (pH 7.8) containing 5 mM EDTA, 0.25 M sucrose, and 0.5 mg of lysozyme per milliliter is placed in a flask and is frozen by swirling in a dry ice–acetone bath. The flask is then thawed in warm water until the ice melts, and the contents are poured into volumes of cold 0.02 M Tris buffer containing 0.5 mM MgCl$_2$ and 0.1 mg/mL of deoxyribonuclease. This material is immediately treated in a laboratory blender for about 20 minutes to shear the cell wall and disperse the cells.

Use of Detergents in Cell Lysis

Detergents are amphiphatic molecules that have both hydrophilic and hydrophobic properties. They provide a gentle means of lysing cells once the integrity of the peptidoglycan (gram-positive bacteria) or the outer membrane (gram-negative bacteria) have been damaged. Detergents are used to solubilize the cytoplasmic membrane (the lipid bilayer) selectively while leaving the outer membrane intact. This causes the cells to lyse. Detergents can also be used to remove membrane contamination from ribosomes, polysomes, and gram-positive cell walls. They are also known to denature proteins. This is one of the most prevalent methods of cell lysis because both chromosomal and plasmid DNA are sensitive to flow-induced stresses encountered during mechanical lysing.

The most commonly used ionic detergents are sodium dodecyl sulfate (SDS), sodium N-lauryl sarcosinate (Sarkosyl), alkyl benzene sulfonates, and quaternary amine salts such as cetyl trimethylammonium bromide (CETAB). Ionic detergents tend to form small micelles (molecular weight around 10,000) and exhibit a rather high critical micelle concentration (CMC). CMC is the critical concentration of the surfactant molecules above which micelle formation takes place. The CMC for SDS in dilute buffers is about 0.2% at room temperature. The nonionic detergents include polyoxyethylene(10), isooctylcyclohexylether (Triton X-100), Nonidet P-40 (NP-40), polyoxyethylene(20) sorbitan monooleate (Tween 80), and octyl glucoside. In general, these detergents are characterized by higher molecular weight (50,000 or greater) and lower CMC (0.1% or less).

An alkali-detergent solution is used during the recovery of plasmid DNA. The alkaline pH causes the chromosomal DNA to be irreversibly denatured while the plasmid is reversibly denatured. The mixture is subsequently neutralized by the addition of a suitable reagent. At neutral pH, the plasmid DNA renatures and remains in solution while the denatured chromosomal DNA precipitates, forming a complex network with other materials, such as proteins and cell debris. The precipitated material flocculates and transforms into a porous gel over a period of 1 to 2 hours. The gel slowly floats to the surface, leaving behind in the solution the plasmid DNA and the fine particulates.

Although detergents provide a fast and gentle means of lysing cells, some of the detergents are also known to interfere with the PCR, even at low concentrations [3,4]. Therefore, any residual detergent not removed during the purification process could coelute with the DNA and inhibit the PCR process. Thus, detergent cleanup may be necessary prior to PCR.

The detergent lysis is relatively fast, and the release of intracellular contents causes dramatic changes in the physical properties, such as viscosity and solution consistency. For example, Levy et al. [5] carried out the lysis of a suspension of *E. coli* C600 with 0.2 M NaOH containing 1% w/v of SDS. The cell suspension and the detergent solution were mixed in a 1:1 volumetric ratio. The reaction was carried out in the narrow gap of a coaxial viscometer. The viscosity increased rapidly with addition of the SDS–NaOH solution to the cell suspension. The maximum viscosity occurred in 100 seconds, suggesting that all the cells had been lysed in that time.

Electroporation

Cell lysis under a high electric field is referred to as *electroporation* [6]. Under these conditions, the cell membrane experiences dramatic changes in permeability to macromolecules. The main applications of the electroporation include the electrotransformation of cells and the electroporative gene transfer by the uptake of foreign DNA or RNA (in plants, animals, bacteria, and yeast). The electric field generates permeable microspores at the cell membrane, so that the nucleic acid can be introduced by electroosmosis or diffusion.

The microspores generated during electroporation are instantly resealable. However, if the electric field is high enough, irreversible mechanical breakdown of the cell membrane occurs as depicted in Figure 8.8. This is a result of imbalance in the osmotic pressure of the cytosol and the external medium that makes the cells swell and eventually break. AC fields as high as 21 kV/cm at 2 MHz and dc fields as high as 10 kV/cm at 100 μs are used in electroporation. The mechanism of electroporation involves the attraction of opposite charges induced on the inner and outer membrane generating compression pressure, which makes the membrane thinner. Once the electric field strength exceeds a critical value, the cell membrane becomes permeable to the medium and the lysis is permanent.

8.3. ISOLATION OF NUCLEIC ACIDS

After cell lysis and nuclease inactivation, cellular debris may be removed by filtration and precipitation. However, other smaller molecules and proteins that remain in the solution need to be separated from the nucleic acids.

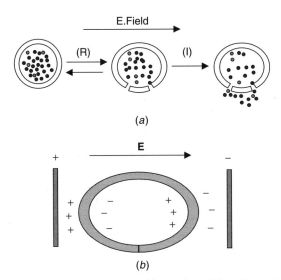

Figure 8.8. (*a*) Reversible (R) and irreversible (I) poration of the cell membrane. (*b*) Charge polarization on the membrane due to the application of electric field.

Some of the commonly used purification methods at this stage are solvent extraction, precipitation, membrane filtration, chromatography, affinity purification, and electrophoresis. Sometimes more than one method is necessary to attain the required level of purity. The challenge is the removal of contaminants that have similar physical–chemical properties as the nucleic acids of interest. Depending on the analysis, these could be lipopolysaccharides, RNA, or chromosomal DNA. While most genomic DNA is denatured and precipitated during lysis, large amounts of RNA and proteins need to be removed in the subsequent steps. Isolation and purification methods are also important in nonanalytical applications, such as in gene therapy and in the production of pharmaceutical-grade plasmids. Here the steps are similar: the isolation of plasmids from the host cells, clarification and concentration of the extract, and finally the purification of the plasmids.

The usefulness of sample cleanup [7] cannot be understated. Figure 8.9 shows an example of the cleanup of a plasmid-containing stream. Figure 8.9*a* is a chromatogram of a pMa-L plasmid standard. The first peak here corresponds to sample buffer (TE), and the plasmid is eluted as a single peak at 5.8 minutes. Figure 8.9*b–d* show chromatograms as the plasmid undergoes different levels of purification. The first peak is from impurities (RNA, proteins, oligoribonucleotides) and the second is the plasmid peak. The first peak is very pronounced in Figure 8.9*b*, right after cell lysis. It decreases in

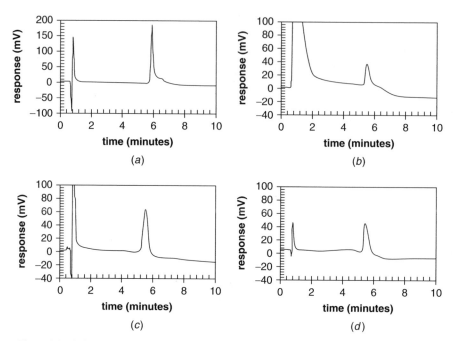

Figure 8.9. Anion-exchange HPLC analysis of a plasmid sample. (*a*) Qiagen-purified plasmid (standard). (*b–d*) Analysis of streams within the process: (*b*) after lysis; (*c*) after clarification/ concentration; (*d*) after ion exchange. (Reprinted with permission from Ref. 7.)

Figure 8.9*c* after clarification/concentration. Further decrease is observed in Figure 8.9*d* after ion-exchange cleanup. The percentage of plasmid peak area in each chromatogram can be used as an estimate of sample purity.

8.3.1. Solvent Extraction and Precipitation

Phenolic extraction of cell lysates is one of the oldest techniques in DNA preparation. Examples of these have been presented in Chapters 6 and 7. Single cells in suspension are lysed with a detergent, and a proteinase enzyme is used to break down the protein molecules. Non–nucleic acid components are then extracted into an organic (phenol–chloroform) solvent, leaving nucleic acids in the aqueous layer. Two volumes of isopropanol are added to the isolated aqueous phase to precipitate the high-molecular-weight nucleic acids as a white mass. These are then treated with DNase-free ribonuclease (RNase) to remove the RNA. This is followed by a second treatment with proteinase, phenol extraction, and isopropanol precipitation. After precipitation, the DNA is separated from the isopropanol by

Figure 8.10. Precipitation of DNA on adding isopropanol.

spooling or centrifugation and is washed twice with ethanol to isolate pure DNA, which by then is a clean, white, fibrous material. This is shown in Figure 8.10.

8.3.2. Membrane Filtration

Membranes with small pores can be used to retain selectively nucleic acids of different sizes. For example, membrane filtration has been used to retain circular double-stranded DNA, which is larger in diameter than linear double-stranded DNA of the same molecular weight [8]. Membrane filters can also filter out alkaline lysates and other cell debris. Martinex et al. [9] reported the use of a poly(ether sulfone) ultrafiltration membrane, pretreated with a solution of linear polyacrylamide (LPA) to minimize the adsorption of DNA sequencing fragments and to eliminate DNA templates (or circular DNA vectors) from the sequencing reaction products. The membranes with a 0.01-μm pore size and with a molecular-weight cutoff (MWCO) of 100,000 in a spin column format (Pall Filtron, North Borough, MA) were found to be efficient in retaining the DNA templates. The membrane pore size was found to be critical in trapping the circular DNA vectors. Saucier and Wang [8] reported the preferential retention of circular double-stranded DNA over the linear double-stranded ones by passage through a cellulose ester membrane. The retention was found to be sensitive to flow rate but insensitive to the membrane pore size in the range of 0.2 to 0.8 μm. Over 80% of circular λ DNA with a molecular weight greater than 30.5×10^6 was retained along with 10% of the linear form. The difference in retention between the nicked circular and the linear form became smaller as the molecular weight decreased. Superhelical λ DNA was retained less than the circular λ DNA with a few single-chain scissions.

Many companies, such as Qiagen (Hilden, Germany), Pall Corporation (East Hills, NY), and Becton Dickinson Inc. (Franklin Lakes, NJ) have proprietary membrane filtration technologies. The membrane filters allow rapid clearing of alkaline lysates without centrifugation. Postlysis centrifugation is one of the most time-consuming steps in plasmid purification, especially when large culture volumes or large numbers of samples are involved. Commercially available filtration membranes remove SDS precipitates and cell debris efficiently following alkaline lysis. Insoluble complexes containing chromosomal DNA, salt, detergent, and proteins formed during the neutralization step are removed without clogging, foaming, or shearing of DNA. The filtration materials used do not bind DNA, thereby avoiding any loss during the filtration step.

A recent development is to combine filtration with solid-phase extraction separation. These filter modules contain a unique silica gel membrane that binds up to 20 µg of DNA in the presence of a high concentration of chaotropic salt and allow eventual elution in a small volume of low-salt buffer. They also contain an asymmetric laminar membrane with a gradation of pore sizes for efficient removal of material precipitated in the lysate. Such membrane filters eliminate time-consuming phenol–chloroform extraction and alcohol precipitation. The impregnation of silica in the membrane matrix also prevents the problems associated with loose resins and slurries. High-purity plasmid DNA eluted from such modules is ready to use and often needs no further precipitation, concentration, or desalting.

8.4. CHROMATOGRAPHIC METHODS FOR THE PURIFICATION OF NUCLEIC ACIDS

It is often desirable to go through a postlysis separation/concentration step prior to chromatography. Concentration methods involve the use of ammonium acetate and polyethylene glycol precipitation to further remove host proteins and small nucleic acids. These methods also reduce the volume of the sample (or the process streams) prior to chromatographic purification. The separation may also involve centrifugation and filtration to remove cell debris.

Chromatography is relatively easy to optimize and scale up, and several plasmid properties, such as charge and size, can be exploited in the design of these separations. Typically, plasmid DNA of 3000-base pair (bp) size has an average length of 10,050 Å (based on 3.35 Å/bp). However, upon super-coiling, plasmids adopt a branched interwined shape and become more

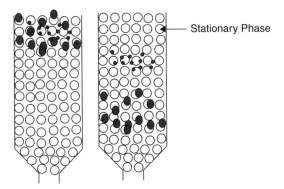

Figure 8.11. Size-exclusion chromatography. The small solid circles are retained in the gel pores represented by the open circles, while the larger molecules pass through.

compact and uniform. Chromatographic methods [10] for purification of nucleic acids include gel-based size exclusion chromatography, ion-exchange chromatography, adsorption chromatography, and solid-phase extraction.

8.4.1. Size-Exclusion Chromatography

In size-exclusion chromatography (SEC), also known as *gel filtration chromatography*, a nonionic, hydrophilic stationary phase is used along with an aqueous mobile phase. When hydrophobic packings are used along with nonaqeous eluents, it is called *gel permeation chromatography*. Large molecules elute first, because they are excluded from the pores in the stationary phase, while smaller molecules are retained in the pore maze (Figure 8.11). The separation in SEC depends on the size and shape of the molecules. Gels of appropriate porosity are selected to exclude pre-selected molecules. They are available in various porosities that can exclude molecules in the range 10,000 to 200,000 Da. The excluded DNAs are easily separated from RNA hydrolysate and smaller molecules (10^2 to 10^3 Da), which are eluted with different retention times. Usually, the purification of nucleic acids from nucleotides works well using an exclusion size of 25,000 Da. Table 8.3 lists the various pore sizes of the gels needed to elute analytes in different molecular weight ranges.

SEC is appropriate for the separation of small linkers from large plasmids or for removing salts and other small molecules from high-molecular-weight material [11]. It is also used for the separation of unincorporated fluorescently tagged nucleotides from the labeling reaction mixture. The main disadvantage is that all SEC materials have an upper limit in fractionating

Table 8.3. Molecular Weight as a Function of Pore Size in Gels

Pore Size (Å)	Globular Molecules (Molecular Weight)	Linear Molecules (Molecular Weight)
60	$5 \times 10^3 - 45 \times 10^3$	$5 \times 10^2 - 1 \times 10^4$
100	$5 \times 10^3 - 16 \times 10^4$	$5 \times 10^2 - 25 \times 10^3$
300	$1 \times 10^4 - 1 \times 10^6$	$2 \times 10^3 - 1 \times 10^5$
500	$4 \times 10^4 - 1 \times 10^7$	$1 \times 10^4 - 35 \times 10^4$
1000	$4 \times 10^5 - 1 \times 10^7$	$4 \times 10^4 - 1 \times 10^6$
4000	NA	$7 \times 10^4 - 1 \times 10^7$

the size of the restriction fragments. Thus, a single gel-packed column cannot be used for the purification of all nucleic acids.

The matrices used in SEC are either polymeric or silica-based particles with a hydrophilic coating. The disadvantage of silica particles is that they tend to retain solutes by adsorption and may catalyze the degradation of solute molecules. To reduce adsorption, the surfaces of these particles are often modified by reaction with organic substituents.

The polymer-based packings are more common. The hydrophilic gels are preferred because they allow the use of aqueous solvents for the elution of nucleic acids. The gels are relatively inert, and the degree of cross-linking determines the average pore size of the gel. Dextran, polyacrylamide, and agarose are the three common cross-linked polymers. Cross-linked dextrans are sold under the trade name *Sephadex* by Amersham Biosciences (Uppsala, Sweden). These beads are classified based on the amount of water retained when swelled in water. Cross-linked Agaorse is sold under the trade name *Sepharose* by Amersham Biosciences (Uppsala, Sweden). Other examples of commercial gel column packings are Fractogel HW 55F (Merck, Whitehouse Station, NJ) and TSK G4000-SW (Phenomenex, Torrance, CA). Table 8.4 lists few commercially available gel filtration columns. Packed gel filtration spin columns, which are disposable and centrifugeable, are also available commercially. The spin columns are faster, and separation takes only a few minutes.

8.4.2. Anion-Exchange Chromatography

The negatively charged nucleic acids are retained on the positively charged stationary phase during anion-exchange chromatography. They are displaced from the resin by a mobile phase of increasing ionic strength. The DNA is adsorbed onto the anion-exchange silica matrix, while RNA, protein, and other cellular components are washed free. As with SEC, silica and

Table 8.4. Characteristics of Commercially Available Gel Filtration Columns

Product Name	Matrix	Avg. Pore Size (μm)	Application
Sephadex G-25	Cross-linked dextran	250	Desalting and buffer exchange
Sephacryl (S-1000SF)	Allyl dextran and/or bisacrylamide	50	Purification of DNA up to 20,000 bp
Superose-6 HR	Highly cross-linked agarose	13	Separation of DNA up to 400 bp
Superdex 200	Composite of dextran and agarose	13	Separation of DNA up to 200 bp

polymeric phases are prevalent in ion-exchange chromatography. Examples include silica particles modified with weak anion-exchange ligands such as DEAE (diethyl amino ethyl) and polymer beads coated with strong ligands such as quaternary amines. Some common resins are listed in Table 8.5.

The anion-exchange resins are based on both porous and nonporous supports. A nonporous microparticulate (<5 μm) anion exchanger having the ability to elute DNA restriction fragments and oligonucleotides up to 12,000 bp has been reported [12]. Porous resins with different pore sizes are available for higher and lower molecular-weight nucleic acids. The pores should be large enough for nucleic acids to penetrate, so that the size exclusion is eliminated. The stationary phase also needs to provide enough surface area for interaction. Consequently, resins with large pores (>4000 Å) are preferred for DNA restriction fragments of high molecular mass. Most common mobile phases for elution from anion-exchange columns are buffers such as Tris–sodium chloride and phosphate buffers with sodium (or potas-

Table 8.5. Characteristics of a Few Common Anion Exchangers Used in DNA Purification

Trade Name	Support Material	Functional Group	Avg. Particle Size (μm)
Nucleogen–DEAE 4000	Coated silica	Diethylamino ethyl	10
Q-Sepharose	6% Cross-linked agarose	Trimethylamino (quaternary amine)	90 (45–165)
ANX Sepharose	4% Cross-linked agarose	Diethylamino propyl	90 (45–165)

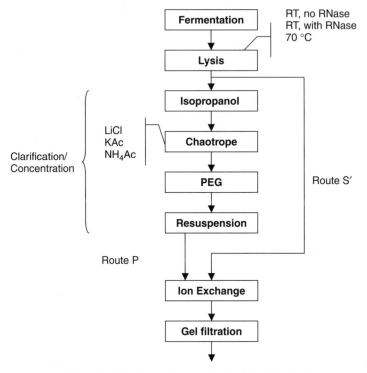

Figure 8.12. Process flow sheet for the purification of plamid DNA showing two alternative purification routes, S' and P. (Reprinted with permission from Ref. 7.)

sium) chloride as the eluting salt. The purified DNA is then eluted in a form that is ready for sequencing. The adsorbents are available in prepackaged kits from a variety of manufacturers and offer greater convenience and higher throughput than both phenol extraction and SEC methods.

A process flow sheet [13] for the purification of supercoiled plasmid DNA for gene therapy applications is shown in Figure 8.12. It is based on alkaline lysis and ion-exchange chromatography. The possibility of bypassing the clarification and concentration steps by performing ion-exchange chromatography right after lysis was investigated. It was found that ion-exchange chromatography by itself was capable of achieving purification levels similar to what was obtained with combined precipitation/clarification (2-propanol precipitation, clarification with chaotropic salt, and polyethylene glycol concentration) and chromatography. This demonstrates the power of chromatographic separation. It was also found that the overall yield of the direct chromatography route was higher (38%) than that of the combined steps (24%). It is an important finding from a high-throughput standpoint, because chromatography is easier to automate than clarification and precipitation.

8.4.3. Solid-Phase Extraction

Solid-phase extraction (SPE) has evolved to be an important sample preparation technique, due to its ease of automation, high analyte recovery, and excellent selectivity. The commercial availability of compact SPE devices with a wide selection of sorbent materials adds to their attraction. A major advantage of SPE is that multiple samples can be prepared in parallel using low volumes of solvents.

SPE has been described in Chapter 2. In principle, SPE is a chromatographic technique. The analyte is selectively adsorbed onto a sorbent phase while unadsorbed species pass through. A wash solution is used to eliminate possible contaminants (phenol–chloroform readily elutes proteins from a nucleic acid mixture) while retaining the analytes of interest. Finally, an eluent (such as Tris–EDTA-based buffer) is used to recover the nucleic acids. In principle the sorbents used in high-performance liquid chromatography (HPLC; anion exchange based) and SEC can be used in SPE. The SPE devices are packed with larger-particle-size sorbents (typically, 30 to 100 μm) with smaller bed lengths, thus requiring lower backpressure. Traditionally, the sorbents have been purer forms of silica oxide free of DNA-binding metallic components. Other silica-based materials, such as glass beads [14], modified siliceous particulates [15], glass wool [16], and diatomaceous earth (99% pure SiO_2) [17] in the presence of a chaotropic reagent such as guadinium thiocyante (GuSCN), have also been used as sorbents to bind DNA. Besides DNA purification, SPE is used for oligonucleotide desalting, primer removal prior to DNA sequencing, purification of crude synthetic oligonucleotides, and oligonucleotide desalting prior to genotyping by mass spectrometry [18]. The process of DNA adsorption on a silica surface (common SPE phase) is not clearly understood. However, results based on Tian et al. [19] and a proposed mechanism for DNA–silica interaction [20] suggest that a high-ionic-strength buffer with pH at or below the pK_a value of the surface silanol groups provide high adsorption efficiency.

The traditional SPE format is that a single disposable cartridge is filled with solid sorbent particles (50 to 500 mg), which are held in place by two polyethylene frits. Some common sorbents are listed in Table 8.6. The typical sorbent mass in a SPE disk ranges from 9 to 15 mg and between 25 and 100 mg in a SPE cartridge. The reduced sorbent mass and the dense packing in a disk allows the use of low solvent volumes. In many cases, elution of analytes can be accomplished in a small enough volume so that direct anal-

Table 8.6. Common Sorbents Used in Chromatography and SPE

Sorbent	Reference
Pure silica	32
Glass beads	14
Glass wool	16
Diatomaceous earth (Celite)	17
DEAE silica (diethylaminoethyl-coated silica)	15
TMA (trimethylamino-coated silica)	
DEAP silica (diethylaminopropyl-coated silica)	

ysis can be carried out without the time-consuming concentration and reconstitution steps.

Solid-Phase Reversible Immobilization

A variation of the solid-phase approach based on magnetic beads is referred to as solid-phase reversible immobilization (SPRI) [21,22]. The procedure is shown in Figure 8.13, where the extraction of genomic DNA from blood using magnetic silica beads is shown. The chosen cells in the blood sample are lysed enzymatically, liberating the nucleic acids into the solution. The nucleic acids are then precipitated by addition of isopropanol. Efficient DNA and RNA isolation from the cell lysate solution relies on the binding of nucleic acid to the surface of paramagnetic beads coated with a sorbent material such as silica or carboxylate coatings. Polyvinyl alcohol–based magnetic (M-PVA) beads are also used [23]. The magnetic beads bind to the DNA in the presence of a high concentration of polyethylene glycol and salt. The paramagnetic beads display magnetic properties when placed in a magnetic field but retain no residual magnetism when removed from it. Magnets are used to immobilize the DNA-bound beads while the solvent (or buffer) is selectively removed. The DNA can then be eluted from the beads using 10 mM, pH 8.0 Tris–EDTA or Tris–acetate buffer. An example of RNA isolation using this technique is presented in Chapter 7. This eliminates traditional solvent extraction and precipitation steps. Advantages of magnetic bead technology include isolation of high-quality nucleic acids, scalability, no centrifugation, ease of buffer exchanges, fast processing, and high reproducibility.

8.4.4. Affinity Purification

Nucleic acids such as RNA and DNA can be separated from a solution using a complementary probe. For example, the strong affinity between

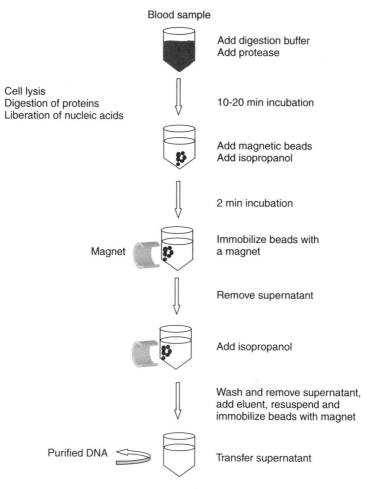

Figure 8.13. Solid-phase reversible immobilization (SPRI): extraction of genomic DNA from blood using magnetic silica beads.

biotin (a naturally occurring vitamin) and streptavidine (bacterial protein) has been used for the purification of nucleic acids [24,25]. This means that the biotinylated probe or the DNA has the biotin incorporated into one end of the DNA molecule by introducing a biotin-labeled nucleotide. The procedure is shown in Figure 8.14. The nucleic acids of interest (amid other cell debris) bind specifically to the biotinylated fragment due to its complementary nature. Streptavidine-coated magnetic particles introduced in the solution bind specifically to the biotin. A magnet can then be used to attract the strepatavidine-coated magnetic particles and thus isolate the nucleic

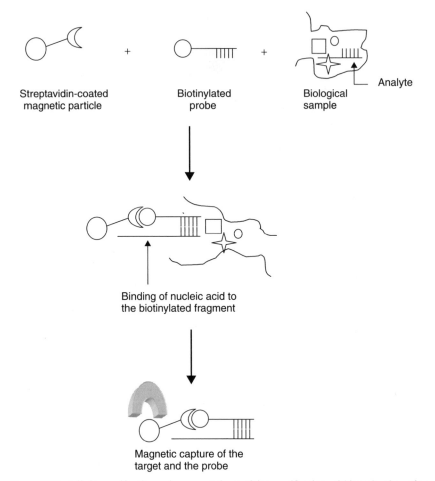

Figure 8.14. Affinity purification using magnetic particles: purification of biomolecules using streptavidine and biotin affinity.

acids bound to the protein. Finally, the DNA is eluted from the biotin–streptavidine complex with a suitable buffer. This solid-phase technique also simplifies nucleic acid purification by incorporating a rapid magnetic separation step. The limitation of this purification method is that the sequence of the nucleic acids to be isolated has to be known beforehand, since the probe has to be the complementary one. An example of RNA isolation using this technique is presented in Chapter 7.

A novel approach to specific binding involves the use of *triple-helix affinity capture* [26,27]. Triple-helix DNA has proven to be a useful approach to

DNA targeting. It was originally developed to produce gene therapy–grade plasmid and can be scaled up to production quantities. The principle can also be adapted to smaller-scale high-throughput analytical systems. This technique uses an affinity matrix coupled to small oligonucleotides that hydrogen bond to the major groove at a unique site on the double-stranded DNA. It is based on the specific binding of pyrimidine oligonucleotides to the purine strand in duplex DNA, forming a local triple helical structure. The pyrimidine oligonucleotides bind in the major groove of DNA parallel to the purine bases in the duplex DNA through the formation of Hoogsteen hydrogen bonds. The binding properties of a matrix can be adjusted by altering wash and rinse buffers or by chemical modification. Triple-helix-mediated capture has been used for the enrichment and screening of recombinant DNA and for the isolation of PCR products and plasmid DNA from a bacterial cell lysate.

8.5. AUTOMATED HIGH-THROUGHPUT DNA PURIFICATION SYSTEMS

SPE sorbents with their fast, efficient DNA purification capabilities have led to the development of automated procedures for high-throughput DNA purification systems [28,29]. The goal here is to prepare many samples simultaneously with as little manual intervention as possible. Multiple samples are processed in parallel in multiwell plates (96 wells or more). The basic steps of DNA purification in a high-throughput workstation are listed in Figure 8.15, with each step being automated. After the cells are lysed, the DNA is adsorbed either on a sorbent bed or in a SPRI bed. Proteins, RNA, and other cellular components are filtered out and washed free in the washing steps. The multiple samples of purified DNA are then eluted simultaneously and are ready for sequencing. The adsorbents are available in prepackaged 96-well (or as necessary) microplate kits from a variety of manufacturers. A 96-well SPE filter plate is shown in Figure 2.18. These kits offer easier automation than traditional phenol extraction or gel chromatography.

The degree of automations has literally revolutionized nucleic acid purification. The core of these systems are the automated liquid-handling workstations that involve the movement of multiple probes in Cartesian axes (x, y, z) over a deck configured with lab ware, such as microplates, tube racks, solvent reservoirs, washbowls, and disposable tips. They have the ability of aspirating and dispensing solvents from a source to a destination. Devices such as vacuum manifolds, heating blocks, and shakers have also been modified to handle 96-well (or more) plates. The multiprobe liquid handlers have a variable tip spacing that allows them to expand their tip-to-

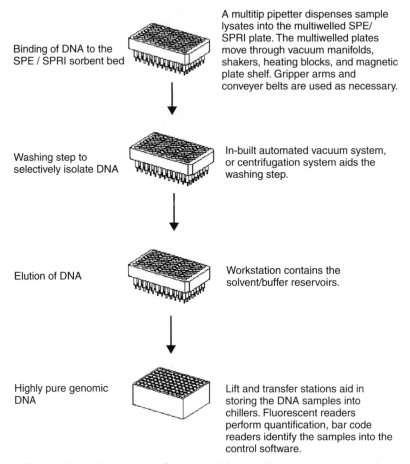

Binding of DNA to the
SPE / SPRI sorbent bed

A multitip pipetter dispenses sample
lysates into the multiwelled SPE/
SPRI plate. The multiwelled plates
move through vacuum manifolds,
shakers, heating blocks, and magnetic
plate shelf. Gripper arms and
conveyer belts are used as necessary.

Washing step to
selectively isolate DNA

In-built automated vacuum system,
or centrifugation system aids the
washing step.

Elution of DNA

Workstation contains the
solvent/buffer reservoirs.

Highly pure genomic
DNA

Lift and transfer stations aid in
storing the DNA samples into
chillers. Fluorescent readers
perform quantification, bar code
readers identify the samples into the
control software.

Figure 8.15. Typical sequence of steps in a high-throughput automated workstation.

tip width to aspirate from various test tube sizes and to reduce the tip spacing width when dispensing into microplate wells. A few workstations are also equipped with a gripper arm that moves microplates around the deck, such as to microplate stackers and fluorescence readers. Some commercial vendors providing such automated instruments are listed in Table 8.7. All of them employ one or more of the techniques mentioned before (i.e., SPE, SPRI, affinity purification, filtration, and centrifugation).

An automated DNA purification system called a PlateTrak system [30] based on SPRI protocol has been developed by PerkinElmer Life Sciences, Inc., in collaboration with the Center for Genome Research at the White-

Table 8.7. Few High Throughput DNA Purification Systems

Manufacturer	Product Name	Purification System
Beckman Coulter Inc. (Fullerton, CA)	Biomek 2000 and Biomek FX	Multiwell SPE plates
CRS Robotics Corp. (Burlington, Ontario, Canada)	CRS DNA purification system	SPRI
Eppendorf-5 Prime Inc. (Boulder, Colorado) (with Zymark Corp, Hopkinton, MA)	PerfectPrep-96 VAC	Alkaline lysis and filtration method
GeneMachines (San Carlos, CA)	RevPrep	Array centrifuge technology
Genovision (Oslo, Norway)	GenoM-96	SPRI
Orochem Technologies Inc. (Westmont, IL)	SpeedPREP	Multiwell SPE plates
Qiagen (Hilden, Germany)	BioRobot 8000	SPE and SPRI
Tecan (Maennedorf, Switzerland)	Genesis	SPE and SPRI

head Institute for Biomedical Research (Cambridge, MA). The workstation, shown in Figure 8.16, is part of a complex system that comprises a multi-position 96- or 384-channel dispensing module, robotic arms, conveyers, lift and transfer stations, recirculating chillers, magnetic plate shelf, and other automated devices. The functionality of the instrument is split into five phases, with phase 1 dedicated to lysis and resuspension procedure. Phases 2 and 3 are designed for DNA purification involving separation of plasmid DNA from genomic DNA and the purification of plamid DNA from the RNA, respectively. Phase 4 consists of a sequencing reaction setup utilizing both dye primer and dye terminator chemistries in the forward and reverse directions. Phase 5 performs the reaction pooling and sequencing reaction cleanup where the 384-well plate is pooled or compressed to a single 96-well purification plate for bead-based cleanup. This is particularly important when utilizing dye primer chemistries for sequencing.

According to the instrument designers, the conveyor-based microplate processing instrument allows a throughput of five hundred 384-well plates, or 200,000 separate DNA preparations per day. They have developed and optimized the procedure for single-stranded DNA isolation, such as M13 phage utilizing iron oxide magnetic particles, and double-stranded plasmid DNA isolation utilizing carboxyl-coated magnetic particles on PlateTrak systems.

The description of a centrifuge [28] used in a high-throughput format marketed by Genomic Solutions (Ann Arbor, MI) is presented in Figure

Figure 8.16. PlateTrak, automated microplate processing system developed by PerkinElmer in collaboration with the Whitehead Institute for Biomedical Research. (Reprinted with permission from Ref. 30.)

8.17. It is designed to isolate plasmid DNA from bacterial cultures using an alkaline lysis protocol with isopropanol precipitation. This workstation is based on an array centrifugation technology system which is an attractive alternative to filter-based automated plasmid purification. Neither filtration nor manual transfers to a centrifuge are required. The workstation comprises 96 separate rotors with microwell plate spacing that functions as both sample wells and miniature centrifuges. The rotors hold as much as 500 μL and generate forces as high as 60,000 rpm per well. The 96 samples are transferred from a standard sample plate and dispensed into individual wells. Then they are spun simultaneously, after which the pipetter removes and discards the supernatant. After a step of resuspending the pellet in the buffer, the pipetter carefully aspirates the buffer containing the DNA and dispenses into an output plate.

The total system includes two 96-channel array centrifuges, a 96-channel pipetter, an eight-reagent bulk dispenser, a wash station, a server arm, four storage cassettes for plates, and control software. According to the manufacturer, the workstation isolates plasmid DNA in less than 40 minutes and operates unattended for up to 8 hours, while purifying over 1100 samples. The estimated cost per sample can be as low as $0.10. Several companies offer similar sample preparation suites comprising modules and protocols that involve extraction, purification, PCR, and other sequencing reaction preparation.

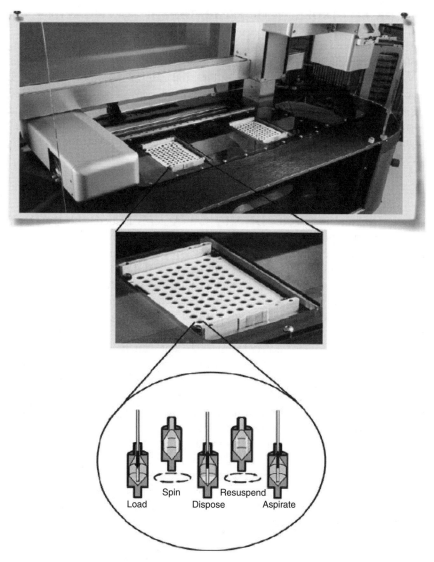

Figure 8.17. RevPrep Orbit workstation from Genomic Solutions (Ann Arbor, MI). (Reprinted with permission from *http://www.genemachines.com/orbit/orbitac.html*)

8.6. ELECTROPHORETIC SEPARATION OF NUCLEIC ACIDS

Electrophoresis is used widely for the separation and purification of macro-molecules, especially proteins and nucleic acids. Separation of these charged species tends to occur due to their differential rate of migration in a buffer across which a dc field is applied. Due to the consistent negative charge imparted by the phosphate backbone, the nucleic acids migrate toward the positive electrode. On the other hand, the proteins can have either a net negative or a net positive charge, which determines the electrode to which they will migrate.

8.6.1. Gel Electrophoresis for Nucleic Acids Purification

Gel electrophoresis can sort DNA by size. A gel is loaded with the DNA fragments and a potential is applied across the gel. As the DNA is negatively charged, it migrates toward the positive electrode. The larger fragments collide with the gel matrix more often and are slowed down, while the smaller fragments move faster. The frictional force of the gel acts as a "molecular sieve" and separates the molecules by size. The rate of migration of the macromolecules depends on the strength of the field, the size/shape of the molecules, and on ionic strength and temperature of the buffer. After staining, the separated macromolecules in each lane can be seen as a series of bands spread from one end of the gel to the other.

Electrophoretic separations of nucleic acids are usually done in agarose gels. The gel is cast in the shape of a thin slab with wells for loading the sample. It is immersed in a buffer medium, which maintains the required pH and provides the ions that carry the electrical current. Staining the gel with the aid of a dye such as ethidium bromide (5 µg/mL) allows detection of the nucleic acids by their fluorescence. During gel electrophoresis, the DNA samples are mixed with a "loading dye" that allows the DNA to be seen as it is being loaded. It also contains glycerol or sucrose to make the sample dense enough to sink to the bottom of the well in the gel. Two types of gel matrices are commonly used: agarose and polyacrylamide.

Characteristics of the Gels

Agarose is a linear polysaccharide obtained from seaweed (average molecular mass about 12,000). It is made up of the basic repeat unit agarobiose, which comprises alternating galactose and 3,6-anhydrogalactose. Agarose is usually used at concentrations between 0.5 and 3%. Agarose gels are formed by suspending dry agarose in an aqueous buffer and then boiling the mixture until a clear solution is formed. This is poured and allowed to cool to room temperature to form a rigid gel. However, it is fragile and can easily be destroyed during handling. The higher the agarose concentration, the stiffer is the gel, leading to a decrease in pore size. It is known that the electrophoretic mobility of a macromolecule is proportional to the volume fraction of the pores that the macromolecule can enter. By using an appropriate concentration of agarose and by applying the right electric field, DNA fragments ranging from 200 to 50,000 bp can be separated. By using a technique called *pulsed field gel electrophoresis*, where the direction of current flow in the electrophoresis chamber is altered periodically, very large fragments of nucleic acids ranging from 50,000 to 5 millon bp can be separated. Agarose gels can be processed faster than polyacrylamide gels; however, the former have a lower resolving power, due to their larger pore size. At the proper agarose concentration, a linear relationship exists between the migration rate of a given DNA fragment and the logarithm of its size (in base pairs). Table 8.8 lists the appropriate agarose concentrations for the separation of DNA fragments of different sizes.

Polyacrylamide is a cross-linked polymer of acrylamide. These gels are more difficult to prepare than agarose. Monomeric acrylamide (which is a known neurotoxin) is polymerized in the presence of free radicals to form polyacrylamide. The free radicals are provided by ammonium persulfate and stabilized by TEMED (N',N',N',N'-tetramethylethylenediamine). The chains of polyacrylamide are cross-linked by the addition of methylenebisacrylamide to form a gel whose porosity is determined by the length of chains and the degree of cross-linking. The chain length is proportional to the acrylamide concentration; usually between 3.5 and 20%. Cross-linking bis-acrylamide is usually added at the ratio 2 g bis/38 g acrylamide.

Polyacrylamide gels are poured between two glass plates held apart by spacers of 0.4 to 1.0 mm, and sealed with tape. Most of the acrylamide

Table 8.8. Concentration of Agarose in Gel for the Separation of DNA Fragments

Percent (w/v) of Agarose in Gel	Range of Linear DNA (bp)
0.3	5000–60,000
0.6	1000–20,000
0.7	800–10,000
0.9	500–7000
1.2	400–6000
1.5	200–3000
2.0	100–2000

**Table 8.9. Concentration of Acrylamide for the
Separation of Different DNA Fragments**

Percentage Acrylamide (w/v) with BIS at 1:20	Effective Range for Separation of Linear DNA (bp)
3.5	1000–2000
5.0	80–500
8.0	60–400
12.0	40–200
15.0	25–150
20.0	6–100

solution is shielded from oxygen so that inhibition of polymerization is confined to the very top portion of the gel. The length of the gel can vary between 10 and 100 cm, depending on the separation required. They are always run vertically with 0.5 M or 1 M TBE as a buffer. These gels separate DNA fragments smaller than 800 bp at high resolution. Thus, they are often the obvious choice in the sequencing of low-molecular-weight fragments. Table 8.9 lists the appropriate acrylamide concentration for the separation of DNA fragments of different sizes.

8.6.2. Techniques for the Isolation of DNA from Gels

After separation by gel electrophoresis, the required band is sliced out of the ethidium-stained gel and can be visualized under an ultraviolet (UV) light. Care is taken to cut out as little of the gel as possible, using a clean, sharp razor blade. The gel slice containing the DNA is then subjected to any of the following isolation techniques.

Electroelution

The block of agarose is placed in a piece of dialysis tubing containing a small amount of electrophoresis buffer. Dialysis tubing is a porous membrane in the form of a tube available in different pore sizes. The ends of the tubing are sealed, and it is placed in an electrophoresis chamber (Figure 8.18). On application of the electric field, the DNA migrates out of the agarose and is trapped within the membrane/bag. Movement of the DNA could be monitored using a transilluminator. When the flow of the current is reversed for a few seconds, the DNA, which is out of the agarose, can be knocked off the side of the tubing into the buffer. The buffer con-

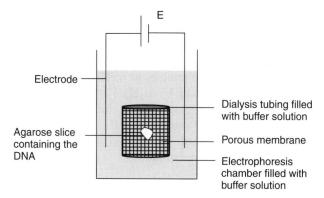

Figure 8.18. Schematic diagram showing electroelution.

taining the DNA is then collected and is precipitated with ethanol. The technique [31] is more time consuming than some of the other methods, but it works well for the recovery of DNA fragments larger than 5 kb.

Binding and Elution from Glass or Silica Particles

DNA binds to diatoms, glass, or silica particles [32] in an environment of high salinity and at neutral to low pH. This phenomenon can be exploited to purify and recover DNA from agarose solutions. Typically, a solution of chaotropic salt (e.g., sodium iodide or guanidium thiocyanate) at a pH of 7.5 or lower containing the slice of agarose is taken. The agarose slice is then melted by incubation at temperatures below 65°C so that the DNA is not denatured. Diatoms, glass, or silica particles are then added to the chaotropic solution and the suspension is mixed to allow adsorption of DNA. The chaotropic agents disrupt the hydrogen bonds of the agarose gel, allowing the DNA to be released into the solution to be adsorbed onto the silica particles. The particles are then recovered from the original liquid, washed by centrifugation, and resuspended in high-salt ethanol buffer. The free particles are pelleted by another centrifugation step, and the DNA containing the supernate is then recovered.

Electrophoresis onto DEAE–Cellulose Membranes

NA45 DEAE anion-exchange membrane is a cellulose support containing diethyl aminoethyl (DEAE) functional groups. At low salt concentrations, DNA binds to DEAE–cellulose membranes [33,34]. Fragments of DNA are electrophoresed in a standard agarose gel until they resolve adequately. A slit is made in the gel slightly ahead of the fragment(s) of interest, the membrane is placed, and electrophoresis is resumed. The fragments migrate and are stuck on the membrane. The membrane is then removed and washed free of agarose in low-salt buffer (150 mM NaCl, 50 mM Tris, 10 mM EDTA). It is then incubated for about 30 minutes at 65°C in a high-salt buffer (1 M NaCl, 50 mM Tris, 10 mM EDTA) to elute the DNA. The progress in binding of the DNA to the membrane, and its elution can be monitored with a UV light or by the ethidium bromide bound to DNA. After elution, the DNA is precipitated with ethanol. However, fragments larger than about 5 kb do not elute well from the membrane.

High-Speed Centrifugation

The agarose gel pieces are subjected to high-speed centrifugation [35] at 12,000 to 14,000 × g (or greater) for 10 minutes at room temperature. The strong centrifugal force compresses the agarose matrix and/or partially destroys it. This releases the DNA, and along with the fluid from the gel piece, it forms the supernatant fluid. On completion of centrifugation, the release of the DNA can be monitored by UV. An orange-red color indicates the presence of DNA in the fluid. Following centrifugation, the supernatant DNA is quickly poured into another tube, because the compressed agarose pellet may swell and reabsorb the DNA.

Low-Melting-Temperature Agarose

The agarose gel piece can be melted by heating to about 65°C [36]. The DNA remains intact and can be extracted with an equal volume of a phenol–chloroform mixture from the molten agarose. This is followed by DNA precipitation with ethanol, and redissolution in a buffer.

8.7. CAPILLARY ELECTROPHORESIS FOR SEQUENCING AND SIZING

In recent years, capillary electrophoresis (CE) [37] has been the technique of choice in the determination of the size and purity of DNA. CE offers some clear advantages over slab gel electrophoresis. These include easier automation, smaller sample volume, and the capability of real-time quantitative monitoring. Thus, CE has been an important tool in the completion of the human genome project. Figure 8.19 is a representation of electrophoresis in a capillary. Small-diameter capillaries are used in CE as the electromigration channels, which vary between 20 and 100 μm in diameter and 20 to 100 cm in length. The narrow diameter allows the application of high voltages and

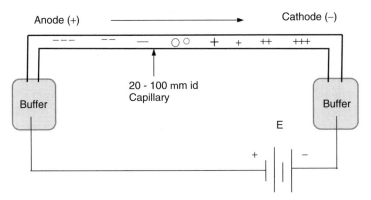

Figure 8.19. Capillary electrophoresis: pictorial representation of electrophoresis in a capillary.

ensures rapid heat dissipation due to the high surface/volume ratio. These lead to high resolution, and nucleic acids differing by a base pair can be separated with ease. Two types of separations have been attempted for the analysis of DNA [38] in capillaries: capillary zone electrophoresis (CZE) and capillary gel electrophoresis (CGE).

Electrophoresis in the CZE mode takes place in an open tube and in a free solution without any separation matrix in the capillary. The separation is based on the mass/charge ratio of the analytes. It is appropriate for the separation of nucleosides and nucleotides. It is not well suited for medium to large oligonucleotides, because their mass/charge ratio tend to be smaller. The use of a separation matrix becomes necessary for these species. Various capillary systems, including bare fused silica capillaries and surface-coated capillaries, have been used in CZE.

Historically, CGE has been translated from the slab format to the capillary format using the same matrices (i.e., cross-linked polyacrylamide and agarose). The gels are prepared in the same manner as slab gels, by adding a catalyst to the monomer solution before it is pumped into the capillary. The polymerization with cross-linking occurs in the capillary. Polyacrylamide gels are stable up to about 450 V/cm. At this field strength, up to 350 bases of DNA can be sequenced. It's been found that Long-Ranger, a modified acrylamide distributed by J. T. Baker (Phillipsburg, NJ) is stable at electric fields as high as 800 V/cm. These gels have a well-defined pore structure, and the life of the gel determines the life of the capillary. Gel degradation by hydrolysis, the small sample size, the tendency to retain high-molecular-weight DNA, and bubble formation at high field strengths (resulting in loss of conductance) are some of the problems associated with these gels. DNA restriction fragments of up to 12,000 bp and polynucleotides up to 500 bases can be separated at high resolution by CGE.

Linear/non-cross-linked polymers are an alternative to cross-linked polymers, where the separation can be carried out in the CGE format. These polymer solutions are made of hydrophilic polymers dissolved in an appropriate buffer. They have relatively lower viscosity than the cross-linked polymers. They can be pumped out of the capillary at the conclusion of a run, thus allowing a fresh separation media to be used for each analysis. In case of cross-linked polymers, once the separation medium has degraded, the entire capillary must be replaced. This can be tedious; for example, it may require realignment of the optical system with the narrow-bore capillary. The linear polymer solutions can withstand temperatures up to 70°C and high field strengths up to 1000 V/cm.

The mechanism of separation with linear polymers is as follows. At a certain polymer concentration known as the *entanglement threshold*, the individual polymer strands begin to interact with each other, leading to a meshlike structure within the capillary. This allows DNA separation to take place. Many of the common polymers are cellulose derivatives, such as hydroxyethyl cellulose, hydroxypropyl cellulose, hydroxypropylmethyl cellulose, and methylcellulose. Other applicable polymers include linear polyacrylamide, polyethylene oxide, agarose, polyvinyl pyrrolidone, and poly-N,N-dimethylacrylamide. High-resolution separation up to 12,000 bp has been reported using entangled polymer solutions.

Capillary gel electrophoresis (CGE) with polymer solutions is about 8 to 10 times faster than slab gel electrophoresis. However, the single-lane nature of CE was unable to compete in throughput with slab gel instruments, which are run in parallel. This led to the development of capillary array electrophoresis (CAE) [38] in 1992. As the name suggests, electrophoresis is performed in an array of capillaries to run multiple samples in parallel. Figure 8.20 shows a microfabricated capillary array system [39] on a glass wafer consisting of 96 channels.

8.8. MICROFABRICATED DEVICES FOR NUCLEIC ACIDS ANALYSIS

Microfluidic concepts can be used to develop an integrated *total chemical analysis system* (TAS) [40], which include sample preparation, separation, and detection. The microminiaturization of a TAS onto a monolithic structure produces a μ-TAS that resembles a small sensor. The first μ-TAS was a micro-gas chromatograph (GC) fabricated on a 5-in. silicon wafer in 1979 by a group at Stanford University [41]. Since then, developments in micromachining has led to the development of microsensors, microreactors,

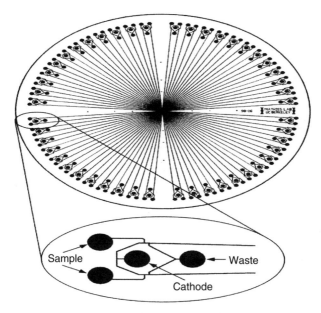

Figure 8.20. Capillary array electrophoresis on a chip: mask pattern for the 96-channel radial capillary array electrophoresis microplate. (Reprinted with permission from Ref. 39.)

and other elements of μ-TAS. The μ-TAS approach promises ultrahigh-throughput analysis with rapid speed, integration of sample preparation with analysis functions, consumption of just picoliters of samples/reagents, and the development of inexpensive disposable devices through mass fabrication.

CE on microchips [42] is one of the most promising technologies in μ-TAS. The CE is amenable to miniaturization because it involves the movement of fluids in a microchannel by electroosmosis. This precludes the need for mechanical pumps and valves, which are difficult to miniaturize and integrate into microdevices. Nucleic acid sizing, genotyping, DNA sequencing, and integrated nucleic acid sample preparation analysis are some of the potential applications for these microdevices. Short oligonucleotides (10 to 25), bases, restriction fragments such as $\Phi x174$ Hae-III DNA [43] r-RNA, and proteins have been separated on microchips. As shown in Figure 8.21, the analysis time of [44] DNA fragments is reduced greatly from the slab gel to the capillary to the microchip format without substantial loss of efficiency. DNA sequencing on a microchip was first demonstrated in 1995 by Woolley and Mathies [45]. Single-base-pair resolution of 150 to 200 bases in 10 to 15 minutes was achieved in a 5-cm-long channel. However,

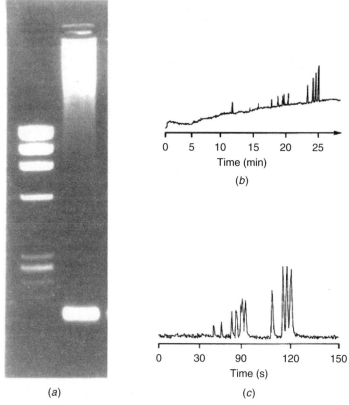

Figure 8.21. Electropherograms of the Hae-III digest of FX-174-RF DNA using (*a*) an agarose slab gel (total running time was approximately 40 min), (*b*) a polyacrylamide-coated capillary, and (*c*) microchips on poly(methyl methacrylate) substrate. The separation buffer for both polyacrylamide-coated capillary and microchips was 1.5% HPMC in TBE buffer (100 mM Tris-borate and 5 mM EDTA, pH 8.2) with 10^{-6} M of TO-PRO-3. (Reprinted with permission from Ref. 44.)

read lengths of over 500 bases were achieved with 99.4% accuracy in about 20 min.

The microchannels are fabricated on silicon or glass wafers using standard lithography and micromachining techniques. Polymeric substrates fabricated via laser ablation, casting, hot embossing, and injection molding have also been used. As shown in Figure 8.22*a*, a microfluidic electrophoretic chip consists of an injection channel connecting the sample reservoir and the sample waste reservoir. The separation channel connects the buffer reservoir and the buffer waste reservoir. In the injection mode, a field

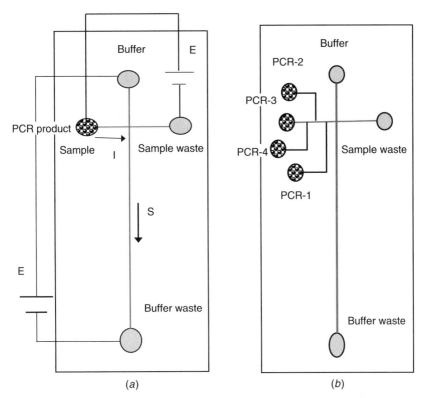

Figure 8.22. Capillary electrophoresis on a chip. (*a*) Schematic of the microchip used for PCR amplification and electrophoresis. The direction of arrows indicate injection (I) and separation (S). (*b*) Electrophoretic microchip with multiple PCR chambers.

is applied between the sample reservoir and the sample waste, thus causing the DNA to migrate to the intersection cross. The small DNA plug at the intersection cross serves as an injection to the separation channel. In the separation mode, a field is applied between the buffer and the buffer waste reservoir. The moving buffer elutes the DNA mixture, which separates as it migrates down the separation channel. Typical dimensions of the electrophoretic channels range from 20 to 100 μm in width, 15 to 100 μm in depth, and 5 to 20 cm in length. The microscale dimensions of the channels and the low thermal mass of the microchip allow rapid dissipation of the heat generated during electromigration. This allows the application of higher electric fields (over 2 kV/cm) for separations. The walls of the channels are modified to minimize the electroosmotic flow so that the separation takes place based on the difference in electrophoretic mobility in the presence of the sieving agent. However, separation is also possible in channels that support electro-osmotic flow (i.e., capillary zone electrophoresis).

The μ-TAS offers some excellent possibilities and is in a state of rapid development. However, several challenges need to be overcome for their successful real-world implementation. For example, detection limits are low due to the small sample size, and the principal detection method so far is laser-induced fluorescence, which offers high sensitivity and low detection limits. Other problems include interfacing microfabricated devices to conventional macro-size instruments and fluid handling.

8.8.1. Sample Preparation on Microchips

Integration of sample preparation and analysis [46] is one of the prime objectives of μ-TAS. PCR on a chip is one of the earliest applications of sample preparation. It has been carried out in the sample reservoir of the electrophoretic chip shown in Figure 8.22a. The nucleotides, primers, and other chemicals are added into the sample reservoir, and the entire device is introduced into a conventional PCR thermal cycler. The PCR products from the sample reservoir are then injected into the separation channel and analyzed. A more complex chip with multiple PCR chambers is shown in Figure 8.22b.

A microfabricated silicon PCR reactor coupled with capillary electrophoresis has been developed in an effort to carry out all the steps on a chip [47]. The device combined the rapid thermal cycling capabilities ($10°C/s$ heating, $2.5°C/s$ cooling) needed to carry out the PCR, followed by high-speed (120-second) DNA separations on a CE chip. The sample was locally heated, causing only a small part of the microchip to be heated instead of the conventional method of loading the entire device onto a PCR thermal cycler. This protected the heat-sensitive parts fabricated on the microchip. The recipe for the PCR remained the same, except that it was carried out in the small-sample reservoir wells of the microchip. The amplified DNA fragments are then subjected to online CE on the same microchip, thus decreasing analysis time. The device was also capable of performing real-time monitoring of the PCR amplification, as shown in the Figure 8.23.

Lagally et al. [48] developed a sophisticated PCR-CE device (Figure 8.24) with microfluidic valves and hydrophobic vents that enable precise handling of submicroliter samples. The sample is loaded from the right by opening the valve using vacuum (30 mmHg) and then forcing the sample under the membrane using pressure (10 to 12 psi). Simultaneously, vacuum is applied to the vent to evacuate air from the chamber. The sample stops at the vent, and the valve is pressure-sealed to enclose the sample. The valve, the vent

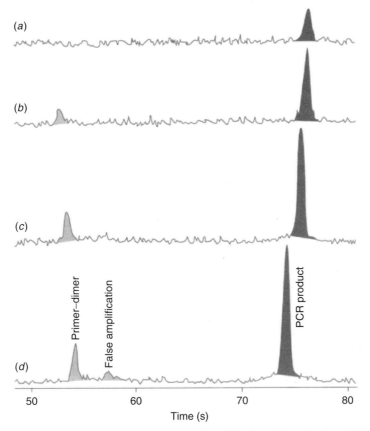

Figure 8.23. Real-time analysis of a β-globin PCR amplification using an integrated PCR-CE microdevice. Chip CE separations of the same sample were performed sequentially in the integrated PCR-CE microdevice after (a) 15, (b) 20, (c) 25, and (d) 30 cycles at 96°C for 30 seconds and 60°C for 30 seconds. (Reprinted with permission from Ref. 47.)

structures, and the ports were formed by drilling holes into the silicon substrate with a diamond-tipped drill bit. The PCR chambers were connected to a common sample bus through a set of valves for sample introduction. Hydrophobic vents at the other end of the PCR chambers were used to locate the sample and to eliminate the gas. They accomplished thermal cycling using a resistive heater, and a miniature thermocouple below the 280-nL PCR chamber was used to provide feedback for temperature control. The PCR chambers were connected directly to the cross-channel of the CE system for product injection and analysis. Two aluminum manifolds,

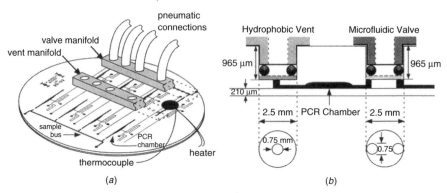

Figure 8.24. Single-molecule DNA amplification and analysis in an integrated microfluidic PCR-CE device. (Reprinted with permission from Ref. 48.)

one each for the vents and valves, were placed onto the respective ports and clamped in place using vacuum. The manifolds were connected to external solenoid valves for pressure and vacuum actuation.

With the need to provide PCR-amplifiable DNA, multiple approaches for incorporation of the extraction protocol onto microchips were examined. Recent development includes the implementation of a solid-phase extraction of DNA on a microchip [49]. The extraction procedure utilized was based on adsorption of the DNA onto bare silica. The silica beads were immobilized into the channel using a sol-gel network. This method made possible the extraction and elution of DNA in a pressure-driven system.

Cell lysis on a chip has been carried out by several different approaches. Detergents have been used to lyse cells on a chip. Thermal lysing on a chip has been carried out by placing the cell in the sample reservoir and then raising the temperature of the chip [50]. A practical approach for microchip applications is lysis by electroporation. Since fluids are moved around on chips by the application of an electric field, its use in cell lysis is an obvious choice.

An example of a microelectroporation device [51,52] fabricated on a silicon substrate is shown in Figure 8.25. It consisted of patterned electrode blocks separated by a 5-µm gap. The blocks of electrodes were separated by parylene. First, the cells and the medium were pumped into the channel. Next, the cells were attracted to the sharp point of the electrode by dielectrophoretic force using ac voltage in the frequency range of a few hundred kilohertz to a few megahertz. Then they were lysed by a pulsed electric field. The electrode was designed to have sharp edges, so that the electric field was concentrated there.

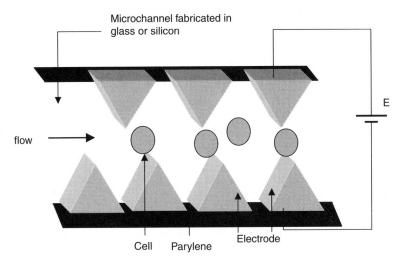

Figure 8.25. Schematic of a microcell lysis device.

ACKNOWLEDGMENTS

The authors wish to acknowledge the Center for Microflow Control and the New Jersey Commission on Science and Technology for financial support.

REFERENCES

1. *Manual of Methods for General Bacteriology*, American Society for Microbiology, Washington, DC, 1981, pp. 52–59.
2. J. R. Norris and D. W. Ribbons, eds., *Methods in Microbiology*, Vol. 5B, Academic Press, New York, 1971, pp. 1–55.
3. A. S. Ochert, M. Slomka, J. Ellis, and C. Teo, in A. Rolfs, I. W. Rolfs, and U. Finckh, eds., *Methods in DNA Amplification*, Plenum Press, New York, 1994, pp. 47–53.
4. A. Rolfs, I. Schuller, U. Finckh, and I. Weber-Rolfs, *PCR: Clinical Diagnostics and Research*, Springer-Verlag, Berlin, 1992, pp. 1–18, 51–58, 79–88.
5. M. S. Levy, L. A. S. Ciccolini, S. S. S. Yim, J. T. Tsai, N. Titchener-Hooker, P. Ayazi Shamlou, and P. Dunnill, *Chem. Eng. Sci.*, **54**, 3171–3178 (1999).
6. J. C. Weaver and Y. A. Chizmadzhev, *Bioelectrochem. Bioenerg.*, **41**, 135–160 (1996).
7. G. N. M. Ferreira, J. M. S. Cabral, and D. M. F. Prazeres, *Biotechnol. Prog.*, **15**(4), 725–731 (1999).

8. J. M. Saucier and J. C. Wang, *Biochemistry*, **12**(14), 2755–2758 (1973).

9. M. C. Ruiz-Martinex, O. Salas-Solano, E. Carrilho, L. Kotler, and B. L. Karger, *Anal. Chem.*, **70**(8), 1516–1527 (1998).

10. R. Hecker and D. Riesner, *J. Chromatogr.*, **418**, 97–114 (1987).

11. M. Polverelli, L. Voituriez, F. Odin, J. F. Mouret, and J. Cadet, *J. Chromatogr.*, **539**(2), 373–381 (1991).

12. C. Sumita, Y. Baba, K. Hide, N. Ishimaru, K. Samata, A. Tanaka, and M. Tsuhako, *J. Chromatogr. A*, **661**(1/2), 297–303 (1994).

13. G. N. M. Ferreira, J. M. S. Cabral, and D. M. F. Prazeres, *Biotechnol. Prog.*, **15**(4), 725–731 (1999).

14. R. Yang, J. Lis, and R. Wu, *Methods Enzymol.*, **68**, 176–182 (1979).

15. M. R. McCormick, *Anal. Biochem.*, **181**, 66–74 (1989).

16. T. M. McNally, *Biotechniques*, **27**, 68–71 (1999).

17. J. M. Carter and D. I. Milton, *Nucleic Acids Res.*, **21**(4), 1044 (1993).

18. M. Gilar, E. S. P. Bouvier, and B. J. Compton, *J. Chromatogr. A*, **909**(2), 111–135 (2001).

19. H. Tian, A. F. R. Huhmer, and J. P. Landers, *Anal. Biochem.*, **283**(2), 175–191 (2000).

20. K. A. Melzak, C. S. Sherwood, R. F. B. Turner, and C. A. Haynes, *J. Colloid Interface Sci.*, **181**(2), 635–644 (1996).

21. M. M. DeAngelis, D. G. Wang, and T. L. Hawkins, *Nucleic Acids Res.*, **23**(22), 4742–4743 (1995).

22. C. J. Elkin, P. M. Richardson, H. M. Fourcade, N. M. Hammon, M. J. Pollard, P. F. Predki, T. Glavina, and T. L. Hawkins, *Genome Res.*, **11**(7), 1269–1274 (2001).

23. J. Oster, J. Parker, and L. Brassard, *J. Magnet. Magnet. Mater.*, **225**(1/2), 145–150 (2001).

24. M. Uhlen, O. Olsvik, and E. Hornes, *Mol. Interact. Biosep.*, 479–485 (1993).

25. X. Tong and L. M. Smith, *Anal. Chem.*, **64**(22), 2672–2677 (1992).

26. J. Huamin and L. M. Smith, *Anal. Chem.*, **65**(10), 1323–1328 (1993).

27. A. F. Johnson, R. Wang, H. Ji, D. Chen, R. A. Guilfoyle, and L. M. Smith, *Anal. Biochem.*, **234**(1), 83–95 (1996).

28. R. E. Majors, *LC-GC North Am.*, **20**(5), 16–28 (2002).

29. K. Wang, L. Gan, C. Boysen, and L. Hood, *Anal. Biochem.*, **226**(1), 85–90 (1995).

30. M. Stevens and K. McKernan, *Automation of DNA Purification Using the PlateTrak*^{TM} *Automated Microplate Processing System*, Application note AN004-CCS, Packard Bioscience Co., Meriden, CT, Nov. 2000.

31. S. R. Pai and R. C. Bird, *Genet. Anal. Tech. Appl.*, **8**(7), 214–216 (1991).

32. W. Mann and J. Jeffery, *Anal. Biochem.*, **178**(1), 82–87 (1989).

33. T. Kaczorowski, M. Sektas, and B. Furmanek, *Biotechniques*, **14**(6), 900 (1993).

34. N. Van Huynh, J. C. Motte, J. F. Pilette, M. Decleire, and C. Colson, *Anal. Biochem.*, **211**(1), 61–65 (1993).

35. W. Wu and M. J. Welsh, *Anal. Biochem.*, **229**(2), 350–352 (1995).

36. R. S. Seelan and L. I. Grossman, *Biotechniques*, **10**(2), 186–188 (1991).

37. J. P. Landers, ed., *Handbook of Capillary Electrophoresis*, CRC Press, Boca Raton, FL, 1997.

38. C. Heller, *Electrophoresis*, **22**(4), 629–643 (2001).

39. Y. Shi, P. C. Simpson, J. R. Scherer, D. Wexler, C. Skibola, M. T. Smith, and R. A. Mathies, *Anal. Chem.*, **71**(23), 5354–5361 (1999).

40. S. Shoji, *Chem. Sensors*, **15** (Suppl. A, Proceedings of the 28th Chemical Sensor Symposium), 34–36 (1999).

41. S. C. Terry, J. H. Jerman, and J. B. Engell, *IEEE Trans. Electron. Devices*, **26**, 1880–1886 (1979).

42. V. Dolnik, S. Liu, and S. Jovanovich, *Electrophoresis*, **21**(1), 41–54 (2000).

43. C. S. Effenhauser, G. J. Bruin, and A. Paulus, *Electrophoresis*, **18**(12/13), 2203–2213 (1997).

44. S. Chen, *LC-GC North Am.*, **20**(2), 164–173 (2002).

45. A. T. Woolley and R. A. Mathies, *Anal. Chem.*, **67**(20), 3676–3680 (1995).

46. M. A. Burns, B. N. Johnson, S. N. Brahmasandra, K. Handique, J. R. Webster, M. Krishnan, T. S. Sammarco, P. M. Man, D. Jones, and Heldsinger, *Science*, **282**(5388), 484–487 (1998).

47. A. T. Woolley, D. Hadley, P. Landre, A. J. deMello, R. A. Mathies, and M. A. Northrup, *Anal. Chem.*, **68**(23), 4081–4086 (1996).

48. E. T. Lagally, I. Medintz, and R. A. Mathies, *Anal. Chem.*, **73**(3), 565–570 (2001).

49. K. A. Wolfe, M. C. Breadmore, J. P. Ferrance, M. E. Power, J. F. Conroy, P. M. Norris, and J. P. Landers, *Electrophoresis*, **23**, 727–733 (2002).

50. L. C. Waters, S. C. Jacobson, N. Kroutchinina, J. Khandurina, R. S. Foote, and J. M. Ramsey, *Anal. Chem.*, **70**(1), 158–162 (1998).

51. S. W. Lee and Y. C. Tai, *Sensors Actuators A*, **73**, 74–79 (1999).

52. Y. Huang and B. Rubinsky, *Sensors Actuators A*, **89**, 242–249 (2001).

CHAPTER

9

SAMPLE PREPARATION FOR MICROSCOPIC AND SPECTROSCOPIC CHARACTERIZATION OF SOLID SURFACES AND FILMS

SHARMILA M. MUKHOPADHYAY

Department of Mechanical and Materials Engineering, Wright State University, Dayton, Ohio

9.1. INTRODUCTION

Characterization of materials in the solid state, often loosely referred to as *materials characterization*, can be a vast and diverse field encompassing many techniques [1–3]. In the last few decades, revolutionary changes in electronic instrumentation have increased the use of highly effective automated instruments for obtaining analytical information on the composition, chemistry, surface, and internal structures of solids at micrometer and nanometer scales. These techniques are based on various underlying principles and cannot be put under one discipline or umbrella. Therefore, it is important first to define the scope of techniques that can be covered in one chapter.

In this chapter we are concerned with the two common categories of materials characterization: microscopy and spectroscopy. Microscopy implies obtaining magnified images to study the morphology, structure, and shape of various features, including grains, phases, embedded phases, embedded particles, and so on. Spectroscopy implies investigation of chemical composition and chemistry of the solid. Within spectroscopy, bulk techniques such as infrared, Raman, and Rutherford backscattering require minimal sample preparation and are not touched upon. Emphasis is placed on the spectroscopy of the outer atomic layers where sample preparation and handling become important.

Within each category, different techniques may have their own restrictions, requirements, and concerns. As the analytical instruments become

Sample Preparation Techniques in Analytical Chemistry, Edited by Somenath Mitra
ISBN 0-471-32845-6   Copyright © 2003 John Wiley & Sons, Inc.

Table 9.1. Common Microscopic Techniques and Sample Preparation Concerns

Optical microscopy (OM) Reflection Transmission Phase contrast Polarized light	Surface and internal microscopy, crystallographic information identification of particulates. Maximum magnification ~1000×. Final sample preparation: Polish and etch one side for reflection modes (Fig. 9.1). Some thinning for transmission mode.
Scanning electron microscopy (SEM)	Surface and internal morphology with 1000 Å or better resolution. Special techniques to characterize semiconductor and magnetic devices. Final sample preparation: Polish and etch (apply coating if required) one side (Fig. 9.1).
Transmission electron microscope (TEM) Scanning transmission electron microscope High-resolution electron microscope Analytical electron microscope	Internal nanostructure. Some case of surface structure if using replicas. Spatial resolution 2–5 Å. Phase determination (often with stained specimens) capability. Crystallographic information from ~4000 Å^2 area. Sample preparation: Very critical. Ultrathin specimens needed (Section 9.3 Table 9.4).

more sophisticated, robust, and user friendly, some stringency of sample specifications can be relaxed, but those fundamental to the analytical process remain. In this chapter, we provide a brief introduction to those sample preparation concerns that every user should be aware of. Tables 9.1 and 9.2 provide a brief summary of the analytical techniques whose sample preparation concerns are covered in this chapter.

9.1.1. Microscopy of Solids

The oldest microscopy technique for materials analysis was optical microscopy. Even to this day, for feature sizes above 1 μm, this is one of the most popular tools. For smaller features, electron microscopy techniques such as scanning electron microscopy (SEM) and transmission electron microscopy (TEM) are the tools of choice. A third family of microscopy includes scanning probe tools such as scanning tunneling microscopy (STM) and atomic force microscopy (AFM). In these relatively recent techniques, sample preparation concerns are of minor importance compared to other problems, such as vibration isolation and processing of atomically sharp probes. Therefore, the latter techniques are not discussed here. This chapter is aimed at introducing the user to general specimen preparation steps involved in optical and electron microscopy [3–7], which to date are the most common

Table 9.2. Common Surface Spectroscopic Techniques and Sample Preparation Concerns

Auger electron spectroscopy (AES)	Elemental analysis of surfaces and films, high resolution (ca. 500 Å) from top $\sim$1- to 20-Å layer. Limited valence-state information. Depth profiling. Sample preparation: Surface cleaning or in situ surface creation.
X-ray photoelectron spectroscopy (XPS)	Elemental analysis of surfaces and films, depth profiling (slow). Reveals detailed chemical state of elements; molecular composition can be deduced from peak sizes and shapes. Sample preparation: Surface cleaning or in situ surface creation.
Secondary-ion mass spectroscopy (SIMS)	Ultrahigh sensitivity in qualitative elemental and molecular compound analysis, isotope analysis, rapid depth profiling of composition, but no chemical information. Spectra interpretation and quantitation difficult. Sample preparation: Minimal (included here for comparison only).
Ion scattering spectroscopy (ISS)	Monolayer or less contaminant can be analyzed in the ppm range. Elemental information. Sample preparation: Surface cleaning or in situ surface creation.
Energy dispersive spectroscopy (EDS)	Qualitative and quantitative elemental analysis and elemental maps inside electron microscope. With Be window detector Na $\rightarrow$ U, with thin window detector C $\rightarrow$ U analyzed. Detection limit $\sim$0.1%. Sample preparation: Same as SEM or TEM (wherever attached).
Wavelength dispersive spectroscopy (WDS)	Qualitative and quantitative elemental analysis inside electron microscope, no elemental mapping. Sharper peaks compared to EDS and no peak overlaps. Detectable elements C $\rightarrow$ U, detection limit $\sim$0.2%. Sample preparation: Same as SEM or TEM (wherever attached).

microscopic techniques used by the scientific community. If one had to identify which technique is most heavily dependent on sample preparation methods (and related facilities and skill), the unanimous answer would be transmission electron microscopy. It is therefore reasonable that the longest section of this chapter is devoted to that technique.

For both optical and electron microscopy, specimen preparation is crucial, the basic concern being that the specimen prepared be a true representative of the sample. The first step obviously is to cut the specimen to size and to grind and polish the surface to expose the feature(s) of interest. These steps are commonly referred to as *metallography* even though they are applicable to all materials, and are discussed in Section 9.2.1. For reflection modes of microscopy, optical and SEM, polishing may need to be followed by etching, as discussed in Section 9.2.2.

In optical microscopy, the probing (or illuminating) beam is light that is either reflected off or transmitted through a specimen before forming its image. The image is formed by contrast between different features of the sample (brightness, phase, color, polarization, fluorescence, etc.) depending on the illuminating source. Magnification is controlled by a system of optical lenses. The limit of resolution (or the maximum magnification that will provide any meaningful contrast) is normally limited by the wavelength of the light used and not by the lens. According to diffraction theory, the closest distance between two points that can be resolved in an image is proportional to the wavelength λ.

The primary difference between optical and electron microscopy is that the latter uses an electron beam as the probe. Since 10- to 500-keV electron beams have much lower wavelengths than light, the resolution is greater. At the same time, the electron beam requires completely different instrumentation (source, collimator, detector, magnification control, etc.). Moreover, electrons are very readily absorbed by matter. Therefore, the entire path of the beam, from source to specimen to detector, has to be in vacuum. From the sample preparation point of view, this is of major significance. For specimens that may change in vacuum, biological tissues, for instance, this can be a major concern, and newly developed accessories such as environmental cells [8] need to be added to the microscope.

For scanning electron microscopy of electrically insulating materials, the surface of the specimen may be electrically isolated when bombarded with electrons. This leads to charge buildup on the specimens that makes imaging or other analysis difficult. To address this issue, special sample coating steps are often required and have been discussed in Section 9.2.3.

When transmission electron microscopy is used, the specimen has to be extremely thin (on the order of 0.1 to 10 μm) for the highly absorbable electrons to penetrate the solid and form an image. Preparing such a thin solid specimen with minimal artifacts is a very complicated problem that makes sample preparation a crucial step in the use of this technique. Therefore, a substantial part of this chapter (Section 9.3) is devoted to specimen thinning issues in TEM.

As the title suggests, in this chapter we stress solid materials and films. Therefore, special concerns related to fluids or biological specimens are not addressed [9]. We cover the most commonly applicable methods that the user can employ in most laboratories with commercially available instrumentation. Also discussed are possible artifacts arising from each preparation step and ways of minimizing or countering them. In addition to the most widely used sample preparation techniques, some newer developments have been touched upon, but these are by no means exhaustive. It must be stressed that despite this being a mature field, many new techniques and variations are being introduced regularly [10] and it is not possible to explain or even list them all. So, some omissions are inevitable.

9.1.2. Spectroscopic Techniques for Solids

Bulk spectroscopic techniques such as x-ray fluorescence and optical and infrared spectroscopies involve minimal sample preparation beyond cutting and mounting the sample. These are discussed in Section 9.2.1. Spectroscopic techniques such as wavelength dispersive spectroscopy (WDS) and energy dispersive spectroscopy (EDS) are performed inside the SEM and TEM during microscopic analysis. Therefore, the sample preparation concerns there are identical to those for SEM and TEM sample preparation as covered in Section 9.3. Some special requirements are to be met for surface spectroscopic techniques because of the vulnerability of this region. These are outlined in Section 9.5.

In recent decades we have seen an explosion of various spectroscopic techniques for analyzing the elemental composition and chemical states of solid surfaces and films. This explosion has stemmed in part from the large number of surface- or interface-related problems seen in integrated-circuit performance, composite reliability, corrosion, nanostructured components, and so on. Instruments themselves can range from stand-alone units to attachments in national synchrotron facilities or multitechnique systems built around special fabrication sites. However, the basic principle of the technique, and therefore the basic concerns with sample preparation, stay the same.

The most commonly used surface spectroscopy techniques for analyzing the composition and chemistry of solid surfaces are x-ray photoelectron spectroscopy (XPS), auger electron spectroscopy (AES), secondary-ion mass spectroscopy (SIMS) and ion scattering spectroscopy (ISS). Of these, the first two are the most popular for quantitative analysis of the outer surface (10 to 20 Å). All of these involve bombarding the surface with a particle probe (electron, photon, or ion) and analyzing the energy of an outgoing particle. In XPS, the probe is an x-ray photon and the detected particle is the photoelectron emitted by it. In AES, the probe is an electron and the signature particle is a lower-energy electron. In SIMS and ISS, both are ions. The relative advantages and disadvantages of these techniques are tabulated in Table 9.3. Most of the sample preparation concerns we discuss in this chapter are pertinent to AES, XPS, and ISS. Since SIMS is a completely destructive technique involving postmortem analysis, sample preparation does not require as much care.

9.2. SAMPLE PREPARATION FOR MICROSCOPIC EVALUATION

See Figure 9.1 for the basic steps in microscopic evaluation.

9.2.1. Sectioning and Polishing

The most obvious requirement, of course, is that the specimen be cut to size. The size depends on the microscope and could range from a few centimeters in a normal SEM to a few inches in a specially designed SEM. In TEM, of course, since the thickness is extremely low and the sample needs to be on a grid or support, the specimen is normally a few millimeters in size. Ductile metals are sometimes rolled into sheets before cutting into the desired size. It needs to be kept in mind that this process itself will lead to defect creation and microstructural changes that need to be annealed out [11]. Some polymers and composites are easily available as sheets anyway, so this step is not of any concern. In the large variety of bulk materials that it is not possible to form into sheets, sectioning the sample to a thin slice is the only way to start.

Sectioning is generally done by saw or cutting wheel. With a regular saw, surface damage can extend 200 μm or more into the sample. This damage depth can be reduced considerably if fine cutting tools are used. This is where a rotating saw with fine blades can help. Diamond-impregnated blades as thin as 10 μm are readily available for this purpose. These wheels have counterbalanced loading to avoid excessive pressure on the sample. Simultaneous lubrication and cooling with water, oil, or alcohol is desirable, and by proper selection of rotational speed, cutting pressure, and saw size, it is possible to get thin (perhaps 100 μm) slices of even the hardest materials, with surface damage extending to less than 1 μm [12].

A still narrower and more precise cut is possible with a wire saw, whose cutting surface is a fine wire wetted with an abrasive-containing liquid. The wire can be made to form a loop running over pulleys or can be a single length running back and forth on an autoreversal system. The main drawback with either of these designs is that the wire gets thin with cutting and might break before the specimen is complete. This is especially true when cutting hard samples. Replacing a broken wire halfway through a cut may make it difficult to resume cutting at exactly the same place.

Table 9.3. Capability Comparison of Common Surface Spectroscopic Techniques That Involve Electron or Ion Detection

Technique	Information Obtained	Elements Detected	Analysis Volume Depth	Analysis Volume Width
Auger electron spectroscopy (AES)	Elemental surface composition, lateral mapping	Li-U	0.5–10 nm	50 nm–30 μm
X-ray photoelectron spectroscopy (XPS)	Elemental surface composition, chemical states and bonding, lateral mapping	Li-U	0.5–10 nm	10 μm–1 mm
Ion scattering spectroscopy (ISS)	Atoms exclusively at outermost monolayer	Li-U	One monolayer	1 mm
Secondary-ion mass spectroscopy (SIMS)	Elemental composition profile, isotope identification	H-U	0.5–500 nm	1 μm–1 mm

Technique	Advantages and Limitations	Sensitivity	Probing Particle	Analyzed Particle
Auger electron spectroscopy (AES)	Fast, semi-quantitative, possible beam damage, very limited chemical information	10^{-3}	1- to 10-keV electrons	1- to 2000-eV electrons
X-ray photoelectron spectroscopy (XPS)	Minimal damage, very sensitive to chemical states, quantitative, depth profiling slow	10^{-3}	X-rays	1- to 1500-eV electrons
Ion scattering spectroscopy (ISS)	Exclusively top monolayer, charging effects and contamination extremely critical	Varies, higher for heavy elements	He$^+$ ion	He$^+$ ion
Secondary ion mass spectroscopy (SIMS)	H–He detection, very high sensitivity, quantification unreliable, destructive	10^{-4}–10^{-8}	0.5- to 10-keV ions (Ar$^+$, O$^+$, etc.)	Secondary ions

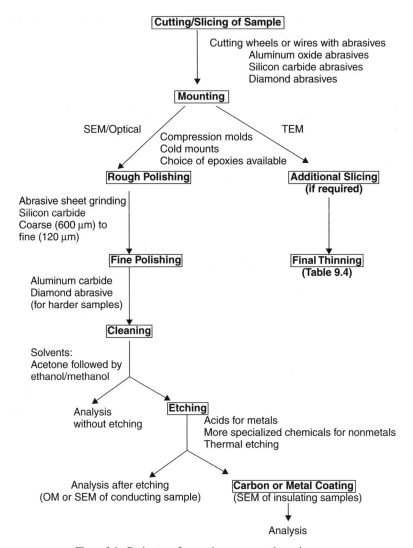

Figure 9.1. Basic steps for specimen preparation-microscopy.

A variation of the wire saw that can cut some specimens without deformation or mechanical damage is the acid string saw [13]. This is a wire saw where the abrasive is replaced by an etching agent and the cut occurs from a chemical reaction rather than mechanical abrasion. This is suitable for metals or other reactive solids that have effective etching solutions. Of course, for chemically inert samples such as some ceramics, this is not an option and slicing has to done mechanically.

After the sample has been sliced, the surface needs to be ground and polished to get a flat face with uniform analysis conditions across the region of interest. This procedure can be tedious and, in some cases, challenging. In most cases, the cut specimen is either compression-molded or cold-mounted in a polymer mold. If this is not possible, the specimen can be glued externally on a metallic mount. The mold (or mount) makes it easy to hold the specimen by hand or machine during polishing. When the specimen is set inside the plastic mold, the edges are protected during polishing. When externally glued on the mount, the edges can be rounded during polishing.

The next step is to grind the surface on abrasive paper or cloth, starting from course grit and using progressively finer and finer grit sizes. A general guideline for simple materials is to start with 50-grit SiC paper and go through three or four levels, finishing with 600 grit. This is followed by finer polish, Al_2O_3 suspension is recommended for most except for very hard surfaces, where diamond paste can be used. These suspensions and pastes are available with abrasives as fine as 0.05-μm particle size [14]. At each step of polishing, deformations introduced during the previous step need to be removed [15]. Since very little material is removed at the finer steps, the preceding step has to be thorough. Polishing wheels on which the abrasive is placed can be rotated at different speeds and the sample (mounted or molded) can be held on it with moderate pressure, either manually or on an automatic arm. Automatic polishers often offer better reproducibility [16]. After the final grinding step, no scratches should be visible on the surface.

9.2.2. Chemical and Thermal Etching

Polished unetched samples can show macroscopic cracks, pits, and so on, but no microstructural details because there is not yet any contrast-producing feature on the surface. These will be revealed by the etching process. The term *etching* is generally used to mean physical or chemical peeling of atomic layers. However, in the context of surface etching for microstructural evaluation, the idea is to expose the lowest-energy surface by chemical or thermal means. This will expose defects such as grain boundaries and bring out the contrast between different phases or different crystallographic orientations that etch at different rates. Specimen etching is a vast and matured area in itself, and several handbooks are available that describe and tabulate recipes for final polishing and etching of specific materials [6,17–19].

A simple example of the importance of the etching process is illustrated in Figure 9.2. The freshly polished surface prior to etching will have no varia-

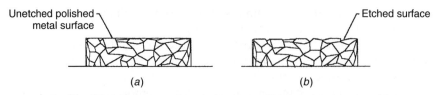

Figure 9.2. Effect of etching on surface profile; the polished unetched surface (*a*) is completely flat with no features to show, whereas the etched surface (*b*) shows the microstructural profile.

tion in contrast across the grain boundary because it is completely flat. But during chemical attack on the surface, the grain boundary region will be eroded faster than the rest of the grain and therefore there will be very fine grooves along the boundary that will be visible under the microscope.

The choice of a chemical etchant is, of course, very dependent on the sample that needs to be etched. As mentioned earlier, a large number of compilations are available in the literature and this is an ever-expanding field in an age of ever-increasing use of new materials. The common thread among all these recipes is that the surface material needs to be chemically attacked so that fresh surface is exposed underneath. For metallic elements and alloys, these are predominantly acid- or peroxide-containing solutions. Aqueous nitric acid (hot or cold) is often the first solution tried. A stronger etchant could be a mixture of nitric, hydrofluoric, and hydrochloric acids. In some cases, methanol is used as a solvent instead of water. Hot orthophosphoric acid can be used in the case of inert oxides. Many electronic materials such as GaAs and recently, superconductors [20] can use halogen in ethanol. The extent of etching needs to be monitored carefully. After sufficient contrast is brought out, the specimen should be rinsed thoroughly in a non-reactive solvent (e.g., acetone, alcohol) to prevent further corrosion. It must be noted that the same ingredient that is used for limited surface etching in optical microscopy or SEM is often used in a different consistency and potency for sample thinning that is crucial for transmission electron microscopy. Therefore, more details of chemical etching and polishing are given in Section 9.3.3.

If the material is so inert chemically that no corrosive etchant is available, allowing the surface to relax at a high-enough temperature (in the range where substantial diffusion is possible) will have a similar effect. Diffusion of atoms will tend to bring the surface to its equilibrium or quasi-equilibrium state [21,22], which often leads to phenomena such as faceting of certain planes and grain boundary grooving. These processes will lead to contrast between different areas of the sample.

9.2.3. Sample Coating Techniques

In the SEM, electrically nonconducting specimens can absorb electrons and accumulate a net negative charge that repels the following electron beam, thereby degrading the image [21]. To a certain extent, lowering the accelerating voltage or reducing the spot size can reduce this artifact, but that would limit the instrument capability considerably. The best way to counter this is to coat the specimen with a thin conducting film. In the past, organic antistatic agents have been tried, but the best method is to deposit a thin film (tens of nanometers) of a metal or carbon [6]. This step, although not mandatory, is also used in some TEM studies to enhance electronic contrast.

It needs to be pointed out that inside most electron microscopes, spectroscopy is also performed. The electron beam used for imaging can excite x-ray fluorescence, especially in the heavy elements of the sample, and the energies of these photons can be analyzed to identify these elements. For this type of analysis (energy dispersive spectroscopy being the most common configuration), the x-ray signal from the coating element needs to be kept in mind. Carbon is the most benign because it gives an almost undetectable signal. Metal coatings such as gold will give their characteristic signal and the investigator needs to check in advance whether this will interfere with any peaks from the specimen. The most common techniques of sample coating are thermal evaporation and sputter coating.

Thermal Evaporation

Thermal evaporation involves passing a current through a refractory filament that holds the evaporation source. This source can be a metal such as gold or palladium, or pure carbon. The assembly is placed in an evacuated chamber containing the sample (Figure 9.3). The filament is resistively heated by passing high current through it, and this in turn heats the evaporation source. As the vaporization temperature of this source is reached, a stream of atoms is released in the chamber. This stream of metal or carbon atoms will coat every object in its line of sight, including the sample. A common step used to ensure uniform coating is to rotate and tilt the sample stage during evaporation. This technique is sometimes called *rotary evaporation*. The reverse trick can be used in special circumstances to create the opposite effect: nonuniform coating for shadowing purposes. If this is desired, the sample is held stationary at an oblique angle to the evaporation beam so that surface features sticking out produce shadows on the deposited coating. This artifact would highlight such features.

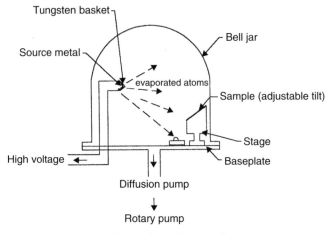

Figure 9.3. Simple thermal evaporation system.

Sputter Coating

Sputter coating involves erosion of atoms from a suitable target by energetic particles and subsequent deposition of these atoms on the sample. It requires lower vacuum than thermal evaporation coaters and does not depend on line-of-sight phenomena to coat the target. Sputter coaters are classified into five types depending on how the energetic particles are produced: plasma, ion beams, radio frequency, penning, and magnetron sputtering. Detailed designs and principles of each type are available in several books and monographs [24]. At this time, the use of sputter deposition is not confined to basic metal/carbon coating for microscopic purposes. Sputter technique is used today to deposit complex compounds in electronic devices, and many sophisticated sputtering systems and targets are available commercially. The most basic type that can commonly be used for SEM (Figure 9.4) consists of an evacuated bell jar containing a cathode made of the target material (the material with which the sample needs to be coated), an anode, and the sample stage. Inert gas (Ar, N) is bled into the chamber and energized by the creation of glow discharge. This kicks off target atoms which are deflected in all directions by collision with the gas atoms and are eventually deposited on cold surfaces, including the sample. The overall drift is toward the anode, but the random motion of individual metal atoms makes the deposition multidirectional in the surface scale, and even rough surfaces can be uniformly coated.

Artifacts of Coating

Some artifacts may be caused by surface deposition of which the user should be aware. One possible problem that can arise in either of the two techniques

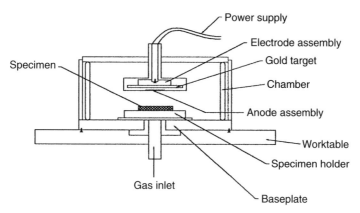

Figure 9.4. Commonly used sputter-coating arrangement.

is hydrocarbon contamination. Vacuum pump oils and improperly cleaned starting sample are common sources. This may produce uneven coating or, in extreme cases, cracks or discontinuities in the coating. Another artifact is thermal damage, which sometimes manifests as pitting or local melting of the film. This can be avoided by increasing the source–sample distance in thermal evaporators or by using lower plasma currents and voltages in sputter systems. Modern deposition chamber designs have reduced this problem to a great extent, and only very sensitive samples require cooling stages. An artifact that may be quite pronounced in thermal evaporators and much less troublesome in sputter coaters is distortion of a rough surface profile. Features that rise above the surface shadow the region behind it and can be exaggerated, whereas pits or grooves that are below the surface level are shielded and do not obtain a coating. This can be avoided by rotating and tilting during the deposition process. A problem that can arise in poorly designed sputter coating systems, but not in a thermal evaporator, is surface etching of the specimen itself. Sometimes a material from a chamber component other than the target material may be sputtered onto the sample. But these problems can easily be recognized and corrected by chamber modifications.

9.3. SPECIMEN THINNING FOR TEM ANALYSIS

As mentioned earlier, once a TEM sample is cut into a thin roughly uniform slice, it needs to be thinned extensively in regions where it will be electron transparent. In extremely rare cases of synthetic materials, the specimen itself can be prepared as a thin film. This is often the technique used to make test specimens for calibrating the instrument [7]. In such specimens, sample thinning is not an issue. But in the vast majority of TEM studies, the starting material is much larger and a slice from it is cut out which eventually needs to be thinned down to an acceptable thickness.

The maximum thickness allowable depends on the electron scattering factor of the material. A general rule of thumb is that the higher the atomic number of the elements in the sample, the greater the electron scattering factor and the thinner the specimen needs to be. Therefore, under identical conditions, an aluminum (Al) sample could be more than 10 times thicker than a uranium (U) sample to provide the same TEM picture quality. For amorphous samples under 100-kV electrons, a few hundred nanometers of Al and a few tens of nanometers of U are often the limits for regular TEM analysis. Higher accelerating voltages can tolerate thicker specimens. When the sample is crystalline, the thickness requirement need not be as stringent. Bragg reflection at certain orientations allows an anomalous thickness of the material to be penetrated [7,11]. So a curved specimen can have certain regions with enhanced transparency (regions that have the correct orientation for Bragg's diffraction condition). The alternative approach used in all TEM systems today is to have a tilting stage. Here, the specimen can be tilted so that any particular area can be put in a Bragg's or anomalous absorption condition. Modern TEMs also have image-intensifying devices for low-intensity operation which can "see through" slightly thicker samples.

All things considered, specimen thickness is still a crucial issue in TEM, and all thinning techniques are geared toward creating foils or regions in foils that are 0.1 to 10 μm in thickness. Often, it is convenient to keep thinning a region until the sample is perforated near the center, with a ring of thicker specimen outside to provide support. The edge of the perforation will probably have thin regions suitable for analysis. This can be accomplished by starting with a "dimpled" sample. This means that the sample is cut such that a small region near the center has a smaller cross section that its surrounding (Figure 9.5a). Dimpling can be accomplished by any of the modern machining and micromachining tools, such as spark machining, ultrasonic drilling, photolithography, and jet drilling. The most common route is to use a mechanical dimpler, which could be as simple as a 1-mm rod tool. A second option is to start with a wedge-shaped sample (Figure 9.5b) supported at the thick end. After final thinning, the tip of the wedge will have thin regions of acceptable transparency. A third approach, more commonly used in conjunction with chemical or electropolishing, is the window technique. Here, the specimen is protected on the outer edges by a chemically inert lacquer that can be painted on to form a frame. Subsequent thinning will allow only the unprotected window to be thinned down (Figure

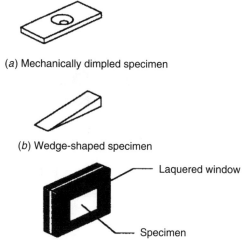

(a) Mechanically dimpled specimen

(b) Wedge-shaped specimen

Laquered window

Specimen

(c) Specimen with window

Figure 9.5. Specimen geometries prior to the final thinning step: (a) dimpled specimen; (b) wedge sample; (c) lacquered window for chemical or electropolishing.

9.5c). A specimen that is cut into one of the shapes above can subsequently be thinned down to electron transparencies. The most commonly used methods for final thinning can be categorized as described below.

9.3.1. Ion Milling

Ion milling involves bombarding the specimen at an oblique angle with a beam of inert gas ions (such as Ar) so that surface atoms are stripped off. The scientific principle behind ion-beam thinning and semiquantitative treatments of the thinning process are available in many books [25]. In general, ion bombardment is a very versatile process that can be used in several ways. When low-energy (1 to 5 keV) ions are used at oblique incidence to the surface, the erosion or sputtering rate can be very slow. This layer-by-layer erosion at the atomic scale, used extensively for cleaning and contamination removal of surfaces, is discussed in Section 1.5. At higher voltages and medium beam currents (typically, 5 to 10 keV voltage and 200 $\mu A/cm^2$ current density of Ar^+ beams), ion bombardment can be used for macroscopic thinning of TEM specimens at a reasonable rate.

A schematic of the experimental setup is shown in Figure 9.6. (This particular apparatus has two chambers, so that two samples can be thinned simultaneously, but it is also common to have a single-chamber ion mill.)

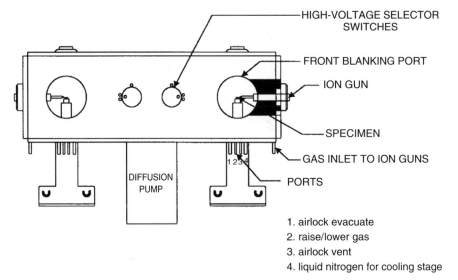

Figure 9.6. Ion milling apparatus.

The basic requirement is a diffusion-pumped chamber attached to an ion gun. The ion gun is filled with high-purity inert gas such as Ar. This gas is accelerated between two electrodes with a high potential difference. This ionizes the gas and a beam of focused and collimated gas particles is aimed at the specimen surface. Modern electronics allows very precise manipulation of the ion beam in several ways. Most ion milling machines are single-beam systems where one surface of the specimen is thinned. Alternatively, double-sided machines are also available where there are two ion beams focused on either side of the same specimen that is milled from both the top and bottom surfaces.

The primary advantage of ion milling is that it is universally applicable to any solid material. The major possible artifact that needs to be understood and monitored by the investigator is beam-induced damage [10,26–28]. There are many aspects to changes in the near-surface region caused by ion beams. Some changes are related to very superficial surface bonding and compositional changes that may not be of much concern in TEM. But other "deeper" changes that can influence TEM studies are structural and compositional alterations. Figure 9.7 shows a ripple pattern on a carbon fiber that has been ion milled for TEM observation. The exact mechanisms that lead to such an alteration are not always clear, but beam-induced roughness is often to blame.

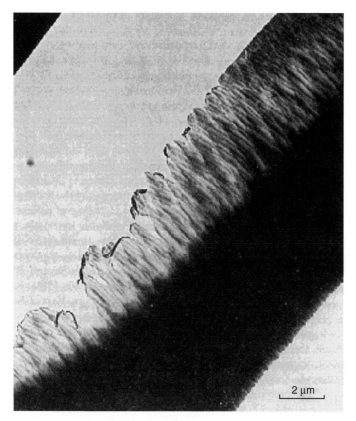

Figure 9.7. Carbon fiber ion-milled for TEM analysis; the ripple pattern is a common observation attributed to the ion milling process. (Adapted from Ref. 11.)

Many artifacts are further aggravated by sample heating because the ion milling process can cause a substantial increase in temperature in some materials. As an example, temperature increases of 100 to 370°C have been reported for semiconductor materials under normal conditions [26]. These effects can be minimized by (1) keeping the ion current density low, (2) using a lower incident angle, or (3) using a heat sink. The latter option is available in most new machines where the specimen can be mounted on a "cold stage" that has liquid nitrogen circulating through it.

9.3.2. Reactive Ion Techniques

Reactive ion techniques are relatively recent and popular modifications of the traditional ion milling technique described earlier. Here, a reactive gas is used to supplement or replace the inert-gas ions. This approach is becoming widely available because reactive ions (mainly halogen-containing gases) are being used extensively by the semiconductor industry for cleaning and patterning very large scale integrated (VLSI) device materials.

In reactive ion-beam etching (RIBE), the inert gas is replaced completely by a chemically reactive gas, so the sample is bombarded with a stream of ions that have a strong interaction with the substrate, and material removal can be very rapid. However, instrument corrosion can be a major concern. The ion gun, milling chamber, and pumping system are all exposed to large quantities of reactive gases and are prone to degradation.

This problem is reduced in the chemically assisted ion-beam etching (CAIBE) approach, which is a compromise between RIBE and inert ion milling. In this technique, a reactive gas is kept in contact with the area as it is being milled with inert Ar ions. For several compounds that produce undesirable artifacts with inert ion milling, RIBE or the gentler CAIBE can be useful alternative [30] dry milling procedures. Figure 9.8 shows such an example in a compound semiconductor (InP). Regular ion milling produces islands of metallic indium due to preferential sputtering of P. This artifact is eliminated completely when iodine-assisted CAIBE is used.

9.3.3. Chemical Polishing and Electropolishing

Chemical polishing and electropolishing were the most commonly used techniques in the past when metals were the materials most commonly studied in TEM [11]. The idea is to corrode the material rapidly and wash away the corrosion products so that it keeps getting thinner. The main difference between these polishing steps and the surface etching step discussed in Section 9.2 is that here, rapid and uniform material removal is the prime concern, whereas in chemical etching case, the goal was to expose low-energy surface configurations in order to enhance contrast.

The key again is selection of the proper chemicals. Here, three functions are required of the polishing chemical: an oxidizing (corroding) agent, a depassivator that constantly dissolves the stable or passivating layer formed near the surface, and a viscous component that lingers near the surface to provide macroscopic polishing. For standard metals (and recently, for other materials) tabulated recipes are available in the literature [11]. The easiest method of chemical polishing would be to dip the sample in the chemical using tweezers or a clamp. Slight heat may be applied if required. Since the goal of a final thinning step is to cause perforation, a weak zone may be created by using a dimpled specimen or a window sample and dipping it halfway into the reactive chemical. Attack occurs most rapidly at the solution surface, starting the perforation at the center in that level. Since chemi-

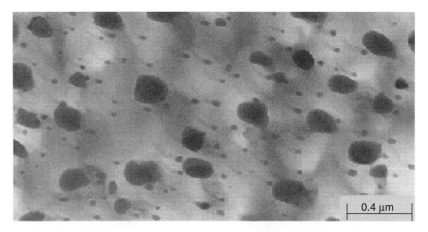

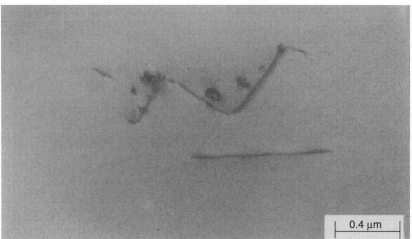

Figure 9.8. Influence of reactive gases on ion milling of delicate materials. The top figure shows an InP specimen after regular Ar-ion etching. Islands of metallic indium are formed by this process. The bottom figure shows same material thinned by iodine-jet-assisted ion etching (CAIBE). The islands are not formed and actual nanostructural features can now be studied. (Adapted from Ref. 30.)

cal polishing uses primarily strong corrosives at high temperatures, it is difficult to control the final stages of thinning once perforation begins. This can be eased in case of conductive specimens by using an electric field to control the potency of the chemical (electroplishing).

The term *electropolishing* is used when an electric potential is applied through the chemical solution using the specimen as the anode. A simple

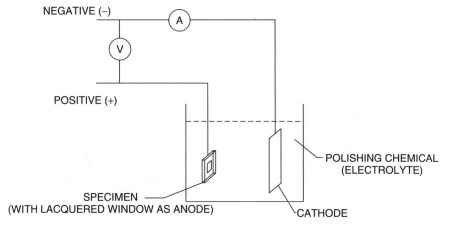

Figure 9.9. Schematic of an electropolishing unit.

schematic is shown in Figure 9.9. At low voltages, current through the electrolytic cell increases linearly with voltage and slow surface etching occurs. At higher voltages, where the current–voltage plot indicates uniform current, steady removal of material occurs at the anode. This voltage range is preferred for thinning purposes. Each sample–electrolyte system is calibrated for optimum conditions, and a large number of studies are summarized in handbooks and textbooks [11].

An important variation of electropolishing is the jet polishing technique. In this method, the electrolyte is introduced as a jet through a nozzle. The jet can be directed parallel or perpendicular to the sample, depending on what flow pattern is desired. Parameters such as sample visibility and thinning geometry are taken into consideration in different designs for commercial jet polishing systems.

Needless to say, all wet chemical techniques should be followed by thorough and repeated washing and drying after processing. Residues from insufficient cleaning can be a major problem not only for surface spectroscopy techniques (discussed later), but also for TEM analysis, where every "speck" of solvent residue is considerably magnified (Figure 9.10).

9.3.4. Tripod Polishing

It is possible to prepare thin foils from hard materials by mechanical methods alone. This is especially useful for modern nonmetallic electronic materials such as compound semiconductors and multication oxides. These materials are not easily polished chemically, and ion beams can cause unequal

Figure 9.10. Importance of specimen cleaning after chemical or electrochemical processing: The image on the left was inadequately washed, and the image on the right was taken after thorough washing. (Adapted from Ref. 11.)

sputtering of different elements, thereby changing the material [10,28,29]. Modern mechanical polishing setups such as the tripod polisher [31] can be especially useful for these samples. This setup (Figure 9.11) allows lapping of the material with a progressively increasing wedge angle so that the final specimen is thin enough for electron transmission at one end. One side of the sample is polished by a conventional technique to the finest final polish available (0.05-μm alumina, if possible). The specimen is then glued on the polished side to a platform that is held by three micrometers (forming a tripod). The micrometer heights can be adjusted individually so that the exposed side of the specimen faces the polishing wheel at any desired angle. The idea is to keep lapping off this side with a gradually increasing angle with respect to the other side so that the final shape is a wedge. This is a delicate operation, especially in the final stages when the sample is very small and fragile. But with some experience, this often becomes the quickest and least damaging thinning route for complex compounds.

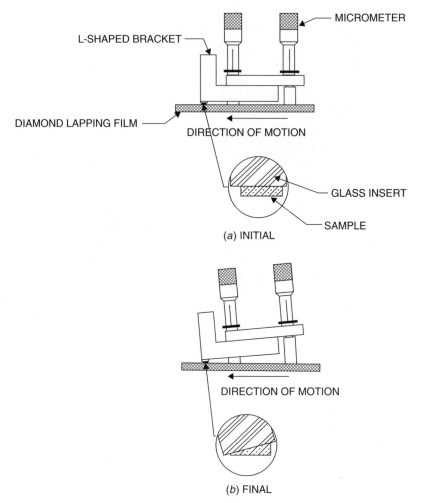

Figure 9.11. Modern tripod polisher.

9.3.5. Ultramicrotomy

Ultramicrotomy was one of the oldest sample preparation techniques used for soft biological specimens. With the improvement in instrumentation capabilities, this approach is making a comeback into the mainstream engineering materials, especially polymers. It involves directly sectioning an extremely thin sample using an ultramicrotome and dropping it in a liquid, where it will float and can latter be retrieved. A schematic of the ultra-

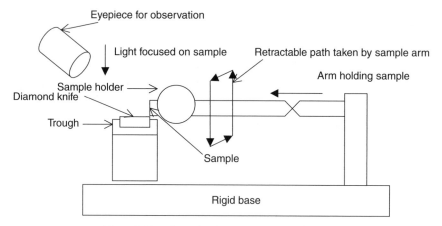

Figure 9.12. Schematic of a modern ultramicrotome.

microtome is shown in Figure 9.12. Samples processed in this way are different from those obtained by other techniques discussed so far. The earlier techniques resulted in thin wedges or perforated foils that were supported by thicker parts of the specimen. Here, the entire sample is a thin piece that has to be self-supporting and also retrievable from the liquid into which it is dropped. It must be noted that except for strong bulk materials strong enough to withstand the cutting force and remain rigid, most samples require embedding, special trimming, and specimen holding arrangements. It is therefore a slightly more complicated method of sample preparation, but works very well in some cases. Some recent articles [32] give detailed description of accessories and recent variations used by investigators. Figure 9.13 is an example of how an ultramicrotome section can reveal features distributed over a large area.

9.3.6. Special Techniques and Variations

Since the consumers of the TEM technique come from a wide variety of backgrounds, interesting variations of sample preparation are introduced all the time. Some examples of unusual approaches are as follows [8]:

- Modern lithography techniques can be used to make many sub-micrometer windows on the sample. The sample can then be thinned to obtain many small transparent regions. The advantage is that if lithography facilities are available, several regions of the specimen can be analyzed simultaneously for statistical sampling.

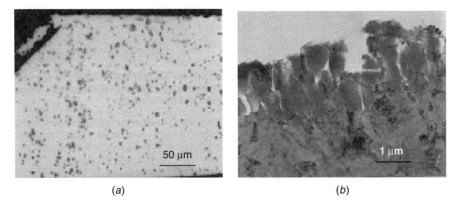

(a)   (b)

Figure 9.13. Correlation of ultramicrotome specimens with more traditional images. Image (*a*) is from a routinely sectioned and polished specimen and image (*b*) is from an ultramicrotomed specimen of the same sample. This provides a relatively large area of electron-transparent region so that details of the grains can be studied. (Adapted from Ref. 32.)

- The conventional dimpling machine has recently been modified to perform chemical polishing with a reactive etchant [33].
- Some crystalline materials can have two cleavage planes that form a thin wedge. They can therefore be fractured along these planes to form wedges with electron-transparent regions. This technique, called *wedge cleaving*, can only be applied to specific crystals.
- A focused ion beam (FIB) can be used instead of a conventional ion mill to mill a sample. In such cases, especially targeted regions of a sample can be thinned for observation in the TEM. This technique requires expensive instrumentation but is becoming extremely popular in the age of VLSI devices and nanostructured components, where precise thinning of specific areas is necessary.

9.4. SUMMARY: SAMPLE PREPARATION FOR MICROSCOPY

In summary, sample preparation is an essential part of microscopy and there are many techniques (and variations) that can be used. The approaches very commonly used to prepare specimens for analysis are as follows: The sample needs to be cut to size using one of the slicing methods outlined. The cut sample is either set in a mold or mounted externally on a polishing mount. This step is followed by a series of coarser to finer grinding on SiC grit paper. For optical microscopy and SEM, subsequent fine polish is done using diamond-abrasive paste or alumina suspension. Polished samples are then cleaned thoroughly and etched chemically or thermally to reveal surface contrast.

For TEM analysis, the cutting and grinding steps are similar except that samples are cut as small as one can handle. Subsequently, the ground sample is dimpled, wedged, or lacquered to provide a thin region supported by a thicker rim. It is then processed further using one of the final thinning techniques until some electron transparent regions are obtained. Table 9.4 summarizes some options, and provides guidelines for the new user. After this step, the very delicate sample is retrieved, cleaned, and placed in the grid or glued to the special holder suitable for TEM.

Table 9.4. Summary of Some Final Thinning Techniques for TEM[a]

Technique	Advantages	Disadvantages
Ion-beam thinning	Universally applicable; good for two-phase materials and chemically resistant materials; large thin areas; reproducible	Slow, ion-beam damage and structural alterations often possible
Chemical thinning	Quick	Not easy to control; chemical recipes for new materials often not available
Electropolishing	Quick and controllable	Applicable to electrical conductors only
Mechanical polishing (tripod technique or similar setup)	Fairly simple; no chemical or ion-beam concerns	Only for very hard materials or too much damage; slow and tedious; needs practice
Ultramicrotomy	Large thin areas that may not require additional thinning	High amount of deformation; not suitable for hard materials; slow; often irreproducible; needs practice
Special method: cleavage	Quick and easy	Very limited applicability (only materials that have clear cleavage planes); may introduce damage

[a]It must be noted that this is a very vast field, and many techniques, patents, and variations are used for specific applications.

9.5. SAMPLE PREPARATION FOR SURFACE SPECTROSCOPY

See Figure 9.14 for the basic steps in surface spectroscopy.

Special Constraints for Surface Spectroscopy

As discussed earlier, bulk spectroscopic techniques do not require much sample preparation and are not included here. Surface spectroscopic techniques have special concerns. Since the surface is the outer skin of the solid, it is the most dynamic and sensitive region. It can change constantly by two types of mechanisms: (1) exchanging atoms, ions, or molecules with the environment: (2) restructuring and redistributing atoms with the bulk. The first mechanism (exchange with environment) results in impurity adsorption, vaporization, and corrosion. The second process results in segregation, relaxation, and restructuring of the surface. Because of the evolutionary nature of this region, the major sample preparation concern is to make sure that the required surface (and not a contaminated or altered one) is the one that is exposed to the probe and getting analyzed. In other words, preserving the test surface or cleaning it with minimal alterations is the major sample preparation challenge.

The other feature specific to surface spectroscopy techniques is that they require ultrahigh vacuum (10^{-8} to 10^{-11} torr) since they involve detection of charged particles (Table 9.3). Therefore, the investigator needs to be aware if their sample is prone to degradation or alteration in vacuum. This is especially true of biosolids that prefer a liquid environment or even complex compounds that may have volatile components. In some cases, surface spectroscopy is still performed on such solids taking the vacuum-related artifacts into account. In other cases, differentially pumped sample holders might be designed which can keep the test surface at somewhat higher pressure than ultrahigh vacuum, but the range of allowable environments is not very large. Owing to the extremely low penetration depth of low-energy electrons, the extent of pressure and atmospheric manipulation possible for successful electron spectroscopy of vacuum-sensitive samples is extremely limited, even to this day.

From a sample preparation point of view, it must be remembered that several of the methods may require processing in vacuum, which implies remote sample handling and manipulation from outside the test chamber. There is a wide variety of intricate commercial instrumentation available for this step, and most designs allow additional customization, depending on vacuum chamber configuration.

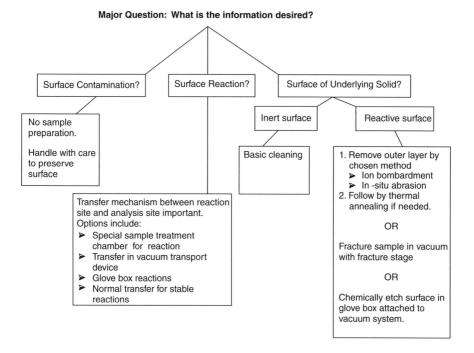

Major Question: What is the information desired?

Surface Contamination? | Surface Reaction? | Surface of Underlying Solid?

No sample preparation.

Handle with care to preserve surface

Inert surface | Reactive surface

Basic cleaning

Transfer mechanism between reaction site and analysis site important. Options include:
- ➤ Special sample treatment chamber for reaction
- ➤ Transfer in vacuum transport device
- ➤ Glove box reactions
- ➤ Normal transfer for stable reactions

1. Remove outer layer by chosen method
   - ➤ Ion bombardment
   - ➤ In-situ abrasion
2. Follow by thermal annealing if needed.

OR

Fracture sample in vacuum with fracture stage

OR

Chemically etch surface in glove box attached to vacuum system.

General Considerations: Sample Handling in Surface Spectroscopy

- ➤ Do not touch with bare hands. Use clean tweezers or lint- and dust-free gloves
- ➤ Avoid cuttingsamples. If it cannot be avoided, try clean diamond saw without cutting fluids.
- ➤ Avoid solvents if possible. If sample is dirty or has been handled before, use solvents or soap and water, but give a final rinse with a solvent that gives minimal residue, such as methanol or ethanol, then blow dry completely.
- ➤ Store in clean containers (preferably glass or metal if long-term storage is required)
- ➤ For storage and transport, mount samples such that surface of interest does not touch the container.
- ➤ Avoid using adhesive tape for long-term mounting (convenient for quick mounting and analysis).
- ➤ For new materials, pre-sputter analysis of surface is recommended even if ion beam sputtering or in-situ abrasion may be necessary sample preparation steps. There may be artifacts introduced by these steps which need to be identified.

Figure 9.14. Specimen preparation/handling for surface spectroscopy.

Sample Handling and Storage Requirements

It cannot be overemphasized that these techniques study the top 1 to 20 nm of the surface, which is extremely prone to contamination. Therefore, sample handling and storage become serious concerns for these techniques,

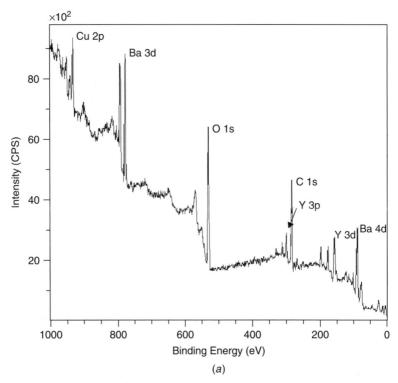

Figure 9.15. Influence of sample cleaning on XPS scans taken on a thin-film superconductor. (*a*) Survey scan from an as-received surface. (*b*) Survey scan from surface after ion-beam (sputter) cleaning. Note the reduction in the C1s peak after cleaning. (*c*) Comparative Ba3d scans from both cases. Note the change in shape and size as the surface contaminant layers (probably containing carbonates and hydroxides of Ba in addition to other components) are removed. The peak shapes and intensities of other cations change, too. Initial data represent the composition and chemistry of the contaminant layer, whereas that from sputtered sample represents those of the pure underlying superconductor (possibly with sputter-induced changes that need to be accounted for).

more so in some samples than others. The general rule of thumb is that high-surface-energy materials (such as metals, especially the reactive ones) are always coated with atmospheric reaction products, whereas low-energy surfaces (such as Teflon) are relatively stable. The stable group can be analyzed directly on introduction into the vacuum chamber. But a vast majority of solids fall under the former group and need to be treated in vacuum by one of the in situ methods outlined below (unless, of course, one is interested in the analysis of the atmospheric contaminant itself).

Figure 9.15 illustrates this point from XPS data taken on a complex oxide

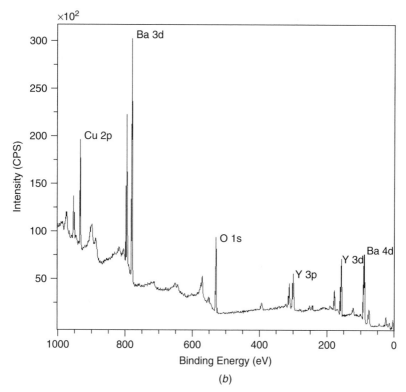

Figure 9.15. *(Continued)*

(thin-film superconductor). Figure 9.13*a* was taken on the as-received sample that was carefully handled and stored in a desiccator immediately after fabrication. Figure 9.15*b* was taken from the same sample after it was sputter cleaned as described in Section 9.5.1. Carbon is detected in Figure 9.15*a* as indicated by the C1s photoelectron peak. In addition, the shapes and sizes of component peaks can be substantially different, as is apparent in Figure 9.15*c*, which is the Ba3d peak. The shape change indicates that the binding environment of the detected atom is different in the as-received surface and the cleaned surface. Therefore, data prior to sample processing would be useful in identifying initial surface contaminants, whereas the data after sputter cleaning would be required for the actual composition and chemistry of the solid. Therefore, the investigator should be clear about what information is needed before processing the sample for analysis.

In all situations, grease-free, powder-free gloves and/or clean dry tweezers are essential for handling. Any grease or oil from human skin and other

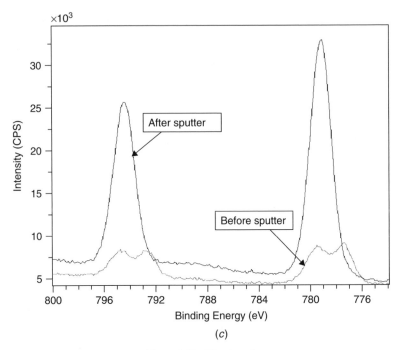

Figure 9.15. *(Continued)*

sources can vaporize in the chamber and degrade the vacuum in addition to contaminating the test surface. In general, storage in a desiccator or a partially evacuated chamber is recommended. It is sometimes necessary to leave the sample in a vacuum chamber overnight to desorb atmospheric contaminants. If the sample is mounted with adhesive tape or silver paint for analysis, care must be taken to check the vacuum compatibility of the adhesive as well as the solvent/sample compatibility. Some solvents can diffuse along the sides of the sample and leave a film of contaminant on the analysis surface.

If a surface-sensitive solid is processed in one site and needs to be transported to the analysis site without exposure to the atmosphere, a "vacuum briefcase" or special transportation module needs to be used. This would consist of a small portable vacuum chamber that is capable of attaching and transferring samples between processing and analysis stations. Understandably, designs of such instruments are system specific and often complicated. Most manufacturers of vacuum and surface analysis systems can offer customized options for specific systems.

9.5.1. Ion Bombardment

Ion bombardment is the most common treatment used for surface cleaning inside vacuum and almost a standard attachment in most surface analysis instruments. It requires a controlled gas inlet and an ion gun. The former requires a source of high-purity noble gas (normally Ar), a regulated line between a high-pressure gas container, and an ultrahigh-voltage (UHV) system followed by a precision leak valve that allows extremely controlled introduction of gas into the chamber. The ion gun ionizes the neutral gas atoms introduced and accelerates them to a specific energy. Several designs are available, some with additional capability that can focus, raster, and manipulate the outgoing beam in several ways. The basic principle is to shoot noble gas ions (normally, 0.5- to 5.0-keV Ar^+ ions) at the surface. This results in atoms from the surface being eroded away by energy exchange with this beam. It can be regarded as a slower and more controllable version of the ion milling process used for thinning TEM specimens (Section 9.3.1). In this process, also termed *sputtering*, the rate of material removal is determined by using standards having known thickness. Of course, the sputter rate of each solid will be different other factors remaining identical, but a commonly used standard is an epitaxially grown oxide film on Si. The parameters of the ion beam (beam voltage, gas flow rate, current densities, etc.) are adjusted in a given instrument to maintain a sputter rate of about 3 to 5 nm/min for SiO_2.

Ion bombardment is a relatively severe treatment and can introduce artifacts in terms of compositional, chemical, and topographic changes. Compositional changes can be caused in compounds where different elements are likely to have different sputtering rates [28,36]. Chemical states of elements can also change. For instance, several electronic oxides are known to show lower oxidation states of cations after sputtering [28]. Any initial irregularity or hard particle on the surface can result in increased roughness after sputtering. Chemical and compositional changes cannot be compensated for and therefore should be taken into account during data analysis. Physical roughness can sometimes be dealt with. In some cases, heating the surface after sputtering (annealing) can soothe out surface irregularities. However, all samples cannot tolerate high temperatures. In rare instances, the sample is rotated during sputtering or, alternatively, two or more guns are used to sputter at different angles. These options can reduce the extent of topographic roughness caused by sputtering but add substantially to the cost of the machine.

Despite these artifacts, sputtering is the most versatile, robust, and universal surface cleaning tool used in electron spectroscopy. It can also be used in conjunction with the analysis tool to perform what is commonly referred to as *depth profiling* of the specimen. Depth profiling involves bombarding a specific area of the specimen surface with Ar^+ ions and analyzing the freshly exposed surface after each bombardment. This sputter analysis cycle is repeated several (10 to 100 is typical) times to obtain compositional and chemical information of the solid as a function of depth from the surface. This combination of sample preparation and analysis capabilities makes this tool very popular in surface spectroscopic systems.

9.5.2. Sample Heating

Some stable surfaces that tend to absorb only loosely bound surface contaminants can be cleaned by heating alone. Refractory metals and silicon surfaces can be cleaned sufficiently by *flash heating*, which implies heating them to a very high temperature for a very short time whereby surface oxides become unstable and vaporize in vacuum [34]. Heating of a specimen can be as simple as passing current through the sample holder (many labs build this in-house) to sophisticated heating/cooling stages available commercially that can have programmable heaters for heating and liquid nitrogen pumps for cooling on the same device. It must be noted that while heating alone can clean only few types of solid surfaces, heating in conjunction with ion beams can be adapted to preparing a wide variety of materials.

9.5.3. In Situ Abrasion and Scraping

In situ abrasion and scraping is a specialized method for cleaning relatively soft solid surfaces. A razor blade or a grinding tool (brush, abrasive grinder, etc.) is attached at an appropriate angle to rotating or sliding shafts inside the vacuum system. The surface can thus be scrubbed while inside the chamber prior to analysis. Several types of UHV abrading tools are available commercially, the choice depending on the sample to be cleaned. Needless to say, the cleanliness and purity of the scraping surface are important. Moreover, care should be taken not to use the same scraper on different surfaces without in-between cleaning, as this will result in cross-contamination between samples.

9.5.4. In Situ Cleavage or Fracture Stage

A specialized method for sample preparation is to fracture or cleave the sample inside the vacuum system, thus creating a fresh surface for immediate analysis. Some crystalline materials (semiconductors, anisotropic structures such as graphite, etc.) have preferred cleavage planes that can be sectioned inside the chamber using a blade or chisel (operated through bellows from outside). Other materials can be introduced with a notch or weak spot in a specially designed fracture stage so that the sample is broken inside the chamber and the newly exposed surface placed in analysis position. This type of sample processing is especially useful in studies where one needs to investigate failure mechanisms (intergranular, intragranular, along specific phase boundaries, etc.). In situ fracture attachments can be obtained in several complicated designs and are beyond the scope of this chapter. Some specific examples can be seen in the references cited or manufacturer brochures [34,37].

9.5.5. Sample Preparation/Treatment Options for In Situ Reaction Studies

A large (and ever-expanding) field where surface spectroscopic techniques are used include in situ study of reaction chemistry, film growth, and so on. In these studies it is difficult to argue where sample preparation ends and sample treatment (a part of the actual experiment) starts. Such studies are almost always conducted in a system that has a sample preparation/ treatment chamber attached to the analysis chamber. The initial steps, of course, would be to clean the surface by sputtering, heating, scraping, and so on. This can be followed by deposition of solids or exposure to gases/ plasmas at specific temperatures and pressures. The former (deposition of solids) is in itself a large field of investigation and can be very simple or very complicated. A simple step may involve thermally heating a metal-coated filament, whereas a complex deposition may require a multimillion-dollar deposition system attached to the spectroscopic chamber. Exposure to gases is a relatively common surface preparation option that involves one or more high-purity gas containers, central manifold, and precision leak valve(s). In some ways, these requirements are similar to those for ion-beam sputtering and are often easy to install.

9.6. SUMMARY: SAMPLE PREPARATION FOR SURFACE SPECTROSCOPY

The major challenge of sample preparation for surface spectroscopy involves producing a clean, pristine surface that is well characterized and reproducible. The suitable cleaning technique will depend on several factors, such as chemical affinities, composition, geometry, vacuum tolerance, and so on. The most commonly used technique is ion-beam etching or sputtering. This step can be accompanied by or followed up with heat treatments in vacuum. Other special treatments include in-vacuum scraping, abrasion, and fracturing. Treatment with other gases can be used in rare occasions for specific applications. Most sample preparation processes here involve specialized ultrahigh-vacuum instrumentations.

ACKNOWLEDGMENTS

The author acknowledges funding from the National Science Foundation, Ohio Board of Regents, AFRL, and National Aeronautics and Space Administration, for financial support. Special thanks go to two of her students, N. Mahadev and P. Joshi, for help with the figures and tables.

REFERENCES

1. C. R. Brundle, C. A. Evans, Jr., S. Wilson, and L. E. Fitzpatrick, eds., *Encyclopedia of Materials Characterization: Surfaces, Interfaces, Thin Films*, Butterworth-Heinemann, Woburn, MA, 1992.

2. L. C. Fieldman and J. W. Mayer, *Fundamentals of Surface and Thin Film Analysis*, Prentice Hall, Englewood Cliff, NJ, 1986.

3. J. B. Wacthman, *Characterization of Materials*, Butterworth-Heinemann, Woburn, MA, 1985.

4. P. J. Duke, *Modern Microscopies: Techniques and Applications*, Plenum Press, New York, 1990.

5. M. R. Louthan, Jr., Optical metallography, in *Metals Handbook*, 9th ed., Vol. 10, *Materials Characterization*, American Society for Metals, Metals Park, OH, 1986.

6. B. L. Gabriel, *SEM: A Users Manual for Materials Science*, ASM International, Metals Park, OH, 1985.

7. J. W. Edington, *Practical Electron Microscopy in Materials Science*, Van Nostrand Reinhold, New York, 1976.

8. P. J. R. Uwins, *Mater. Forum*, **18**, 51–75 (1994).

9. P. B. Bell and B. SafiejkoMroczka, *Int. J. Imag. Syst. Technol.*, **8**(3), 225–239 (1997).

10. See, for example, the three-volume symposia proceedings entitled *Specimen Preparation for Transmission Electron Microscopy of Materials*, Vols. I–III, Materials Research Society, Warrendale, PA, 1987, 1990, and 1993.

11. P. J. Goodhew, Specimen preparation in materials science, in *Practical Methods in Electron Microscopy*, North-Holland, New York, 1973.

12. A. Szirmae and R. M. Fisher, *Specimen Damage during Cutting and Grinding*, ASTM Technical Publication 372, 1963, p. 3.

13. R. W. Armstrong and R. A. Rapp, *Rev. Sci. Instrum.*, **29**, 433 (1958).

14. Buehler Ltd., *Met. Dig.*, **20**(2), 1 (1981).

15. J. H. Richardson, Sample preparation methods for microstructure analysis, in

J. C. McCall and W. M. Mueller, eds., *Microstructural Analysis: Tools and Techniques*, Plenum Press, New York, 1974.

16. Stuers, Inc., *Stuers Metallographic News*, special issue on sample preparation, Structure 4.1, 1982.

17. American Society for Metals, *Metals Handbook*, 8th ed., Vol. 7, *Atlas of Microstructure of Industrial Alloys*, ASM, Metals Park, OH, 1972.

18. B. J. Kestel, *Polishing Methods for Metallic and Ceramic TEM Specimens*, ANL-80-120, Argonne National Laboratory, Argonne, IL, 1981.

19. U. Linde and W. U. Kopp, *Structures*, **2**, 9 (1981).

20. S. M. Mukhopadhyay and C. Wei, *Physica C*, **295**, 263–270 (1998).

21. W. D. Kingery, H. K. Bowen, and D. R. Uhlman, *Introduction to Ceramics*, Wiley, New York, 1976.

22. D. A. Porter and K. E. Easterling, *Phase Transformation in Metals and Alloys*, Chapman & Hall, New York, 1996.

23. T. J. Shaffnor and J. W. S. Hearle, *ITRI/SEM*, **1**, 61 (1976).

24. P. Echlin, *ITRI/SEM*, p. 217 (1975); also *SEM Inc.*, **1**, 79 (1981).

25. G. Betz and G. K. Wehner, Sputtering of multicomponent materials, in *Sputtering by Particle Bombardment*, Vol. II, Springer-Verlag, New York, 1983, p. 11.

26. M. J. Kim and R. W. Carpenter, *Ultramicroscopy*, **21**, 327–334 (1987).

27. D. Bahnck and R. Hull, Experimental measurement of transmission electron microscope specimen temperature during ion milling, in R. Anderson, ed., *Specimen Preparation for TEM*, Vol. II, MRS Vol. 199, Materials Research Society, Warrendale, PA, 1990.

28. S. M. Mukhopadhyay and T. C. S. Chen, *J. Appl. Phys.*, **74**(2), 872–876 (1993).

29. M. St. Louis-Weber, V. P. Dravid, and U. Balachandran, *Physica C*, **243**, 3–4 (1995).

30. R. Alani, J. Jones, and P. Swann, CAIBE: a new technique for TEM specimen preparation for materials, in R. Anderson, ed., *MRS Symposium Proceedings*, Vol. 199, 1990, p. 85.

31. J. P. Benedict, S. J. Klepeis, W. G. Vandygrift, and R. Anderson, *A Method of Precision Specimen Preparation for Both SEM and TEM Analysis*, ESMA Bulletin, 19.2, Nov. 1989.

32. T. F. Malis and D. Steele, Ultramicrotomy for materials science, in R. Anderson, ed., *MRS Symposium Proceedings*, 1990, p. 3.

33. Bellcore, U.S. patent 4,885,051.

34. D. Briggs and M. P. Seah, eds., *Practical Surface Analysis*, Vol. 1, Wiley, New York, 1990.

35. K. Kiss, Problem solving with microbeam analysis, in *Studies in Analytical Chemistry*, Vol. 7, Academiai Kiado, Budapest, 1988.

36. S. M. Mukhopadhyay and T. C. S. Chen, *J. Mater. Res.*, **8**(8), 1958–1963, Aug. 1993.

37. See product catalogs of vacuum equipment manufacturers such as Physical Electronics, Varian, Huntington, Kratos, and others.

CHAPTER

10

SURFACE ENHANCEMENT BY SAMPLE AND SUBSTRATE PREPARATION TECHNIQUES IN RAMAN AND INFRARED SPECTROSCOPY

ZAFAR IQBAL

Department of Chemistry and Environmental Science, New Jersey Institute of Technology, Newark, New Jersey

10.1. INTRODUCTION

Sampling in surface-enhanced Raman and infrared spectroscopy is intimately linked to the optical enhancement induced by arrays and fractals of "hot" metal particles, primarily of silver and gold. The key to both techniques is preparation of the metal particles either in a suspension or as architectures on the surface of substrates. We will therefore detail the preparation and self-assembly methods used to obtain films, sols, and arrayed architectures coupled with the methods of adsorbing the species of interest on them to obtain optimal enhancement of the Raman and infrared signatures. Surface-enhanced Raman spectroscopy (SERS) has been more widely used and studied because of the relative ease of the sampling process and the ready availability of lasers in the visible range of the optical spectrum. Surface-enhanced infrared spectroscopy (SEIRA) using attenuated total reflection coupled to Fourier transform infrared spectroscopy, on the other hand, is an attractive alternative to SERS but has yet to be widely applied in analytical chemistry.

To aid the general reader, short descriptions of the fundamentals of modern Raman scattering and attenuated total reflection (ATR) infrared spectroscopy are provided. This is followed for each spectroscopy by brief introductions to the enhancement mechanism involved.

10.1.1. Raman Effect

In the Raman effect, incident radiation is inelastically scattered from a sample and shifted in frequency by the energy of its characteristic molecu-

Sample Preparation Techniques in Analytical Chemistry, Edited by Somenath Mitra
ISBN 0-471-32845-6 Copyright © 2003 John Wiley & Sons, Inc.

lar vibrations. Since its discovery in 1927, the Raman effect has attracted attention from the point of view of basic research as well as a powerful spectroscopic technique with many applications in chemical analysis. The advent of laser sources with monochromatic photons at high flux densities was a major development in the history of Raman spectroscopy and has resulted in dramatically improved scattering signals. For general overviews of modern Raman spectroscopy, the reader is referred to Refs. 1 and 2.

In addition to spontaneous or incoherent Raman scattering, the development of lasers also opened the field of stimulated or coherent Raman scattering where molecular vibrations are coherently excited. Whereas the intensity of spontaneous Raman scattering depends linearly on the number of molecules being probed, nonlinear Raman scattering associated with stimulated or coherent excitation is proportional to the square of the number of molecules probed [3,4]. Coherent Raman techniques can therefore provide interesting new opportunities such as the vibrational imaging of biological samples [5], but have yet to be advanced to the level of ultrasensitive single-molecule detection.

Modern Raman spectroscopy utilizes laser photons over a wide range of frequencies from the near-ultraviolet to the near-infrared region of the optical spectrum. This allows for the selection of optimum excitation conditions for each sample. For example, by choosing wavelengths that excite appropriate electronic transitions, selected components of a molecule can be studied [6]. The extension of excitation wavelengths to the near-infrared (NIR) region, where background fluorescence is reduced and photo-induced degradation from the sample is diminished, has allowed the detection and study of a range of biological molecular systems. High-intensity (NIR) diode lasers are now easily available, making this region attractive for compact, low-cost Raman instrumentation. Coupled with this has been the development of low-noise, high-quantum-efficiency multichannel detectors [charge-coupled-device (CCD) arrays], which when combined with high-throughput single-stage spectrometers and holographic laser rejection filters has led to high-sensitivity NIR Raman systems [7].

As with optical spectroscopy, the Raman effect can be applied non-invasively in a wide range of environments. In contrast with infrared spectroscopy, Raman measurements do not require complicated sampling techniques. In addition, optical fiber probes can be used for bringing the laser light to the sample and transporting scattered light to the spectrometer, thus allowing remote detection of Raman spectra.

The spatial and temporal resolution of Raman scattering are determined by the spot size and pulse length, respectively, of the exciting laser. Femtoliter volumes (ca. 1 μm^3) can be observed using a confocal lens microscope, enabling spatially resolved measurements in biological cells [8]. Techniques such as confocal scanning Fourier transform Raman microscopy [9] allow high-resolution imaging of samples. Recently, near-field Raman spectroscopy measurements have been made that overcome the diffraction limit and allow volumes significantly smaller than the cube of the wavelength of the exciting light [10,11]. In the time domain, Raman spectra can be measured on the picosecond time scale, providing information on short-lived species such as excited states and reaction intermediates [12].

The key advantage of Raman spectroscopy (and this is also largely true for infrared spectroscopy) is its high degree of specificity, which arises from its correlation with the molecular structure of the sample. The Raman spectrum is obtained as a highly specific fingerprint, allowing direct identification of the sample in much more detail than can be achieved by techniques such as fluorescence. Recently, sophisticated data analysis based on multivariate techniques have made it possible to exploit the full information content of Raman spectra to obtain the chemical structure and composition of very complex systems such as biological molecules [13].

10.1.2. Fundamentals of Surface-Enhanced Raman Spectroscopy

The extremely small cross sections for conventional Raman scattering, typically 10^{-30} to 10^{-25} cm^2/molecule has in the past precluded the use of this technique for single-molecule detection and identification. Until recently, optical trace detection with single molecule sensitivity has been achieved mainly using laser-induced fluorescence [14]. The fluorescence method provides ultrahigh sensitivity, but the amount of molecular information, particularly at room temperature, is very limited. Therefore, about 50 years after the discovery of the Raman effect, the novel phenomenon of dramatic Raman signal enhancement from molecules assembled on metallic nanostructures, known as surface-enhanced Raman spectroscopy or SERS, has led to ultrasensitive single-molecule detection.

Jeanmaire and Van Duyne [15] and Albrecht and Creighton [16] concluded that the strong Raman signals measured from electrochemically adsorbed pyridine on a roughened silver electrode are caused by an intrinsic enhancement of the Raman effect and cannot be explained by an increase in number of molecules adsorbed per unit area on the high-surface-area electrode. Within a few years, strongly enhanced Raman spectra were obtained for many different molecules adsorbed on SERS-active substrates. These SERS-active substrates are various metallic structures with sizes on the order of tens of nanometers. The most common types of SERS substrates exhibiting the largest effects are colloidal silver or gold nanoparticles in the size range of 10 to 150 nm.

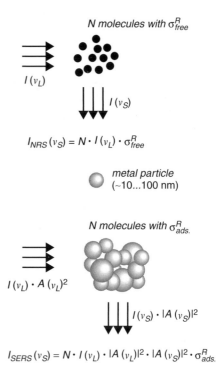

Figure 10.1. Comparison of normal (top) and surface-enhanced (bottom) Raman scattering. The top panel shows the conversion of incident laser light of intensity $I(\nu_L)$ into Stokes scattered light I_{NRS}, which is proportional to the Raman cross section σ^R_{free} and the number of target molecules N in the probed volume. In the bottom panel σ^R_{ads} describes the increased Raman cross section of the adsorbed molecule due to chemical enhancement; $A(\nu_L)$ and $A(\nu_S)$ are the field enhancement factors at the laser and Stokes frequency, respectively, and N' is the number of molecules involved in the SERS process. (With permission from Ref. 17.)

It is generally agreed that more than one effect contributes to the observed enhancement of many orders of magnitude of the Raman signal. A schematic of the normal and surface-enhanced Raman scattering process is shown in Figure 10.1. In normal Raman scattering, the total Stokes Raman signal I_{NRS} is proportional to the Raman cross section σ^R_{free}, the excitation laser intensity $I(\nu_L)$ and the number of molecules N in the probed volume (cf. Figure 10.1, top). Because Raman cross sections are extremely small, approximately 10^8 molecules are typically required to generate a measurable, conventional Raman signal. In a SERS experiment (Figure 10.1, bottom), the molecules are adsorbed on a metallic nanostructure, which can be in the form of a colloid, a nanostructured thin film, an array of metallic spheres, or a grating. Alternatively, the molecules can be electrochemically deposited on a roughened metal electrode. The SERS Raman signal I_{SERS} is proportional to the Raman cross section of the adsorbed molecule σ^R_{ads}, the intensity of the incident laser beam $I(\nu_L)$ and the number of molecules involved in the SERS process N'. N' can be smaller than the number of molecules in the probed volume N.

The enhancement mechanisms are roughly associated with either electromagnetic field enhancement or chemical first-layer effects. The electromagnetic enhancement arises from enhanced local optical fields at the metal surface due to the excitation of electromagnetic resonances that are also called *surface plasmon resonances*. Because the excitation field, as well as the Raman scattered field, contributes to this enhancement, the SERS signal is proportional to the fourth power of the field enhancement. Maximum values for electromagnetic enhancement are on the order of 10^6 to 10^7 for isolated particles of metals. Closely spaced interacting particles appear to provide extra field enhancement, particularly near the gap between two particles in close proximity. SERS enhancement factors up to 10^8 are achieved under these conditions. Theory also predicts strong enhancement of electromagnetic fields for sharp features and large curvature regions, which may exist on silver and gold nanostructures. In many experiments, SERS substrates consist of a collection of nanoparticles exhibiting fractal properties, such as colloidal clusters formed by aggregation of colloidal particles or metal island films. In these structures, the excitation is not distributed uniformly over the entire cluster but is spatially localized in "hot" sites. A typical collection of metal particles used in SERS experiments is shown in Figure 10.2.

Chemical enhancement effects include enhancement mechanisms of the Raman signal that are related to specific interactions involving electronic coupling between the molecule and metal. Roughness, resulting in nanostructuring, appears to play an important role by providing pathways for the "hot" electrons to be transported from the metal to the molecules. The magnitude of chemical enhancement has been estimated to be on the order of 10 to 100.

In the middle 1980s, despite a poor understanding of the effect, SERS generated growing interest as an analytical tool for trace analysis. The ability of SERS to detect substances down to the picogram detection limits was demonstrated [19] for a variety of molecules of environmental, technical, biomedical, and pharmaceutical interest. For example, the separation and highly specific determination of adenine, guanine, hypoxanthine, and xanthine was performed using liquid chromatography in combination with SERS [20]. The key reason for these advancements was the quenching of fluorescence due to additional new relaxation channels to the metal surface for the electronic excitation. This allowed the observation of high-quality vibrational spectra over wide frequency ranges from minimum amounts of substances.

The size of the enhancement factor or the effective SERS cross section is a key question for the application of SERS as a tool for ultrasensitive detection. The effective cross section must be high enough to provide a detectable Raman signal from a few molecules. In the early SERS experiments, Van

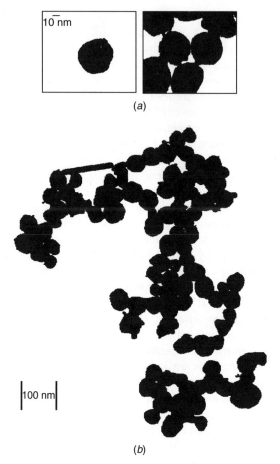

Figure 10.2. Electron micrographs of typical colloidal gold and silver particle structures used in SERS experiments. (*a*) Colloidal gold particles in the isolated and aggregated stage after addition of NaCl. (*b*) Typical colloidal silver clusters exhibiting strong SERS enhancement. (With permission from Refs. 17 and 18.)

Duyne and co-workers estimated enhancement factors on the order of 10^5 to 10^6 for pyridine on rough silver electrodes. The value was obtained from a comparison between surface-enhanced and normal "bulk" Raman signals from pyridine by taking into account the different number of molecules on the electrode and in solution. The size of the enhancement was found to correlate with the electrode roughness, indicating that enhancement occurs via a strong electromagnetic field. On the other hand, the dependence of the enhancement on the electrode potential suggested that chemical enhancement also plays a role.

For excitation wavelengths that are in resonance with an optical transition of the target molecule, surface-enhanced resonance Raman scattering (SERRS) has been observed. In such experiments, total enhancements on the order of 10^{10} have been obtained for molecules such as rhodamine 6G on colloidal silver and excited under molecular resonance conditions. A method of estimating the SERRS enhancement factor involves comparing the intensity of the nonenhanced methanol (e.g., in 5 M solution) Raman line to the enhanced Raman lines of rhodamine 6G (in 8×10^{-11} M) and taking into account the different concentrations of both compounds, provide a total enhancement factor of 5×10^{11}. The problem with this method of estimating the enhancement factor is in the assumption that nearly all molecules in the SERRS sample contribute in a similar way. To avoid this problem, a different approach in which Stokes and anti-Stokes Raman data are used to extract the effective SERS cross section independent of whether the process is resonant or nonresonant. In conventional Raman scattering, the Stokes to anti-Stokes intensity ratio is determined by a Boltzmann thermal population. However, a very strong SERS process can sizably populate the first excited vibrational level in excess of the Boltzmann population. This vibrational population pumping is reflected in deviation of the anti-Stokes/Stokes signal ratio from the Boltzmann population and allows an estimate of the effective SERS cross sections. The excited vibrational level is populated by Stokes scattering and depopulated by anti-Stokes scattering and a spontaneous decay process whose lifetime is given by τ_1. Assuming steady state and weak saturation, a simple theoretical estimate for the anti-Stokes to Stokes signal ratio I_{aS}^{SERS}/I_S^{SERS} can be derived [17] from the relationship

$$\frac{I_{aS}^{SERS}}{I_S^{SERS}} = \sigma_{sers}\tau_1 n_i + e^{-h\nu_m/kt}$$

where the first term on the right-hand side of the equation describes the SERS population of the first excited vibrational state in excess of the Boltzmann population. In conventional Raman scattering this term can be neglected relative to the normal population. To account for the significant deviation of the anti-Stokes/Stokes ratios obtained, the product of the cross section and vibrational lifetime $\sigma_{sers}\tau_1(\nu_m)$ must be approximately 10^{-27} cm^2·s. Assuming vibrational lifetimes on the order of 10 ps, the Raman cross section is estimated to be at least 10^{-16} cm^2/molecule. To make the large cross sections inferred from vibrational pumping consistent with the level of the observed SERS Stokes signal, the number of molecules involved must be very small. This number was experimentally shown to be between 10^{-13} and

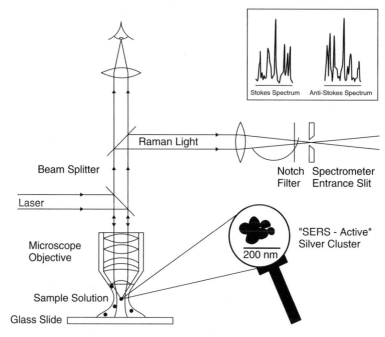

Figure 10.3. Schematic experimental set-up for single-molecule SERS. Insert (top) shows a typical Stokes and anti-Stokes Raman spectrum. Insert (bottom) shows an electron microscope image of SERS-active colloidal clusters. (With permission from Ref. 21.)

10^{-10} M in concentration. This is in the range required for single-molecule detection. These sensitivity levels have been obtained on colloidal clusters at near-infrared excitation. Figure 10.3 is a schematic representation of a single-molecule experiment performed in a gold or silver colloidal solution. The analyte is provided as a solution at concentrations smaller than 10^{-11} M. Table 10.1 lists the anti-Stokes/Stokes intensity ratios for crystal violet (CV) at 1174 cm^{-1} using 830-nm near-infrared radiation well away from the resonance absorption of CV with a power of 10^6 W/cm^2 [34]. CV is attached to various colloidal clusters as indicated in the table. Raman cross sections of 10^{-16} cm^2/molecule or an enhancement factor of 10^{14} can be inferred from the data.

10.1.3. Attenuated Total Reflection Infrared Spectroscopy

Attenuated total reflection infrared (IR) spectra are obtained by pressing the sample against an internal reflection element (IRE) [e.g., zinc selenide (ZnSe) or germanium (Ge)]. IR radiation is focused onto the end of the IRE.

Table 10.1. Anti-Stokes to Stokes SERS Intensity Ratios at 1174 cm^{-1} for Crystal Violet (CV) Attached to Silver Clusters at Various Locations[a]

Sample Location	Anti-Stokes (cps)[b]	Stokes (cps)[b]	Ratio
CV-1	37	780	4.8×10^{-2}
CV-2	53	1100	4.8×10^{-2}
CV-3	28	545	5.1×10^{-2}
CV-4	132	2550	5.2×10^{-2}
CV-5	50	1000	5.0×10^{-2}
CV-6	51	1055	4.8×10^{-2}
Toluene	10.4	1920	5.4×10^{-3}

Source: Ref. 34.
[a]The anti-Stokes to Stokes ratio for toluene at 1211 cm^{-1} establishes the Boltzmann population.
[b]cps, counts per second.

Light enters the IRE and reflects down the length of the crystal. At each internal reflection, the IR radiation actually penetrates a short distance (~ 1 μm) from the surface of the IRE into the sample as shown in Figure 10.4 and enables one to obtain infrared spectra of samples placed in contact with the IRE.

10.1.4. Fundamentals of Surface-Enhanced Infrared Spectroscopy

Modern infrared spectroscopy is performed using Fourier transform interferometry [22]. Hartstein and co-workers [23] were the first to show that

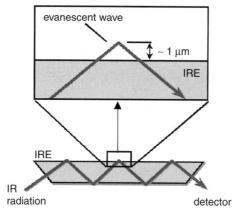

Figure 10.4. Total internal reflection at the interface of an internal reflection element (IRE). Depth of penetration of the evanescent wave is approximately 1 μm. The top picture depicts the evanescent beam in more detail. The sample is coated on both sides of the IRE.

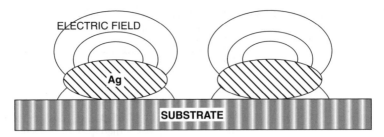

Figure 10.5. Schematic model for electromagnetic enhancement in SEIRA.

infrared absorption from molecular monolayers can be enhanced by a factor of 20 with thin metal overlayers or underlayers using the attenuated total reflection (ATR) technique. The total enhancement, including contributions from the ATR geometry, is 10^4. The effect, analogous to SERS, has been attributed to collective electron or plasmon resonances arising from the island, nanostructured nature of the films. This is depicted schematically in Figure 10.5.

Figure 10.6 demonstrates the SEIRA spectra of the C–H modes of a monolayer of 4-nitrobenzoic acid using ATR infrared techniques. The pres-

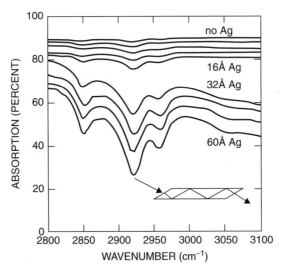

Figure 10.6. Absorption of the C–H modes of a monolayer of 4-nitrobenzoic acid using SEIRA techniques. The curves are for increasing thicknesses of a silver overlayer. The inset shows the path of the infrared beam through the silicon total-internal-reflection plate. The sample is deposited on the two sides of the plate. (With permission from Ref. 23.)

ence in SEIRA spectra of some absorption bands and the absence of others, as compared to transmission spectra, has led to investigations into the possible applicability of surface selection rules similar to those observed in reflection–absorption spectroscopy. SEIRA spectroscopy is therefore useful for in situ observations of metal surface reactions and very promising as a trace analytical technique.

10.2. SAMPLE PREPARATION FOR SERS

10.2.1. Electrochemical Techniques

Bulk gold, silver, or copper electrode surfaces can be roughened in a conventional three-electrode electrochemical cell where the SERS electrode is the working electrode. The electrode is roughened in the nanometer scale by 20 oxidation/reduction cycles (ORCs) between -0.600 and $+1.200$ V at a scan rate of 0.500 V/s with pauses of 8 s at -0.600 and $+1.200$ V. The ORC cycle influences the magnitude of the enhancement both by cleaning the surface and by producing surface roughness. A clean surface allows the adsorbate to interact strongly with the silver surface. After the ORC cycle the electrode is removed from the roughening cell and rinsed with doubly distilled water to ensure the removal of residual chloride ions. The experimental variables involved in the ORC have been investigated extensively. Factors that affect enhancement include the number of ORC cycles, the amount of charge passed during the ORC, the concentration of solute, the electrode potential, and the nature of the electrolyte. In addition, illumination of the electrode during the ORC procedure produces a significant increase in SERS intensity. Therefore, there are many variables in electrochemical preparation techniques and each needs to be carefully controlled to result in a surface that gives highly reproducible enhancement factors for a given adsorbate.

In a typical experiment described by Garrell et al. [24] to detect C_{60} by SERS, 1 μL of 1.1×10^{-6} M solution of fullerene in CCl_4 was deposited on an electrode surface by dipping. This corresponds to a surface coverage of one or two monolayers on a 4-mm-diameter smooth surface. Unbiased SERS data were collected with the electrode immersed in N_2-purged water, which serves as a heat sink to minimize laser heating of the adsorbed layer. Biased spectra of the adsorbed fullerene were obtained in 0.1 M KCl as the electrolyte at $+0.200$ and -0.600 V. Applying a more negative potential causes the three highest-frequency lines of C_{60} to shift to lower values, due to metal–substrate back-donation associated with the chemical mechanism of SERS discussed in Section 10.1.2. No evidence of amorphous carbon was found on the substrate, indicating that coadsorption of a carbon layer was not necessary for inducing SERS in C_{60}.

SERS due to pyridine on Au electrode surfaces appears to arise from the adsorption of pyridine in or on surface carbon present after the oxidation–reduction cycle [25,26]. Anodically roughened Ag electrode surfaces, which were subsequently cathodically cleaned, exhibited no SERS from pyridine. This confirms that the SERS-active phase is carbon–pyridine and not pyridine alone. In ultrahigh vacuum, SERS can be induced in pyridine by coadsorbing pyridine with CO [27]. The effect depends on the type of silver surface and involves shifts in the peak positions and intensities of some of the vibrational modes. SERS peaks were not observed at 2100 cm^{-1} at the position of the C–O stretching mode of CO. A possible interpretation is that surface complexes are formed between pyridine and CO molecules at the active or hot sites on the silver surface.

A combination of electrochemical methods and SERS is used to detect chlorinated hydrocarbons in aqueous solutions [28]. Electrochemistry prepares the surface of a copper electrode for SERS and concomitantly concentrates the analyte on the surface of the electrode, possibly by electrophoretic processes. Detection sensitivity of <1 ppm for trichloroethylene, for example, was achieved.

10.2.2. Vapor Deposition and Chemical Preparation Techniques

The preparation of rough silver films by vapor deposition results in reproducible and stable surfaces for SERS. For example, deposition of 20-nm Ag films onto Teflon, polystyrene, or latex spheres [29,30] has been performed. These substrates produced strong SERS intensities for various organic adsorbates and good reproducibility between multiple runs. However, vapor deposition can be slow and needs access to a vacuum system. There are also some variables that need to be controlled, such as the film thickness, deposition temperature, and use of annealing procedures. Moreover, unless the experiment is performed under vacuum, the film is exposed to the atmosphere after deposition. Even a brief exposure to the atmosphere results in contamination of the surface and the formation of an inactive oxide layer.

Chemically deposited films on frosted glass slides provide a more facile and reproducible approach to SERS substrates. One such approach [31], which has proven to be very successful, involves the initial preparation of Tollen's reagent. The reagent is prepared by adding about 10 drops of fresh 5% sodium hydroxide solution to 10 mL of 2 to 3% silver nitrate solution, whereupon a dark-brown AgOH precipitate is formed. This step is followed by the dropwise addition of concentrated NH$_4$OH, at which point the precipitate redisssoves. Tollen's reagent is then placed in an ice bath. Frosted glass slides cleaned with nitric acid and washed with distilled water were placed in the Tollen's agent and 3 mL of a 10% aqueous solution of D-glucose was added to reduce the Ag^+ ions to Ag. The Tollen's reagent together with the glass slides were taken out of the ice bath and placed in a water bath maintained at 55°C for about 1 minute followed by sonication for another minute, washing in distilled water, and storage in water for several hours prior to exposure to the analyte. For the SERS experiment, the slides were briefly air-dried and then dipped into the analyte solution (e.g., 10^{-3} to 10^{-7} M solutions of 4,4'-bipyridine) for at least 30 minutes. Alternatively, the analyte is adsorbed electrochemically on the silver substrate at a potential of -500 mV versus SSCE (saturated NaCl calomel electrode). The normalized SERS intensity increases with decreased silver nitrate concentration, which scales directly with the thickness of the silver films. The thickness, in turn, influences the morphology of the films. Scanning electron microscope images of the chemically deposited films show that although the particles are not optimally spherical as required by electromagnetic theory, they do conform better to the optimum theoretical shape.

10.2.3. Colloidal Sol Techniques

Colloidal suspensions of silver or gold particles in water can be prepared typically by the reduction of a silver salt (e.g., silver nitrate) using a reducing agent such as D-glucose or a citrate. A novel technique [32] involves the laser ablation of silver foil in water using a 355-nm laser with a pulse energy of about 50 mJ and a 10-Hz repetition rate.

A common practice in SERS studies has been to activate the colloid by electrolyte-induced aggregation. The activated colloid contains large clusters, which are believed to be more efficient for SERS. Nie and Emory [33] showed the existence of Raman enhancement on the order of 10^{14} to 10^{15} for rhodamine 6G molecules on colloidal silver particles under resonant Raman conditions that allowed detection of single molecules. By screening a large number of individual particles immobilized on a glass slide from a colloidal suspension, the authors found a small number of nanoparticles that exhibited unusually high enhancement efficiencies. These particles, shown in Figure 10.7, were labeled "hot" particles. To screen for these particles in a heterogeneous Ag colloid, an aliquot of the colloid was incubated with rhodamine 6G for about 3 hours at room temperature. The particles were then immobilized on polylysine-coated glass because of the interactions between the negative charges on the particles and positive charges on the surface. Other methods using organosilane and thiol compounds are also available for immobilizing and dispersing colloidal particles on surfaces.

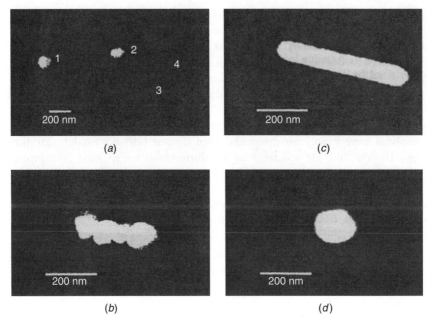

Figure 10.7. Tapping mode atomic force microscope images of typical "hot" particles. (*a*) Four single particles. Only particles 1 and 2 were highly efficient for SERS. (*b*) Close-up image of a "hot" aggregate particle containing four linearly arranged particles. (*c*) Close-up image of a rod-shaped "hot" particle. (*d*) Close-up image of a faceted "hot" particle. (With permission from Ref. 33.)

Kneipp et al. [34] showed that enhancement is independent of cluster sizes ranging from 100 nm to 20 μm. The data and the electron microscope images of the SERS particles are depicted in Figure 10.8 together with the nonresonance SERS spectrum of 10^{-6} M crystal violet. SERS enhancement is estimated to be on the order of 10^6 for the spatially isolated cluster and up to 10^8 for the colloidal clusters. The isolated silver clusters were made by the laser ablation technique mentioned earlier.

Using near-IR excitation at 830 nm these authors found that on the order of some hundreds of adenine molecules without any fluorescence labeling could readily be detected, as shown by the data in Figure 10.9. Note the extremely large anti-Stokes intensities in Figure 10.9 due to vibrational pumping, from which enhancement factors of 10^{14} can be inferred. Due to very similar cluster formation of colloidal silver and gold and very similar dielectric constants in the near IR, gold should also be a very good material for near-IR SERS and might provide some advantages, due to its chemical nobility.

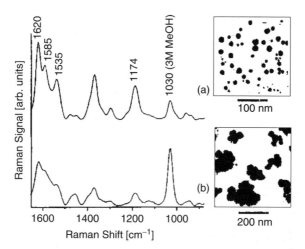

Figure 10.8. SERS spectra of (*a*) 10^{-6} *M* crystal violet on isolated silver spheres and (*b*) 10^{-8} *M* crystal violet on small colloidal silver clusters. The insets show electron micrographs of the SERS-active architectures. (With permission from Ref. 34.)

Resonance SERS spectra of single hemoglobin molecules have been observed [35]. Figure 10.10 shows the resonance SERS spectrum of single-molecule hemoglobin on silver nanoparticles made from a silver hydrosol. The hydrosol of colloidal particles at a concentration of approximately 35 p*M* was prepared by a citrate reduction process. The sol was incubated together with a 10 p*M* solution of adult human hemoglobin to obtain an average of 0.3 hemoglobin molecule per Ag particle. Dispersed hemoglobin–Ag aggregates were immobilized on glass or Si surfaces with a polymer coating.

10.2.4. Nanoparticle Arrays and Gratings

Lithographic techniques have been employed for the production of more uniform and controllable SERS substrates. One of the most interesting possibilities using the lithographic approach is to achieve a controlled electromagnetic coupling between metal particles. Gunnarsson et al. [36] were able to fabricate electromagnetically coupled arrays (Figure 10.11) on silicon by electron beam lithography. A single- or double-layer resist was spin-coated on a clean silicon wafer with (100) surface covered by the native oxide, and the pattern was defined by electron beam lithography. Removal of the exposed resist was performed with a developer, followed by vapor deposition of a 30- to 40-nm-thick silver layer. After this the unexposed resist was dis-

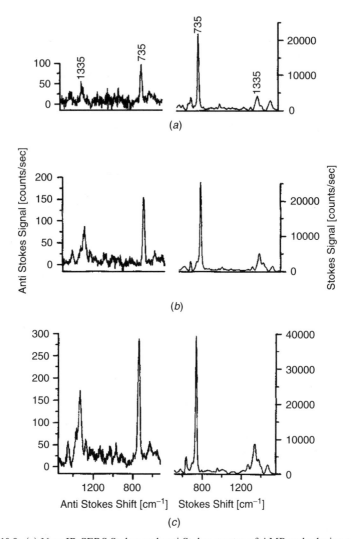

Figure 10.9. (*a*) Near-IR SERS Stokes and anti-Stokes spectra of AMP and adenine measured from 100- to 150-nm silver clusters (*b*) and from an 8-μm cluster (*c*). All spectra were collected in 1 s. (With permission from Ref. 34.)

solved. The latter step also removes the unwanted Ag areas through the lift-off process.

The adsorbates were applied through incubation of the cleaned SERS substrates in a 10 n*M* solution of Rhodamine 6G in water. SERS intensities

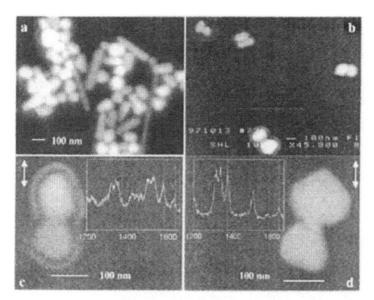

Figure 10.10. SEM micrographs of Ag nanoparticles. The images show (*a*) overview of the particles' shapes and sizes, (*b*) Ag particle dimers after incubation in hemoglobin solution, and (*c*), (*d*) "hot" dimers and corresponding single-molecule SERS spectra. The double arrows in (*c*) and (*d*) indicate the polarization of the incident laser radiation. (With permission from Ref. 35.)

were found to increase with decreasing interparticle separation and found to be comparable to intensities of Rhodamine 6G obtained on Ag nanoparticles from hydrosols similar to that shown in Figure 10.10.

An alternative, more facile approach is to use as the SERS substrate a three-dimensionally ordered array of silica spheres coated with silver or gold. This has been fabricated by Grebel et al. [37] to obtain a grating from which angular dependent SERS can be detected from C_{60}. The ordered structure self-assembles as a coating on a quartz substrate from a suspension of silica spheres in the presence of surfactant. It is then annealed at 600°C prior to deposition of silver or gold layers on top of the silica spheres by evaporation techniques. The adsorbate is added by dipping the substrate in a solution of C_{60} in toluene. Atomic force microscope and scanning electron microscope images of these substrates are depicted in Figure 10.12. Although SERS spectra are obtained at key critical incident angles, no detailed comparison of the SERS efficiencies with other substrates have yet been made.

A similar approach to substrate preparation was adopted by Tessier et al. [38]. Concentrated gold nanoparticles (25 nm) and latex microspheres (630 nm) were mixed together and deposited on a microscope slide. A sec-

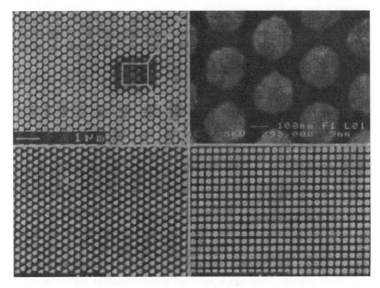

Figure 10.11. Scanning electron micrographs of SERS substrates obtained by electron beam lithography. Examples consist of circular (top left and right), triangular (bottom left), and square (bottom right) 30-nm-thick silver particles on silicon wafer. The predefined particle length scale, defined as the diameter in the case of circular particles and the edge length in the case of triangles or squares, is $D = 200$ nm and predefined interparticle separation distance, defined as the minimum edge-to-edge distance, is $d = 100$ nm. (With permission from Ref. 36.)

ond slide was used to drag a meniscus of the colloidal suspension along the lower slide, depositing a film of latex particles. Thick multilayer latex crystals grew on drying of the films, due to a combination of the increasing latex volume fraction and convective two-dimensional assembly. The gold particles were trapped in the interstitial voids of the latex and eventually assembled around the bottom of the latex particles. The latex–gold composite was then immersed in toluene to dissolve away the latex template, leaving behind the gold structure immobilized on the glass slide. The scanning electron image showing hexagonally ordered pores is shown in Figure 10.13.

The SERS spectra from an organic analyte deposited on the substrate were found to show a enhancement factor of 10^4, which is comparable to the enhancement obtained from two-dimensional silver gratings produced by electron beam lithography by Kahl et al. [39]. Like the self-assembled three-dimensional opal gratings, the templated gratings provide clear practical advantages over ordered substrates produced by more complex and expensive methods. The results reported are summarized in Table 10.2. From the

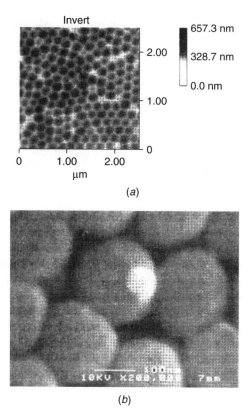

(a)

(b)

Figure 10.12. Atomic force microscope (a) and field-emission scanning electron microscope (b) images of an ordered array of 200-nm silver-coated silica spheres. (With permission from Ref. 37.)

enhancement factors reported it appears that SERS technology will be utilized increasingly to detect organic species (such as explosive vapors) and biological molecules with a high degree of specificity. The potential of this technique has been demonstrated by single-molecule detection of a number of molecular species (See Section 10.4).

10.3. SAMPLE PREPARATION FOR SEIRA

Samples for SEIRA are typically in the form of molecular monolayers deposited on both sides of the internal reflection element or ATR plate, which is usually a silicon crystal. The silicon substrate is cut in the paral-

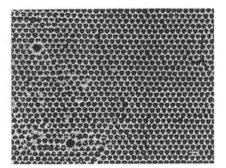

Figure 10.13. Scanning electron microscope image of porous structure after dissolving away the latex from the latex–gold composite. The scale bar at the bottom right of the image corresponds to 1 μm. (With permission from Ref. 38.)

lelepiped geometry illustrated in Figure 10.6. The substrates are cleaned with organic solvents and finally by HF to expose the bare silicon surface. The monolayer deposition is either followed or preceded by the evaporation of a thin metal (typically, silver or gold). The metal overlayers have to be deposited at room temperature to prevent decomposition of the analyte layer.

The internal angle of incidence is generally chosen to be 20° so that it is close to the critical angle of total internal reflection. The infrared absorption of each of the monolayers is measured both with and without the metal layers. The ATR technique is used only over the frequency range for which the substrate is transparent. For silicon substrates, this limits the applicable frequency ranges to between 4000 and 1700 cm^{-1} and below 420 cm^{-1}.

Table 10.2. SERS Results

Nanoparticle	Substrate	Analyte	Enhancement
Silver, gold	Electrode	Pyridine	10^6–10^7
Silver	Film	4,4'-Bipyridine	10^5
Gold, silver	Colloid	Rhodamine-6G	10^{14}–10^{15}
Silver	Isolated spheres	Crystal violet	10^6–10^8
Silver, gold	Isolated spheres	Adenine	10^{14}
Silver	Isolated particles	Hemoglobin	10^{14}
Gold/glass	Templated array	*trans*-1,2-Bis(pyridyl)-ethylene	10^5
Silver/silicon	Nanolithographed 2D array	Rhodamine-6G	10^4

10.4. POTENTIAL APPLICATIONS

SERS and SERRS, in particular, are well positioned for applications in the area of highly sensitive and specific biological and chemical detection. This is due primarily to emerging advances in nanotechnology and the development of miniature laser sources and light detection techniques. Two recent reports clearly point to the feasibility of developing sensors based on the surface-enhanced Raman effect.

In one report, Cao et al. [40] have developed a highly sensitive biosensing method based on SERRS that they hope will make other methods obsolete. The SERRS-based technique, for example, can detect DNA or RNA at a concentration of about 20 femtomolar—about 1 part in 3 trillion in an aqueous solution. This is hundreds of times better than most other methods and is similar to the single-molecule detection levels achieved by Nie and Emory [33] for rhodamine 6G. Conventional methods of detecting genetic material rely on the polymerase chain reaction (PCR) to boost sensitivity, but PCR limits the speed and increases false-positive rates. The SERRS technique involves making an open-faced chemical sandwich, as depicted schematically in Figure 10.14. On one side is a silicon chip covered with alkylthiol-capped DNA strands designed to capture a fragment of genetic material from a biological agent, such as anthrax or HIV. On the other side is a set of 13-nm-sized gold nanoparticles in a solution, each also covered with DNA strands that are complementary to a target's genetic material. The DNA strands on the gold particles are labeled in the experiments performed with six resonant-Raman active dyes (Cy3 is shown in Figure 10.14). When the chip is exposed to the target material, the target strands bind to the complementary DNA strands on the chip with a bit of each target strand jutting above the forest of DNAs. When this is soaked in the solution of modified gold particles, DNAs attached to the gold couple to the loose ends of the target strands. The target strands flag their presence by this process. Raman detection at femtomolar sensitivity levels is then carried out using a fiber optic scanning Raman spectrometer after exposing the chip–gold sandwich to silver hydroquinone to form a silver layer to further enhance the SERRS effect. The spectrum obtained (Figure 10.14) is exclusively from the Cy3 dye and can be used as a spectroscopic fingerprint to exclusively monitor the presence of a specific target oligonucleotide. Dyes other than Cy3 can be used to form a large number of probes with distinct SERRS signals for multiplexed detection (Figure 10.15). This technique is potentially ripe for the marketplace. Mirkin (who is the lead author in Ref. 40) has founded a company, Nanosphere, in Northbrook, Illinois, to exploit the commercial feasibility of this technology.

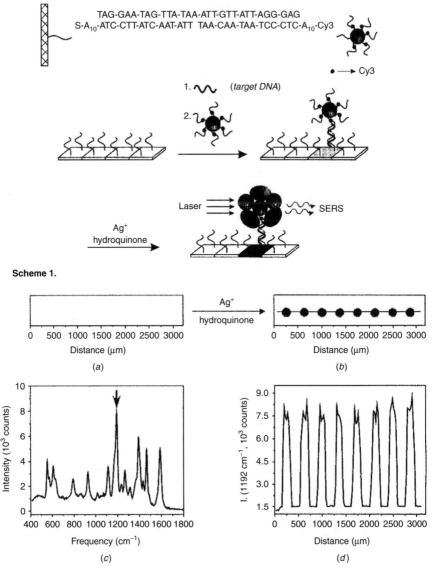

TAG-GAA-TAG-TTA-TAA-ATT-GTT-ATT-AGG-GAG
S-A_{10}-ATC-CTT-ATC-AAT-ATT TAA-CAA-TAA-TCC-CTC-A_{10}-Cy3

• ⟶ Cy3

1. ∿ (target DNA)

2.

Laser ⟹ ∿ SERS

Ag^+
hydroquinone ⟶

Scheme 1.

Ag^+ ⟶
hydroquinone

0   500 1000 1500 2000 2500 3000
Distance (μm)

(a)

0   500 1000 1500 2000 2500 3000
Distance (μm)

(b)

Intensity (10^3 counts)

400 600 800 1000 1200 1400 1600 1800
Frequency (cm^{-1})

(c)

I. (1192 cm^{-1}, 10^3 counts)

0   500 1000 1500 2000 2500 3000
Distance (μm)

(d)

Figure 10.14. Scheme 1 depicts the formation of the three-component sandwich assay discussed in the text. (*a*) and (*b*) show flatbed scanner images of microarrays treated with gold nano-particles before and after silver enhancement, respectively. (*c*) shows a typical Raman spectrum acquired from one of the silver spots. (*d*) shows a profile of the Raman intensity at 1192 cm^{-1} as a function of position on the chip; the laser beam from the Raman instrument is moved over the chip from left to right as defined by the line in (*b*). (With permission from Ref. 40.)

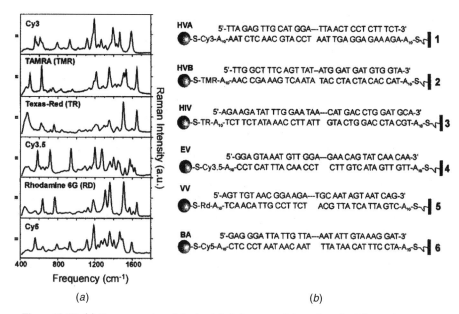

Figure 10.15. (*a*) Raman spectra of six dye-labeled nanoparticle probes after silver enhancement on a chip. (*b*) Six DNA sandwich assays with corresponding target systems. A$_{10}$ is an oligonucleotide tether with 10 adenosine units. (With permission from Ref. 40.)

In the second report, McHugh and co-workers [41] have used a similar SERRS strategy to detect the presence of the high explosive RDX at enhanced sensitivity levels. RDX is present in Semtex, which is widely used in terrorist activities. In 1988 about four pounds of it were detonated on board Pan Am flight 103 over Scotland, destroying the plane and scattering debris onto the town of Lockerbie, killing a total of 270 people. RDX has a small but finite vapor pressure at ambient temperatures, which could in principle alert security systems. There are already several techniques for detecting explosives, but they are not always sensitive enough, especially if the explosives are well wrapped. Sniffer dogs are often the most effective option.

SERRS can be highly sensitive, specific, cheap, and easy to use, but in order for it to detect RDX at near-single-molecule levels, RDX has to be modified. This is achieved by treating RDX with an amalgam of sodium in mercury to form hydrazine, which can then be linked to resonant Raman-active dyes that show up strongly in SERRS. So a Semtex sensor would contain a sodium amalgam to convert RDX to a detectable substance on the spot. Preliminary experiments suggest that such a device could sense just a few trillionths of a gram of RDX.

ACKNOWLEDGMENTS

The author would like to acknowledge support from the Department of the Army by grant DAAE30-02-C-1139.

REFERENCES

1. B. Schrader, ed., *Infrared and Raman Spectroscopy: Methods and Applications*, Wiley, Chichester, West Sussex, England, 1995.

2. J. J. Laserna, *Modern Techniques in Raman Spectroscopy*, Wiley, Chichester, West Sussex, England, 1996.

3. N. Bloembergen, *Pure Appl. Chem.*, **59**, 1229 (1987).

4. A. B. Harvey, ed., *Chemical Applications of Non-linear Raman Spectroscopy*, Academic Press, New York, 1981.

5. A. Zumbusch, G. R. Holtom, and X. S. Hie, *Phys. Rev. Lett.*, **82**, 4142 (1999).

6. S. A. Asher, C. H. Munro, and Z. Chi, *Laser Focus World*, **33**, 99 (1997).

7. R. L. McCreery, in J. J. Laserna, ed., *Modern Techniques in Raman Spectroscopy*, Wiley, Chichester, West Sussex, England, 1996.

8. G. J. Puppels, F. F. M. D. Mul, C. Otto, J. Greve, M. Robert-Nicoud, D. J. Arndt-Jovin, and T. Jovin, *Nature*, **347**, 301 (1990).

9. C. J. H. Brenan and W. Hunter, *Appl. Opt.*, **33**, 7520 (1994).

10. D. A. Smith, S. Webster, M. Ayad, S. D. Evans, D. Fogherty, and D. Batchelder, *Ultramicroscopy*, **61**, 247 (1995).

11. C. L. Jahncke, H. D. Hallen, and M. A. Paesler, *J. Raman Spectrosc.*, **27**, 579 (1996).

12. G. Walker and R. Hochstrasser, in A. B. Myers and T. R. Rizzo, eds., *Laser Techniques in Chemistry*, Wiley, Chichester, West Sussex, England, 1995.

13. L. D. Fisher and G. V. Belle, *Biostatistics: A Methodology for the Health Sciences*, Wiley, Chichester, West Sussex, England, 1996.

14. M. E. Moerner, *Science*, **265**, 46 (1994).

15. D. L. Jeanmaire and R. P. Van Duyne, *J. Electroanal. Chem.*, **84**, 1 (1977).

16. M. G. Albrecht and J. A. Creighton, *J. Am. Chem. Soc.*, **99**, 5215 (1977).

17. K. Kneipp, H. Kneipp, I. Itzkan, R. R. Dasari, and M. S. Feld, *Chem. Rev.*, **99**, 2957 (1999).

18. K. Kneipp, H. Kneipp, R. Manoharan, E. B. Hanlon, I. Itzkan, R. R. Dasari, and M. S. Feld, *Appl. Spectrosc.*, **52**, 1493 (1998).

19. R. L. Garrell, *Anal. Chem.*, **61**, 401A (1989).

20. R. Sheng, F. Ni, and T. M. Cotton, *Anal. Chem.*, **63**, 437 (1991).

21. K. Kneipp, H. Kneipp, R. Manoharan, I. Itzkan, R. R. Dasari, and M. S. Feld, *Bioimaging*, **6**, 104 (1998).

22. P. R. Griffiths and J. A. de Haseth, *Fourier Transform Infrared Spectroscopy*, Wiley, Chichester, West Sussex, England, 1986.

23. A. Hartstein, J. R. Kirtley, and J. C. Tsang, *Phys. Rev. Lett.*, **45**, 201 (1980).

24. R. L. Garrell, T. M. Herne, C. A. Szafranski, F. Diederich, F. Ettl, and R. L. Whetten, *J. Am. Chem. Soc.*, **113**, 6302 (1991).

25. T. P. Mernagh, R. P. Cooney, and J. A. Spink, *J. Raman Spectrosc.*, **16**, 57 (1985).

26. T. P. Mernagh and R. P. Cooney, *J. Raman Spectrosc.*, **14**, 138 (1983).

27. H. Seki, *J. Vac. Sci. Technol.*, **20**, 584 (1982).

28. J. M. E. Storey, R. D. Shelton, T. E. Barber, and E. A. Wachter, *Appl. Spectrosc.*, **48**, 1265 (1994).

29. T. Vo-Dinh, M. Y. K. Hiromoto, G. M. Begun, and R. L. Moody, *Anal. Chem.*, **56**, 1667 (1984).

30. P. D. Enlow, M. Buncick, R. J. Warmack, and T. Vo-Dinh, *Anal. Chem.*, **58**, 1119 (1986).

31. F. Ni and T. M. Cotton, *Anal. Chem.*, **58**, 3159 (1986).

32. J. Nedderson, G. Chumanov, and T. M. Cotton, *Appl. Spectrosc.*, **47**, 1959 (1993).

33. S. Nie and S. R. Emory, *Science*, **275**, 1102 (1997).

34. K. Kneipp, H. Kneipp, V. B. Kartha, R. Manoharan, G. Deinum, I. Itzkan, R. R. Dasari, and M. S. Feld, *Phys. Rev.*, **E57**, R6281 (1998).

35. H. Xu, E. J. Bjerneld, M. Käll, and L. Börjesson, *Phys. Rev. Lett.* **83**, 4357 (1999).

36. L. Gunnarsson, E. J. Bjermeld, H. Xu, S. Petronis, B. Saemo, and M. Käll, *Appl. Phys. Lett.*, **78**, 802 (2001).

37. H. Grebel, Z. Iqbal, and A. Lan, *Appl. Phys. Lett.*, **79**, 3194 (2001).

38. P. M. Tessier, O. D. Velev, A. T. Kalambur, J. F. Rabolt, A. M. Lenhoff, and E. W. Kaler, *J. Am. Chem. Soc.*, **122**, 9554 (2000).

39. M. Kahl, E. Voges, S. Kostrewa, C. Viets, and W. Hill, *Sensors Actuators A Phys.*, **51**, 285 (1998).

40. Y. C. Cao, R. Jin, and C. A. Mirkin, *Science*, **297**, 1536 (2002).

41. C. J. McHugh, W. E. Smith, R. Lacey, and D. Graham, *Chem. Commun.*, 2514 (2002).

INDEX

Sample Preparation Techniques in Analytical Chemistry, Edited by Somenath Mitra
ISBN 0-471-32845-6   Copyright © 2003 John Wiley & Sons, Inc.

439

CHEMICAL ANALYSIS

A SERIES OF MONOGRAPHS ON
ANALYTICAL CHEMISTRY AND ITS APPLICATIONS

J. D. Winefordner, *Series Editor*